Flood Risk and Resilience

Flood Risk and Resilience

Editors

Guangtao Fu
Mónica Rivas Casado
Fanlin Meng
Roy Kalawsky

MDPI • Basel • Beijing • Wuhan • Barcelona • Belgrade • Manchester • Tokyo • Cluj • Tianjin

Editors
Guangtao Fu
University of Exeter
UK

Mónica Rivas Casado
Cranfield University
UK

Fanlin Meng
University of Exeter
UK

Roy Kalawsky
Loughborough University
UK

Editorial Office
MDPI
St. Alban-Anlage 66
4052 Basel, Switzerland

This is a reprint of articles from the Special Issue published online in the open access journal *Water* (ISSN 2073-4441) (available at: https://www.mdpi.com/journal/water/special_issues/Flood_Risk_Resilience).

For citation purposes, cite each article independently as indicated on the article page online and as indicated below:

LastName, A.A.; LastName, B.B.; LastName, C.C. Article Title. *Journal Name* **Year**, *Article Number*, Page Range.

ISBN 978-3-03936-890-7 (Hbk)
ISBN 978-3-03936-891-4 (PDF)

Contents

About the Editors

Guangtao Fu, Professor of Water Intelligence at the University of Exeter, has a research focus on developing and applying new computer models, data analytics and artificial intelligence tools to tackle urban water challenges, including water supply resilience, network leakage, urban flooding and urban wastewater management. He is a Royal Society Industry Fellow and a Turing Fellow at the Alan Turing Institute. He has authored 130 papers in international peer-reviewed journals and conference papers, and has received several awards, including the 2014 'Quentin Martin Best Practice Oriented Paper' and the 2018 'Best Research-Oriented Paper' awards from the American Society of Civil Engineers.

Mónica Rivas Casado, senior lecturer in integrated environmental monitoring at Cranfield University, has expertise in the application of statistics to environmental data. Her academic career has been built around the integration of emerging technologies, advanced statistics and environmental engineering for the design of robust monitoring strategies. Mónica has an MSc in Environmental Water Management and a Ph.D. in applied geostatistics from Cranfield. She is a Chartered Environmentalist (CEnv), a Chartered Scientist (CSci), a Chartered Forestry Engineer and a Fellow of the Higher Education Academy (FHEA).

Fanlin Meng is a research fellow at the University of Exeter. Her research interests include resilient infrastructure, anomaly detection and water policy by the modelling, optimization and control of urban water systems. She has worked on six research projects funded by UK EPSRC and EU Horizon 2020 and FP7 (EU). She has authored over 20 journal papers in *Environmental Science and Technology*, *Water Research*, etc.

Roy Kalawsky (Ph.D. (Hull 1991), MSc (Hull 1984), BSc (Hull 1978), C.Eng, FIET, FRSA) is the Royal Academy of Engineering/Airbus Research Chair in Digital and Data Engineering Information Systems. He is Director of the Advanced VR Research Centre in the Wolfson School of Mechanical, Electrical and Manufacturing Engineering at Loughborough University, UK. He has extensive industrial and academic experience in systems engineering spanning over 32 years. He spent over 17 years working for BAE Systems and was responsible for Advanced Crew Station research across the Military Aircraft Division. He joined Loughborough University in 1995 and established the Advanced VR Research Centre, to specialize in advanced systems, modelling and simulation, digital-twins, synthetic environments, augmented/virtual reality and advanced visual analytics.

Editorial

Towards Integrated Flood Risk and Resilience Management

Guangtao Fu [1], Fanlin Meng [1,*], Mónica Rivas Casado [2] and Roy S. Kalawsky [3]

[1] Centre for Water Systems, College of Engineering, Mathematics and Physical Sciences, University of Exeter, North Park Road, Devon, Exeter EX4 4QF, UK; G.Fu@exeter.ac.uk

[2] School of Water, Energy and Environment, Cranfield University, College Road, Cranfield, Bedfordshire MK430AL, UK; m.rivas-casado@cranfield.ac.uk

[3] Wolfson School of Mechanical, Electrical and Manufacturing Engineering, Advanced VR Research Centre, Loughborough University, Loughborough LE11 3TU, UK; R.S.Kalawsky@lboro.ac.uk

* Correspondence: M.Fanlin@exeter.ac.uk

Received: 4 June 2020; Accepted: 15 June 2020; Published: 23 June 2020

Abstract: Flood resilience is an emerging concept for tackling extreme weathers and minimizing the associated adverse impacts. There is a significant knowledge gap in the study of resilience concepts, assessment frameworks and measures, and management strategies. This editorial introduces the latest advances in flood risk and resilience management, which are published in 11 papers in the Special Issue. A synthesis of these papers is provided in the following themes: hazard and risk analysis, flood behaviour analysis, assessment frameworks and metrics, and intervention strategies. The contributions are discussed in the broader context of the field of flood risk and resilience management and future research directions are identified for sustainable flood management.

Keywords: flood risk; flood resilience; machine learning; management strategy; metrics

1. Introduction

Flooding has been widely recognized as a global threat due to the extent and magnitude of damage it poses around the world each year. Globally, flooding affected 2.3 billion people with an estimated economic loss of USD 662 billion from 1995 to 2015 [1]. Flooding can occur from fluvial, pluvial, coastal or groundwater sources, and the economic costs and disruption to communities are expected to increase as a result of urbanization, economic growth and climate change [2,3]. For example, flood risk is the top risk for the UK among 60 risks arising from climate change; 2.6 million people are projected to live in areas of significant risk (i.e., 1 in 75 or more annual change of flooding) by the 2050s under a 2 °C scenario and 3.3 million under a 4 °C scenario with according to the UK Climate Change Risk Assessment. Flood risk management has proven an effective and successful approach to assess risks and support informed decisions on flood measures and thus reduce economic losses and social-environmental damage.

Risk assessment is now a well-established paradigm in flood management in many countries worldwide. In general, flood risk is perceived as the magnitude of loss, calculated as a function of consequence and probability of flood events. The flood risk assessment process can be represented using the Source-Pathway-Receptor-Consequence conceptual model [4], involving the following key components: (1) understanding the frequency, magnitude and location of one or more hazards (such as storms or cyclones) that can lead to flooding, (2) identifying the route that hazards take to reach the receptors, (3) assessing the vulnerability of the receptors, i.e., people, assets and environment, which could be directly or indirectly affected by flooding, and (4) quantifying the damages that occur to those receptors. Scientific advances have been made in all the above components, such as understanding the impacts of climate change on rainfall [5], representation of rainfall variable dependency [6], development of hydrodynamic or data-driven flood models for flood extent and depth estimation [7,8], as well as flood damage data [9]. However, major research challenges remain in many aspects, in particular, in improving accurate representation and modelling of future uncertainties, capture of system interdependency, and understanding the impacts of human behaviours and stakeholder interactions.

Flood resilience has been gradually recognised as a key aspect in flood management. The term of resilience originated from the field of ecology [10] and was then introduced as a property of a water system which has the ability to prepare for and adapt to changing conditions and absorb, respond to, and recover rapidly from disruptions [11,12]. Conceptual frameworks and metrics were proposed for quantitative and qualitative assessment of flood resilience. On one hand, multi-criteria approaches are commonly used for multi-level and -system assessments; for example, considering physical, and economic and social indicators in failure and recovery phases [13]. On the other hand, performance-based metrics are defined to measure the capacity of a water system in response to specific extreme events in terms of failure duration and magnitude [14,15]. Key to these assessments is the concept of rapid recovery which is contrasted to traditional risk concepts and measures.

To tackle the huge challenges in reducing flood consequences and improving flood emergence planning and preparedness, a Special Issue on flood risk and resilience management was proposed to review the latest developments in the field of flood management. The special issue consists of 11 papers and focuses on the following themes: hazard and risk analysis, flood behavior analysis, assessment frameworks and metrics, and intervention strategies, as introduced in Section 2. This Special Issue will help researchers and practical engineers understand the current challenges in flood management and develop an effective intervention strategy based on the current state-of-the-art knowledge and technologies to tackle these challenges.

2. Overview of the Special Issue

This special issue reports some key advances in flood risk and resilience management. The 11 articles compiled in this issue provide the readers an overview of modelling, optimization, and analytical tool studies for building flood resilience as described below.

(1)   Hazard and Risk Analysis

A matrix-based multi-risk assessment methodology is proposed in Barria et al. [16] to assess the risk of natural hazards such as floods and tsunami. The proposed method can support urban planning and flood mitigation. Stakeholders are closely engaged in the development of the risk assessment, which is an important aspect for enhancing local flood resilience.

Change in land use and rapid urbanization are widely considered as contributory factors for urban flooding. Areu-Rangel et al. [17] performs a quantitative analysis for Villahermosa, Tabasco (Mexico), which shows that the change in land use can increase flood depth by 7% to 22% and urban growth (until 2050) can raise inundation level by up to 0.7 m. The conclusions from this case study reinforced our current understanding that the current way of urban development is lack of sustainability and resilience from the perspective of flood management.

(2)   Flood Behavior Analysis

Our current flood frequency analysis is usually based on stationarity that the probability distributions derived from the historical data is applicable to the future. This can yield misleading results as the temporal and spatial distribution of flood is constantly changing especially under climate change and intensified human activities. Zhang et al. [18] proposes a nonstationary flood frequency analysis for a more accurate representation of flooding. Saravi et al. [19] uses Machine Learning techniques to study the impact of different types of flood based on a big database in the U.S.A. The findings are useful in guiding planning and management to enhance flood resilience.

(3)   Analytical Frameworks and Indices

Fenner et al. [20] presents the latest research outputs from the Urban Flood Resilience research project, where methodologies and tools are developed to facilitate transformative change against extreme rainfall events driven by climate change and rapid urbanization. A roadmap is proposed in this project for urban drainage adaptation over the next 40 years. A Natural Capital Planning Tool is also developed to calculate the theoretical minimum and maximum possible scores of a given site with respect to the natural capital and associated multiple benefits, which is useful in guiding the selection, planning and design of various blue-green infrastructures.

Chen and Leandro [21] propose a novel time-varying index to assess flood resilience at household level during and after a flooding event, based on physical factors, and social and economic factors, respectively. The proposed assessment method is tested on a real-life case in Munich, Germany.

Cascading failures caused by flooding are not uncommon, e.g., critical infrastructures such as a water supply network can be knocked down by a flooding event. Joannou et al. [22] proposes a systems-of-systems approach to identify how resilience can be improved to enhance the performance of a water supply system during times of flooding.

(4)   Intervention Strategies

The reasonable determination of floodway is key to strike a balance between enhancing resilience towards riverine flooding and maximizing the area of land for human activities. Cho et al. [23] provides a contribution to the literature by the study of optimization of floodway using advanced modeling and optimization tools.

Flood control material and emergency logistics play an important role in enhancing flood resilience by providing resources to prepare for, respond to, and recover from flooding. Wang et al. [24] develops an allocation model to maximize the retrieval efficiency and shelf stability.

Unmanned Aircraft Systems (UAS) are increasingly used by emergency responders to acquire core information pre-, during- and post-events. Salmoral et al. [25] develops a guideline on the use of UAS to maximize its benefits for responding to flood and enhancing system resilience that is transferable to multiple countries.

The public reception of flood risk and resilience is of vital importance to the delivery of flood resilience studies, which is surveyed and analysed in Wang et al. [26] by a case study in Jingdezhen, China. Results show that gender, age, education level, experience and knowledge of flooding, income level, and the attitude/level of trust in the governance have influences on public risk perception of flooding. The findings are useful for developing targeted activities to actively engage public in building flood resilience.

3. Conclusions

The research articles in this Special Issue addressed the challenges in flood management and proposed new methods, models and tools for understanding and improve flood resilience in the following four themes: hazard and risk analysis, flood behavior analysis, assessment frameworks and metrics, intervention strategies. Their contributions are discussed in the broader context of the field of flood management and help move towards integrate risk and resilience management.

Research challenges in achieving sustainable flood management remain in many aspects, including developing fast, accurate, high-resolution flood models, characterizing various uncertainties including deep uncertainty, developing integrated risk and resilience frameworks and effective metrics, understanding the relationships between flood risk and resilience, and developing adaptive management strategies with innovative technologies including machine learning technologies.

Funding: This research received no external funding.

Acknowledgments: We acknowledge the financial support from the EPSRC Building Resilience into Risk Management project (EP/N010329/1).

References

1. UNISDR. The human cost of weather-related disasters 1995–2015. 2015. Available online: https://www.unisdr.org/2015/docs/climatechange/COP21_WeatherDisastersReport_2015_FINAL.pdf (accessed on 19 May 2020).
2. Djordevic, S.; Butler, D.; Gourbesville, P.; Mark, O.; Pasche, E. New policies to deal with climate change and other drivers impacting on resilience to flooding in urban areas: The CORFU approach. *Environ. Sci. Policy* **2011**, *14*, 864–873. [CrossRef]
3. Liu, H.; Wang, Y.; Zhang, C.; Chen, A.; Fu, G. Assessing real options in urban surface water flood risk management under climate change. *Nat. Hazard.* **2018**, *94*, 1–18. [CrossRef]
4. Narayan, S.; Nicholls, R.; Clarke, D.; Hanson, S.; Reeve, S.; Horrillo-Caraballo, J.; Cozannet, G.; Hisseld, F.; Kowalska, B.; Parda, R.; et al. The SPR systems model as a conceptual foundation for rapid integrated risk appraisals: Lessons from Europe. *Coast. Eng. J.* **2014**, *87*, 15–31. [CrossRef]
5. Pachauri, R.K.; Allen, M.R.; Barros, V.R.; Broome, J.; Cramer, W.; Christ, R.; Church, J.A.; Dahe, Q.; Dasgupta, P.; Dubash, N.K.; et al. *Synthesis Report. Fifth Assessment Report of the Intergovernmental Panel on Climate Change*; Pachauri, R.K., Meyer, L.A., Eds.; IPCC: Geneva, Switzerland, 2014; pp. 1–151.
6. Fu, G.; Butler, D. Copula-based frequency analysis of overflow and flooding in urban drainage systems. *J. Hydrol.* **2014**, *510*, 49–58. [CrossRef]
7. Guidolin, M.; Chen, A.S.; Ghimire, B.; Keedwell, E.C.; Djordjević, S.; Savić, D.A. A weighted cellular automata 2D inundation model for rapid flood analysis. *Environ. Modell. Softw.* **2016**, *84*, 378–394. [CrossRef]
8. Glenis, V.; Kutija, V.; Kilsby, C. A fully hydrodynamic urban flood modelling system representing buildings, green space and interventions. *Environ. Modell. Softw.* **2018**, *109*, 272–292. [CrossRef]
9. Penning-Rowsell, E.; Priest, S.; Parker, D.; Morris, J.; Tunstall, S.; Viavattene, C.; Chatterton, J.; Owen, D. *Flood and Coastal Erosion Risk Management: A Manual for Economic Appraisal*; Routledge, Taylor & Francis: London, UK, 2013.
10. Holling, C. Resilience and the stability of ecological systems. *Annu. Rev. Ecol. Syst.* **1973**, *4*, 1–23. [CrossRef]
11. Butler, D.; Ward, S.; Sweetapple, C.; Astaraie-Imani, M.; Diao, K.; Farmani, R.; Fu, G. Reliable, resilient and sustainable water management: The Safe & SuRe approach. *Glob. Chall.* **2016**, *1*, 63–77.
12. Linkov, I.; Bridges, T.; Creutzig, F.; Decker, J.; Fox-Lent, C.; Kröger, W.; Lambert, J.; Levermann, A.; Montreuil, B.; Nathwani, J.; et al. Changing the resilience paradigm. *Nat. Clim. Chang.* **2014**, *4*, 407–409. [CrossRef]
13. Kotzee, I.; Reyers, B. Piloting a social-ecological index for measuring flood resilience: A composite index approach. *Ecol. Indicat.* **2016**, *60*, 45–53. [CrossRef]
14. Mugume, S.N.; Gomez, D.E.; Fu, G.; Farmani, R.; Butler, D. A global analysis approach for investigating structural resilience in urban drainage systems. *Water Res.* **2015**, *81*, 15–26. [CrossRef] [PubMed]
15. Wang, Y.; Meng, F.; Liu, H.; Zhang, C.; Fu, G. Assessing catchment scale flood resilience of urban areas using a grid cell based metric. *Water Res.* **2019**, *163*, 114852. [CrossRef] [PubMed]
16. Barría, P.; Cruzat, M.; Cienfuegos, R.; Gironás, J.; Escauriaza, C.; Bonilla, C.; Moris, R.; Ledezma, C.; Guerra, M.; Rodríguez, R.; et al. From multi-risk evaluation to resilience planning: The case of central chilean coastal cities. *Water* **2019**, *11*, 572. [CrossRef]
17. Areu-Rangel, O.; Cea, L.; Bonasia, R.; Espinosa-Echavarria, V. Impact of urban growth and changes in land use on river flood hazard in Villahermosa, Tabasco (Mexico). *Water* **2019**, *11*, 304. [CrossRef]
18. Zhang, T.; Wang, Y.; Wang, B.; Tan, S.; Feng, P. Nonstationary flood frequency analysis using univariate and bivariate time-varying models based on GAMLSS. *Water* **2018**, *10*, 819. [CrossRef]
19. Saravi, S.; Kalawsky, R.; Joannou, D.; Rivas-Casado, M.; Fu, G.; Meng, F. Use of artificial intelligence to improve resilience and preparedness against adverse flood events. *Water* **2019**, *11*, 973. [CrossRef]
20. Fenner, R.; O'Donnell, E.; Sangaralingam, A.; Dawson, D.; Kapetas, L.; Krivtsov, V.; Ncube, S.; Vercruysse, K. Achieving urban flood resilience in an uncertain future. *Water* **2019**, *11*, 1082. [CrossRef]
21. Chen, K.; Leandro, J. A conceptual time-varying flood resilience index for urban areas: Munich city. *Water* **2019**, *11*, 830. [CrossRef]
22. Joannou, D.; Kalawsky, R.; Saravi, S.; Rivas-Casado, M.; Fu, G.; Meng, F. A model-based engineering methodology and architecture for resilience in systems-of-systems: A case of water supply resilience to flooding. *Water* **2019**, *11*, 496. [CrossRef]

23. Cho, H.; Yee, T.; Heo, J. Automated floodway determination using particle swarm optimization. _Water_ **2018**, _10_, 1420. [CrossRef]
24. Wang, W.; Yang, J.; Huang, L.; Proverbs, D.; Wei, J. Intelligent storage location allocation with multiple objectives for flood control materials. _Water_ **2019**, _11_, 1537. [CrossRef]
25. Salmoral, G.; Casado, M.; Muthusamy, M.; Butler, D.; Menon, P.; Leinster, P. Guidelines for the use of unmanned aerial systems in flood emergency response. _Water_ **2020**, _12_, 521. [CrossRef]
26. Wang, Z.; Wang, H.; Huang, J.; Kang, J.; Han, D. Analysis of the public flood risk perception in a flood-prone city: The case of Jingdezhen city in China. _Water_ **2018**, _10_, 1577. [CrossRef]

Article

From Multi-Risk Evaluation to Resilience Planning: The Case of Central Chilean Coastal Cities

Pilar Barría [1,2,*], María Luisa Cruzat [3], Rodrigo Cienfuegos [1,3], Jorge Gironás [1,3,4,5], Cristián Escauriaza [1,3], Carlos Bonilla [3,4], Roberto Moris [6], Christian Ledezma [7], Maricarmen Guerra [3,8], Raimundo Rodríguez [9] and Alma Torres [6]

[1] Centro de Investigación para la Gestión Integrada de Desastres Naturales, Conicyt/Fondap/15110017, Santiago 7820436, Chile; racienfu@ing.puc.cl (R.C.); jgironas@ing.puc.cl (J.G.); cescauri@ing.puc.cl (C.E.)
[2] Departamento de Gestión Forestal y su Medio Ambiente, Facultad de Ciencias Forestales y de la Conservación de la Naturaleza, Universidad de Chile, Santiago 8820808, Chile
[3] Departamento de Ingeniería Hidráulica y Ambiental, Pontificia Universidad Católica de Chile, Santiago 7820436, Chile; mluisacruzats@gmail.com (M.L.C.); cbonilla@ing.puc.cl (C.B.); mnguerra@uc.cl (M.G.)
[4] Centro de Desarrollo Urbano Sustentable, CONICYT/FONDAP/15110020, Santiago 7530092, Chile
[5] Centro Interdisciplinario de Cambio Global, Pontificia Universidad Católica de Chile, Santiago 7820436, Chile
[6] Instituto de Estudios Urbanos y Territoriales, Pontificia Universidad Católica de Chile, Santiago 7530091, Chile; rmoris@uc.cl (R.M.); adtorres@uc.cl (A.T.)
[7] Departamento de Ingeniería Estructural y Geotécnica, Pontificia Universidad Católica de Chile, Santiago 7820436, Chile; ledezma@ing.puc.cl
[8] Applied Physics Laboratory, University of Washington, Seattle, WA 98105, USA
[9] Edic Ingenieros, Santiago 7560873, Chile; raimundorodriguezt@gmail.com
* Correspondence: pbarria@uchile.cl; Tel.: +56-2-2978-5945

Received: 1 February 2019; Accepted: 5 March 2019; Published: 19 March 2019

Abstract: Multi-hazard evaluations are fundamental inputs for disaster risk management plans and the implementation of resilient urban environments, adapted to extreme natural events. Risk assessments from natural hazards have been typically restricted to the analysis of single hazards or focused on the vulnerability of specific targets, which might result in an underestimation of the risk level. This study presents a practical and effective methodology applied to two Chilean coastal cities to characterize risk in data-poor regions, which integrates multi-hazard and multi-vulnerability analyses through physically-based models and easily accessible data. A matrix approach was used to cross the degree of exposure to floods, landslides, tsunamis, and earthquakes hazards, and two dimensions of vulnerability (physical, socio-economical). This information is used to provide the guidelines to lead the development of resilience thinking and disaster risk management in Chile years after the major and destructive 2010 Mw8.8 earthquake.

Keywords: multi-risk matrix; resilience; flood risk; multi-hazard; risk reduction

---

1. Introduction

Deficient urban planning together with a lack of integrated disaster risk management plans may result in important economic and social costs for a community [1]. The continuous population growth, the major growth and urban development trend of coastal cities during the last decades [2], the unregulated urban immigration along with climate change induced extreme hydroclimatic events [3], are likely to strengthen geo-risks derived from natural hazard [1]. Particularly concerning is the case of port cities that concentrate most of the world commercial trade, tourism, and recreation activities [4]. These factors impact what are considered to be the three components interacting to explain disaster risks: the threatening physical phenomena—hazard; the population, infrastructure, economic, and social assets exposed to the hazard—exposure; and the predisposition of society to deal with the event—vulnerability [1,5–7].

The Intergovernmental Panel on Climate Change (IPCC) highlights that suitable methodologies to evaluate the different risk components depend on the local decision-making context, and include qualitative and quantitative approaches [1]. An appropriate risk evaluation methodology for a given region should effectively provide the information about the physical and social-environmental characteristics needed to reduce the risk and to increase resilience through the design and application of disaster risk management plans. Although several studies have investigated disaster risk evaluations worldwide, two limitations are still identified: (1) different hazard events threatening the same region are typically not integrally assessed (e.g., [8,9]), and (2) these studies do not comprehensively contemplate the complexity of all the components that define the risk (e.g., [10,11]), which may result in an underestimation of risk. Thereby, multi-risk approaches, which involve the analysis of multiple hazards and their interactions with exposure and vulnerability are an active area of research and development.

According to [12], multi-risk assessments have been mostly assessed through quantitative tools that aggregate the hazards and vulnerability from a probabilistic perspective. In the frame of the MATRIX project, which aims to set the stages for the multi-risk evaluation for Europe, [13] suggested three tools to address multi-risk assessments: events trees, Bayesian networks and time stepping Monte Carlo simulations. Events trees consider the conditional probability of occurrence of different discrete scenarios of hazards and vulnerabilities, the Bayesian approach analyses the cascade effect of multi-hazard events and time-dependent vulnerability from a probabilistic perspective. Finally, the Monte Carlo approach incorporates the concept of uncertainty in risk evaluation through the generation of multiple random parameters, hazards, and vulnerability scenarios for a given period of time with an associated probability of occurrence. These methodologies require a large and detailed amount of data typically unavailable in many developing countries. Adhering to [14] who stated that multi-risk analyses must acknowledge the vision of stakeholders in the evaluation of all the risk components and the local characteristics of the region under study, this work proposes a simple methodology which fits the needs of regions prone to be affected by multiple natural disasters with insufficient data.

Natural hazards and risk assessment are of extreme relevance for Chile, a country whose average annual economic losses between 1980 and 2011 due to multiple natural disasters have been estimated to represent 1.2% of the gross domestic product, more than for any other country of the G20 group [15]. Moreover, Chile is ranked among the 30 countries under highest water-related risk by 2025, including drought and flood risk [16]. Despite all the territory is highly exposed to multiple natural threats [17], a regulatory framework to support risk management and increase resilience is still under development [11,18,19]. Indeed, the 526 deaths and ~200,000 severely damaged houses caused by the 8.8 Mw Maule earthquake and the subsequent tsunami on 27 February 2010 [20,21], revealed the serious limitations of hazard risk assessments in Chile [22]. Because of these catastrophic losses, and the record of previous hazardous floods and landside episodes in the coastal zone of central Chile [23–25], the Ministry of Housing and Urban Planning (MINVU) and local governments started the development of integrated disaster risk management plans for these areas. These plans must integrate multiple hazards, covering their complex interactions, along with the vulnerability and level of exposure of the community and the infrastructure. Moreover, they must consider the large number of visitors these cities host during the summer season.

This paper presents the development of a multi-risk approach to be applied in urban planning, using two Chilean coastal cities, characterized by low levels of information availability, El Quisco and San Antonio, as a case study. The methods used in the approach are adapted to the local differences of qualitative and quantitative information details. These urban settlements are touristic cities that concentrate a high floating population during summer and are highly exposed to a variety of natural threats that can have large impacts on the local community and economy. This is particularly important, as tourism has experienced continuous growth in the Chilean economy, currently contributing to a 3.3% of the total domestic gross product (GDP), and the number of foreign visitors has doubled between

2010 and 2016 [26]. A matrix-based multi-risk assessment methodology within the Chilean context is proposed, which integrates the risk associated with different natural hazards including floods, landslide, and tsunami processes. Likewise, as earthquakes act as a triggering factor to tsunamis, the earthquake risk has also been estimated. Despite the lack of detailed data, and the differences in qualitative and quantitative information available for the two cities under analysis, disaster risk is addressed comprehensively considering three different components: the physical characteristics of the hazards, the exposure to single and aggregated hazards, and the physical and socio-economic vulnerability of crucial three elements (i.e., housing, urban infrastructure, and facilities). A mitigation plan orientated to increase the resilience in the cities, using integrated quantitative and qualitative tools to support urban planning in the region is designed and proposed.

2. Materials and Methods

2.1. Study Area

El Quisco and San Antonio are nearby coastal cities located 120 km west of Santiago, the capital of Chile (Figure 1). Both cities have a temperate Mediterranean climate, with a mean annual precipitation of about 500 mm [27]. Because of their proximity to Santiago, these cities are of high touristic interest, especially during summer season, when the population of El Quisco increases up to 8 times [28]. On the other hand, San Antonio is the most important port of the country and the economic hub of the province to which it belongs [29]. The majority of the ~90,000 San Antonio and the ~10,000 El Quisco inhabitants [30] belong to low-income groups identified as highly vulnerable to natural hazards [31,32]. Furthermore, the lack of appropriate regulatory and mitigation plans has led to an unregulated continuous urban development, which has increased the exposure levels to natural hazards.

Figure 1. Location of El Quisco and San Antonio. Rivers and streams are presented in blue lines.

Because its location on the subduction area between the Nazca and the South American plates, several strong tsunamis and earthquakes have occurred in the region [33]. The tsunami of 27 February 2010 [20,34] was particularly destructive and produced severe damages (Figure 2). Furthermore, both cities have developed along the coastal floodplains and near rivers or streams, which enhances the risk to floods and landslide events. Because El Quisco is crossed by five ephemeral creeks (i.e., Las Petras, El Batro, Pinomar, Seminario, and Córdova), there are areas in the city prone to be affected by floods and landslides (Figure 1). On the other hand, San Antonio is crossed by Maipo River and El Sauce stream, and limits to the south with the San Juan stream. According to the [23] report, several fluvial floods and landslide events have affected San Antonio precisely because of the proximity of urban areas to these water bodies.

Figure 2. Damage in the San Antonio area caused by the 27 February 2010 tsunami [35].

2.2. Data and Methodology for Multi-Risk Assessment

Following [1,36] recommendations, a scenario-based multi-risk assessment methodology was developed, which integrates hazard and vulnerability degrees (scores) into a risk matrix, designed to meet the local needs using available local data. Then, MINVU and the local governments provided the information needed to design disaster risk management plans through hazard risk zoning. The methodology schematized in Figure 3 considers the analysis of different degrees of risk for each natural hazard, which are finally overlapped to obtain the multi-risk zoning using Geographic Information System (GIS) tools. First, the natural hazards that have historically affected the study region are identified and characterized, while the exposure to each natural hazard is spatially determined and a level of exposure ranging between 0 (low exposure) to 2.5 (high exposure) is assigned. Based on the standard of recent coastal cities vulnerability studies [37–39], and when qualitative information is available, physical and socio-economical dimensions were considered in the vulnerability assessment of the different components analyzed. Then, a level of vulnerability ranging between 0 (no vulnerability) and 4 (high vulnerability) was assigned to every component evaluated using a matrix. When information about materiality is not suitable to assess the physical vulnerability, only socio-economical dimension is considered, which can be adapted and updated when new information is collated for coastal management plans. Finally, the multi-risk zoning is produced by crossing this matrix and the natural hazard exposure assessment. The range of exposure levels (0 to 2.5), and the vulnerability levels (0 to 4) were defined with the local decision-making authority (i.e., the Chilean Secretary of Regional and Administrative Development, hereafter the SUBDERE), to standardize the level under a 0–10 range. The models and the data used to assess the hazard exposure to the different natural phenomena, the vulnerability and risk characterization are presented in the following sections.

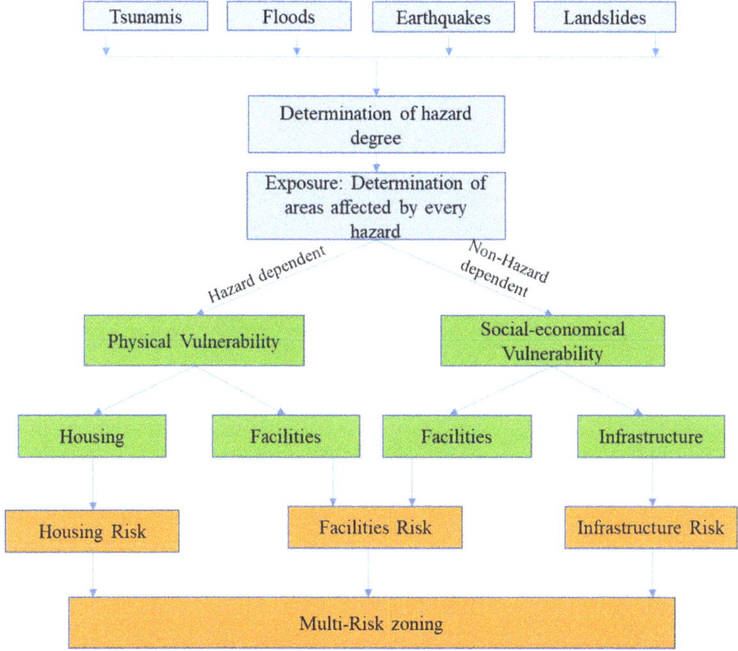

Figure 3. Scheme of the multi-risk assessment methodology. The hazard risk is calculated for each hazard independently, and the overlapping of all the hazards risk determines the multi-risk zoning.

2.2.1. Natural Hazard Characterization

The hazard characterization is developed by constructing scientific physically-based models for each natural phenomenon, extensively used in Chile, to evaluate the exposure to the corresponding hazard (i.e., degree of hazard) considering conservative scenarios and boundary conditions. To standardize the analysis, the degree of hazard is assigned quantitatively using a score or class which was chosen to range from 0 to 2.5, where high (2.5), medium (1–2), and low (0–1.5) degrees of hazard are identified for each natural phenomenon evaluated. Detailed information is provided below.

2.2.2. Flood Hazard

Rainfall-runoff and hydraulic models have commonly been used to characterize the probability of occurrence and the impacts of floods (e.g., [40]). This involves simulating riverine floods associated with precipitation events of different return periods (T), and route them in a hydraulic model. A central assumption then is that T of the flood equals that of the rainfall event that produces it [41]. Based upon the hydraulic design criteria for urban infrastructure specified by the Ministry of Public Works in Chile [23], values of $T = 2, 5, 10, 20, 25, 50,$ and 100-year are considered.

Several rainfall-runoff models developed for ungauged basins extensively used in Chile are implemented to simulate the peak flow discharges for different T. These models include (1) the rational method as defined by [41] using the Kirpich formula for computing the time of concentration, (2) the Snyder synthetic unit hydrograph method [42], (3) the Storm Water Management Model (SWMM; [43]); as well as (4) the Verni-King method [44] and (5) the Modified Verni-King Method [45]. These last two models are local methods defined for Chilean catchments, which consider a non-linear relationship between peak flow and contributing area, similar to that of the Espey 10-Min Synthetic Unit Hydrograph [46] and the Colorado Urban Hydrograph Procedure [47]. Parameters for all these models are obtained from the 30 m digital elevation model [48], local studies [35,45,49] and the literature [41,47]. Local intensity–duration–frequency (IDF) curves and daily precipitation data recorded between 1986 to 2011 at Lagunillas station (33°26′22″ S, 71°27′02″ W) for El Quisco and between 1972–2010 at Punta Panul (33°34′29″ S, 71°37′30″ W) for San Antonio, are utilized to define the rainfall events used in the rainfall-runoff modeling and landslide analysis described later.

The peak flows for the different return periods are propagated using HEC-GeoRAS [50], an ArcGIS® based platform to run the 1D hydraulic model Hec-Ras (Hydrologic Engineering Center River Analysis System; [51]), and display its results. Hec-Ras solves the energy equation (Equation (1)) using an iterative procedure

$$Z_2 + Y_2 + \frac{a_2 V_2^2}{2g} = Z_1 + Y_1 + \frac{a_1 V_1^2}{2g} + h_c \tag{1}$$

where Z is the elevation of the river or channel bed, Y is the water depth, V is the average velocity in the river or the channel, α is the velocity weighting coefficient, and h_c is the energy head loss. Subscript 1 and 2 indicate two sections in the river.

The water surface elevation, the flow velocity and the riverine flooded areas for each stream and return period are obtained from the model. Cross sections measured in the field, contours maps and the 30 m DEM are combined to generate the streambed and floodplain topography, whereas high-tide data from the SHOA (National Hydrographic and Oceanic Service) nautical charts are used to set a conservative downstream boundary condition that maximizes the flooded area (i.e., a water surface elevation of 1.6 m). Finally, the flow discharge record for the period between 1986 and 2011 from the Maipo at Cabimbao gauge (33°43′19″ S, 71°33′18″ W), as well as the topography and other relevant information obtained from local reports [23,29,49,52], are used to calibrate the hydraulic model for the Maipo River.

The level of exposure to flood hazard is categorized based on different scenarios depending on the frequency of the events and following the Organization of American States recommendation [53]. Then, the level of flood hazard exposure was defined as 'low' for a 50-year flood (low frequency), 'medium' for a 20-year flood (medium frequency) in El Quisco (estimated) and 25-year flood in San Antonio (obtained from [23]), and 'high' for a 10-year flood (more frequent).

2.2.3. Landslide Hazard

Landslides are caused by the downward and outward movement of slope-forming materials triggered by the interaction of different conditioning factors. These factors can be the physical characteristics of the terrain (e.g., geomorphology, vegetation, or land use), or external factors such as earthquakes and storms [54], with the last factor being considered in this study. Meteorological variables, particularly precipitation, have the largest impact as a triggering factor of landslides in Chile [55]. The rainfall magnitude and intensity affect the soil moisture before the landslide, increasing the pore pressures and reducing the slope stability. Moreover, precipitation can induce a change in the water table and enhances surface runoff [54].

In this study, landslide hazard is defined by identifying a critical precipitation threshold above which a landslide can occur under a given soil type and topography [56]. This approach allows linking the probability of occurrence of the precipitation event (storm) to the probability of occurrence of the landslide at a specific site. Adapting to the data availability in each city, a physical and a statistical method are used to identify the precipitation threshold in El Quisco and San Antonio respectively. In particular, there was a record of historical landslide events for San Antonio. The physical method used in El Quisco is based on the shallow slope stability (SHALSTAB) model [57], which relates the rainfall rate (P), the pore water pressure (related to W), and the slope stability as

$$W = P\frac{A}{b}T_s sin(\theta) \tag{2}$$

where W (dimensionless) is the wetness, defined as the ratio of the local flux at a given steady state rainfall to that at soil profile saturation. W defines when the saturation overland flow occurs, which depends on the net rainfall rate (*P*, precipitation less evapotranspiration and deep drainage into bedrock), the contributing area (*A*), the contour length (*b*), the soil transmissivity (*Ts*), and the slope θ. The stability in response to a critical precipitation is calculated using the method proposed by [58], which was used to calculate the probability of occurrence of a landslide under different precipitation thresholds.

The statistical method used in San Antonio considers the analysis of the precipitation record and historical landslide events, as suggested by [55]. The method considers the following steps: (1) characterization of historical landslide events by identifying their date, type (debris or mudflow) and triggering factor (precipitation or other factors), (2) analysis of daily precipitation, land use and physical characteristics of the landslide site (conditioning factors), and (3) identification of the magnitude of the triggering precipitation depth and modeling of the conditioning factors that generate the landslides to identify the potentially hazardous areas.

Since the instability of the slope can be the result of either intense precipitation or the saturation of the soil profile because of successive small rainfall events, both the daily rainfall on the day of the event and the cumulative precipitation during the last 20 days are selected as the triggering factors in the statistical method. We also identified the terrain slope and the soil water retention as conditioning factors to landslides. Nineteen different landslide events occurring in San Antonio between 1986 and 2010 are evaluated [23,59,60]. The soil layer depth is obtained from the Chilean Hydrogeologic Map [27], and the land use from the Territorial Master Plan of San Antonio [29].

Both in San Antonio and El Quisco, the level of exposure to landslide hazard is evaluated considering three scenarios based on the return period of the event [61]. Thus, the 50- (less frequent), 20- (medium frequency), and 10- year (more frequent) landslide events are associated with low, medium, and high levels of exposure, respectively.

2.2.4. Tsunami Hazard

Tsunamis represent a threat to coastal cities because the incoming waves have the potential to flood inland areas [62]. A three-step numerical methodology is used to characterize the tsunami hazard, in which (1) the rupture mechanism that originates the tsunami, (2) the regional propagation of the tsunami to the Chilean coast, and (3) the local inundation in the study area, are simulated.

The rupture mechanism was simulated using the Okada model [63], which analytically calculates the surface deformation u_i due to a fault dislocation $\Delta u_j\left(\varepsilon_n,\ \varepsilon_k, \varepsilon_j\right)$ across a three dimensional surface $\sum\left(\varepsilon_n,\ \varepsilon_k \text{ and } \varepsilon_j \text{ coordinates}\right)$ produced by shear and tensile faults F (e.g., an earthquake) on an isotropic homogeneous semi-infinite medium

$$u_i = \frac{1}{F}\int_{\Sigma}\Delta u_j\lambda\delta_{jk}\frac{\partial u_i^n}{\partial\varepsilon_n} + \mu\frac{\partial u_i^j}{\partial\varepsilon_k} + \frac{\partial u_i^k}{\partial\varepsilon_j}v_k d\sum \tag{3}$$

where δ_{jk} is the Kronecker delta, λ and μ are elasticity constants, and v_k is the direction cosine of the normal to the surface element $d\sum$. The parameter values (Table 1) were obtained from a local study developed by [64].

Table 1. Seismic design parameters used in the tsunami modeling, obtained from [65].

Parameter	Tsunami 2010	Scenario 1	Scenario 2
Seismic momentum magnitude (Mw)	8.8	8.6	8.8
Rupture length (km)	550	500	500
Rupture width (km)	100	150	150
Strike	10° N	10° N	10° N
Dip angle	10° to 22°	18°	18°
Slip length (m)	6 to 10	5	10

The regional propagation of the tsunami through the Chilean coast was modeled using the SWAN model (SWAN 41.20AB, Delft University of Technology, Delft, The Netherlands) [66], which solves the non-linear shallow water equations (Equations (4)–(6)) considering Coriolis forces (F_u, F_v), frictional slope effects (Chezy coefficient C) and wind or pressure forces (F_x, F_y)

$$\frac{\partial \eta}{\partial t} + \frac{\partial hu}{\partial x} + \frac{\partial hv}{\partial y} = 0 \tag{4}$$

$$\frac{\partial u}{\partial t} + u\frac{\partial u}{\partial x} + v\frac{\partial u}{\partial y} + g\frac{\partial \eta}{\partial x} = F_v + Fx - g\frac{\overline{uu^2 + v^2}}{C^2 h} \tag{5}$$

$$\frac{\partial v}{\partial t} + u\frac{\partial v}{\partial x} + v\frac{\partial v}{\partial y} + g\frac{\partial \eta}{\partial y} = -F_u + Fy - g\frac{\overline{vu^2 + v^2}}{C^2 h} \tag{6}$$

where η is the vertically averaged flow in terms of horizontal velocity (u,v) and water depth variation (h), and g is the gravity acceleration. The inundation areas at the local level are determined using the nonlinear shallow water model developed by [67] and modified by [68], which propagates rapidly evolving flows over complex topographies using a finite volume approach and a non-orthogonal generalized curvilinear coordinate framework. Topo-bathymetric profiles from the General Bathymetric Chart of the Oceans (GEBCO, http://www.gebco.net), the nautical charts from the SHOA and the study by [65] are used in the regional propagation model.

The regional model for tsunami propagation is coupled with the inundation model by entering the tsunami waves to the local domain using an absorbing-generating boundary condition [69] through a nested grid scheme. To assure numerical convergence, a grid of 0.025° and a time step of 5 s was used in the regional propagation model (i.e., the Courant number is less than 1), and a 30-m resolution grid, restricted by the resolution of the digital elevation model ASTER GDEM (version 2, METI & NASA, USA) [48], is used in the local propagation model. For the whole domain a Manning roughness coefficient of $n = 0.020$ is adopted, representative of smooth natural floodplains [70]. In addition to the information used by the regional propagation model, local topography provided by the cities [29,49] and field data were used in the local inundation model.

We consider inundated areas for one historical and two potential scenarios. The historical scenario used for validation purposes is based on the field measurements taken after the 8.8 Mw Maule earthquake and tsunami [34]. The potential scenarios consider two additional earthquakes of magnitude 8.6 Mw y 8.8 Mw respectively, originated by a rupture located in North-Central Chile, from Santo Domingo to La Serena (33.5° S to 30° S), each with maximum vertical displacements of 1.6 m and 3.2 m, respectively. The regional tsunami model is run in a north–south direction covering ~25° latitude (from Iquique 20° S to Chaitén 45° S), and East-West direction covering 25° longitude of the Chilean coast (~75° W–100° W).

Following [71] recommendations, three scenarios of tsunami degree of hazard are defined: low, medium, and high when the flood depth is less than 0.5 m, between 0.5 and 2 m, and over 2 m respectively.

2.2.5. Earthquake Hazard

Although this work mainly focuses on risks associated with floods, and in order to accomplish the multi-risk assessment for the study area, the earthquake hazard to the extent allowed by the available data and records was also studied. Surface waves from are the main cause of infrastructure damage during large earthquakes. It is well known that the predominant soils and topographic characteristics of an area determine the seismic amplification of these waves [72,73]. As a proxy for seismic amplification, this study considers four soil types (i.e., I, II, III, and IV), where I is a stiff foundation soil, such as rock, characterized by a low seismic amplification, and IV corresponds to a soft deposit, such as soft clays, characterized by a high seismic amplification. Location and information about the soil types in the study area are obtained from [52], while recommendations from Eurocode EN 1998-5:2004 [74] are used to evaluate the topographic effects on the seismic classification. Hence, topographic effects are neglected when the local slope is lower than 15°, whereas the seismic classification of sectors located within 20 m from areas with slopes ranging between 15° and 30° is increased in one level. Finally, the seismic classification of sectors located near topographic slopes larger than 30° is increased in two levels.

This assessment evaluates the exposure to seismic hazard according to the [53] recommendations, considering three scenarios, which assigns a low, medium, and high degree of hazard to areas with low, medium, and high seismic amplification respectively.

2.3. Vulnerability Characterization

Vulnerability refers to the susceptibility of a community or system of being negatively impacted by a hazard [75,76]. Based upon the results of technical studies developed by the UNESCO-RAPCA (Regional Action Program Central America) project, which since 1999 has been working on the use of GIS and remote sensing tools to assess the impact of natural hazard on human and physical infrastructure in Latin America [77], vulnerability of three components were assessed: housing, facilities (i.e., public and private buildings, equipment, and services) and infrastructure. Table 2 lists the elements considered and evaluated in each component. Depending on the component evaluated, the vulnerability of each element is assessed considering a physical (hazard dependent) and/or a socio-economical (non-hazard dependent) dimension. The physical vulnerability refers to the susceptibility of an element to suffer physical damage because of a hazard. On the other hand, although the socio-economic vulnerability is generally defined as the predisposition of social groups in the context of a natural disaster [78,79], the concept has multiple and complex different definitions [76]. In this study, the socio-economic vulnerability refers to the capacity to respond to a hazard given its relevance and hierarchy in the system [75], which are proxies of the intangible losses of elements at risk in society.

Table 2. Elements at risk.

Housing	Facilities	Infrastructure
Houses	Health services	Roads
Residential buildings	Schools	Bridges
	Police stations	Public lighting
	Naval units	Drinking water infrastructure
	Local government facilities	Telecommunication infrastructure
	Harbors	
	Banks	
	Supermarkets	
	Gas stations	
	Sport centers	
	Government offices	
	Churches	
	Childcare facilities	
	Neighborhood council	
	Other services	

The physical dimension was considered for housing in both study cities, and thus the vulnerability is evaluated in terms of the materials, the quality and the height of the construction [77], whereas for the facilities, the socio-economical vulnerability is evaluated in San Antonio and El Quisco. Nevertheless, as of more detailed information about the building materials and constructions of facilities in El Quisco is available, the physical dimension is also considered in the latter. Finally, the socio-economical dimension is used to evaluate the infrastructure vulnerability in both cities, although the physical vulnerability of roads is also assessed, as information about material and the construction is available. Following the local Chilean Secretary of Regional and Administrative Development criteria (Governmental Office in charge of development of risk plans in Chile), a larger weight to the physical dimension is given through a 3 to 2 proportion. This decision was made because, under a natural hazard scenario, the physical vulnerability is considered to be more critical than the socio-economical vulnerability when both are estimated for an element. It is worth noting that, although vulnerability definitions normally acknowledge its multi-dimension and site-specific nature [80], most of the scientific studies explore only one of the dimensions of vulnerability of the elements at risk (either the social or physical dimension). Both together, as it is in the case of our study, are rarely considering in the literature [76].

The level of vulnerability is determined based on qualitative parameters following the criteria defined with the Chilean Secretary of Regional and Administrative Development [81] and MINVU, which were then used to be crossed against hazard exposure degrees. To homogenize the evaluation, values of vulnerability level or classes of vulnerability are assigned as 0 (non-vulnerable), 1 (low), 2 (medium-low), 3 (medium-high), or 4 (high) for each element under the physical and the socio-economical dimension separately. Then, the aggregated vulnerability of the elements evaluated under the physical dimension is calculated as the average of the vulnerabilities to the different hazards. The values used to assess the physical vulnerability to every natural hazard are presented in Table 3.

Table 3. Matrix of natural hazards risk scores arisen from the hazard degree and the vulnerability scores.

	Hazard	Tsunami			Fluvial Flood			Landslide			Seismic Wave Amplification		
		High	Medium	Low	High	Medium	Low	High	Medium	Low	High	Medium	Low
Vulnerability	Score	2.5	2	1.2	2.5	2	0.7	2.5	1	0.4	2.5	2	1.5
High	4	10	8	4.8	10	8	2.8	10	4	1.6	10	8	6
Medium-High	3	7.5	6	3.6	7.5	6	2.1	7.5	3	1.2	7.5	6	4.5
Medium-Low	2	5	4	2.4	5	4	1.4	5	2	0.8	5	4	3
Low	1	2.5	2	1.2	2.5	2	0.7	2.5	1	0.4	2.5	2	1.5
N/V	0	0	0	0	0	0	0	0	0	0	0	0	0

On the other hand, when the socio-economical dimension is evaluated, single vulnerability values for each element are assigned [82]; these values are not hazard dependent and consider the role of the elements during and after a catastrophic event. The function and characteristics of each element are analyzed, such as the number of students enrolled in each school, the capacity and complexity of health services, among others. For instance, as complex health services have larger capacity to provide medical care during natural disasters than ambulatory health care centers, they are characterized by a larger level of vulnerability. Detailed information about the characteristics considered in the evaluation of each element is presented in Table 4.

Table 4. Vulnerability of facilities, evaluated according to the type of facility and the role they have during and after catastrophic disasters (adapted from [82]).

Health Services	Characteristics-Definition	Socio-Economic Vulnerability
Tertiary health care center	High complexity health service, up to 500 beds	4
Secondary health care center	Medium complexity health service, specialist or referral services	3
Urgency primary health care center	Urgent care center	2
Ambulatory primary health care center	Low complexity health service for ambulatory care	1

Educational Services	Characteristics-Definition	Socio-Economic Vulnerability
Educational services T1	More than 500 enrolled students	4
Educational services T2	Between 251 and 500 enrolled students	4
Educational services T3	Between 101 and 250 enrolled students	3
Educational services T4	Between 0 and 100 enrolled students	2

Security services	Characteristics-definition	Socio-economic Vulnerability
Police station T1 ("Comisaría")	High complexity police unit	4
Police station T2 ("Tenencia")	Medium complexity police unit	3
Police station T3 ("Retén")	Low complexity police unit	2

Navy Facilities	Characteristics-Definition	Socio-Economic Vulnerability
Harbormaster facilities	Ultimate navy regional authority	4
Water bailiff facilities	Local navy authorities controlled by the Harbormaster	3
Navy staff offices	Local navy facilities	2

Local Government Facilities	Characteristics-Definition	Socio-Economic Vulnerability
Town hall	Local Government administration office	4
Town hall offices	Various governmental agencies controlled by the Town Hall	2

Table 4. *Cont.*

Small harbours facilities-N° of fisherman	Socio-Economic Vulnerability
>200	4
50–200	3
<50	2

Services- Type of facility	Socio-Economic Vulnerability
Bank	2
Supermarket	2
Gas Station	2
Other services	2

Sport Facilities	Characteristics-Definition	Socio-Economic Vulnerability
Stadium	Large capacity sport facility	4
Fitness center	Medium to low capacity sport facility	2
Football court	Medium to low capacity sport facility	2

Other Facilities	Characteristics-Definition	Socio-Economic Vulnerability
Government offices	Provide shelter and help during and after disasters	2
Churches	Provide shelter during and after disasters	2
Child care facilities	Provide shelter and services during and after disasters	2
Neighborhood council	Provide shelter and services during and after disasters	2
Other services	Provide shelter and services during and after disasters	2

Housing vulnerability to tsunami and fluvial floods was assessed following the [77,81] recommendations. Thereby, higher buildings present lower vulnerability than single-story ones. The level of vulnerability to landslides of building is defined as high for all the constructions. Regarding seismic amplification, the assessment considers lower vulnerabilities for lower buildings, as they typically have larger seismic resistance. Table 5 shows the values used to characterize housing vulnerability.

Table 5. Housing level of vulnerability to natural hazards in San Antonio (SA) and El Quisco (EQ) (I: inundation due to tsunami and fluvial flood, LS: landslide, SA: seismic amplification). Note that values may differ for the same hazard among the locations.

Material	I SA	I EQ	LS	SA	Condition	I SA	I EQ	LS	SA SA	SA EQ	Height	I	LS	SA
Concrete	0	0		0	Good	0	0		1	0	>2 Story	1		4
Bricks	0	1		1	Regular-Good	-	1		-	1	2 Story	2	4	2
Clay	2	2	4	2	Regular	2	2	4	2	2	1 Story	4		1
Wood	3	3		3	Regular Bad	-	3		-	3				
Waste	4	4		4	Bad	4	4		4	4				

Following [81] recommendations to assess infrastructure vulnerability, medium-low and low physical vulnerability are considered for paved roads under different natural hazards, whereas unpaved roads have medium-high physical vulnerability to all the hazards (Table 6). All the infrastructure and facilities elements were considered to have a High physical vulnerability to landslides. Likewise, because road surfaces materials do not provide seismic resistance information, all the infrastructure is evaluated with high physical vulnerability level to this hazard. The socio-economical vulnerability of roads was assessed considering their hierarchy in the city road system, which depends on their size (Table 6).

Table 6. Infrastructure physical and social-economical vulnerability (adapted from [83]).

Road Surface	Physical Vulnerability
Asphalt	1
Gravel	2
Improved dirt	3
Dirt	4
Road Surface	**Physical Vulnerability**
Asphalt	4
Gravel	4
Improved dirt	4
Dirt	4
Type of Road	**Social-Economical Vulnerability**
Paved road	4
Unpaved road	4
Trail	1
Path	1
Type of Road Associated with the Bridge	**Social-Economical Vulnerability**
Paved road	4
Unpaved road	4
Trail	1
Path	1
Lightning Category	**Social-Economical Vulnerability**
Substation	4
Transformer	2
Urban lightning	1
Type of Antenna	**Social-Economical Vulnerability**
Radio antenna	4
Mobile antennas	3
Television antennas	2

Regarding facilities, the socio-economical vulnerability is evaluated for every single element considering its importance or role within the city under a given hazard. Detailed information for each element is presented in Table 4. Finally, the demographic and economic characteristics of the study area are obtained from national and regional studies and entities (i.e., [23,30], the Traffic Headquarters of the Ministry of Public Works, the Communal Planning department and the Education Ministry).

2.4. Risk Characterization

Following the matrix approach, risk is characterized using risk scores calculated as the multiplication of the level of vulnerability and the exposure [6,83] of each evaluated element (i.e., housing, facilities, and infrastructure). Four different risk levels are defined for each hazard in each location: high, medium-high, medium-low, and low, and the scores of these levels are presented in Table 7. Finally, as shown in Figure 3, the integrated multi-risk zoning is defined as the overlapping of the level of risk associated with every hazard, which is presented for each of the elements evaluated separately (i.e., one multi-risk zoning is considered for housing, for facilities, and for infrastructure independently). To be conservative with the results of the risk assessment and following the local decision-making authority (SUBDERE) goal of informing the aggregated risk level (although the hazards do not necessarily have the same probability of occurrence), when overlapping the level of risk associated with each hazard, the highest level of risk predominates for each element.

Table 7. Risk scores for each natural hazard.

Degree of Risk	Risk Score
High	6.1–10
Medium-high	4.1–6
Medium-low	2.1–4
Low	0.1–2
No risk	0
Outside the hazard zones	-

3. Results

3.1. Natural Hazards

3.1.1. Models Validation

The regional tsunami model is validated by the [65] study using the observations of the 27 February 2010 tsunami. The far-field wave simulated by the SWAN model were compared to the DART (Deep-ocean Assessment and Reporting of Tsunamis) buoy measures located in the coast of Perú (Figure 4a). The near-field waves simulated by the SWAN model were validated by comparing the simulated times of arrival of waves to different Chilean coastal cities against observations published in local media (i.e., La Tercera newspaper, 26 March 2010, Figure 4b).

The fluvial flood model for the Maipo river was validated by comparing the simulated flooded area produced on 8 October 2009 by the flow discharge measured at the Cabimbao station, against a digital photography of the river retrieved from Google Earth (Figure 4c) on the same day. Also, other historical floods reported by [23] are used to validate the models for other streams. On the other hand, the SHALSTAB landslide model is validated in San Antonio through its ability to simulate the location of a landslide event in the 'Talud 21 de Mayo' sector that took place on 28 May 2013, which was produced by a 5- to 10-year rainfall event (Figure 4d).

Finally, the seismic wave amplification classification could not be quantitatively validated due to the lack of proper instrumentation in the zone. Nonetheless, because the soil type II representative of a low level of amplification is predominant in both cities under study, significant uncertainty in the risk assessment is not expected.

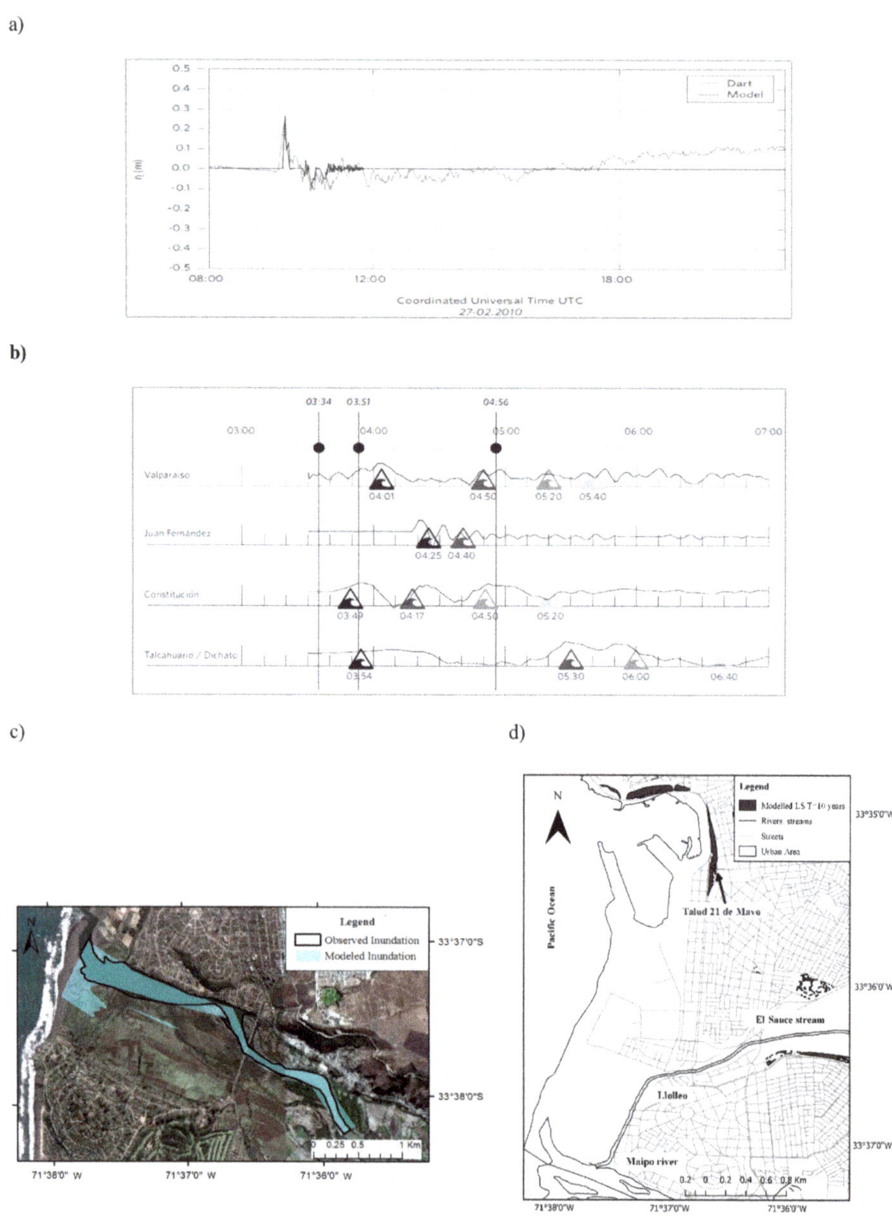

Figure 4. Validation of natural hazard models. (**a**) Validation of the SWAN model by comparing the far-field wave simulation to the DART buoy measures located in the coast of Perú. (**b**) Validation of the near-field SWAN model waves by comparing simulated times of arrival to different Chilean coastal cities against observations published in local media. (**c**) Validation of the flood inundation model using the flooded area retrieved from Google Earth V 6.2 for 8 October 2009. (**d**) Validation of the landslide model through the simulation of the landslide occurred on 28 May 2013 in Talud 21 de Mayo zone under the 5- to 10-year precipitation event.

3.1.2. Natural Hazards Model Results

Results of the regional tsunami propagation model toward the coastline of El Quisco and San Antonio indicate that, for the 8.6 Mw scenario, the first tsunami wave has a maximum amplitude of 1.5 m and reaches the coastline within the first hour of simulation. Simulations under the 8.8 Mw scenarios indicate that the first tsunami wave has a maximum amplitude of 4 m arriving to the Chilean coast within 30 min after the rupture. According to the simulations, inundation extents obtained from the local propagation model over El Quisco are relatively small under both the 8.8 Mw and 8.6 Mw scenarios. Only the beaches, the mouth of streams and some coastal streets are flooded in the model, with inundation depths of ~2 m and ~0.5 m under the 8.8 Mw and 8.6 Mw scenarios respectively (Figure 5a). On the other hand, larger inundations areas are obtained in San Antonio. The harbor, downtown, and some residential sectors located along the streams and the Maipo river are flooded due to the tsunami propagation upstream. As expected, a larger inundation extent is obtained under the 8.8 Mw scenario. The maximum water depths under the 8.8 Mw and 8.6 Mw scenarios are 5 m and 4 m respectively along the coast of San Antonio (Figure 5b).

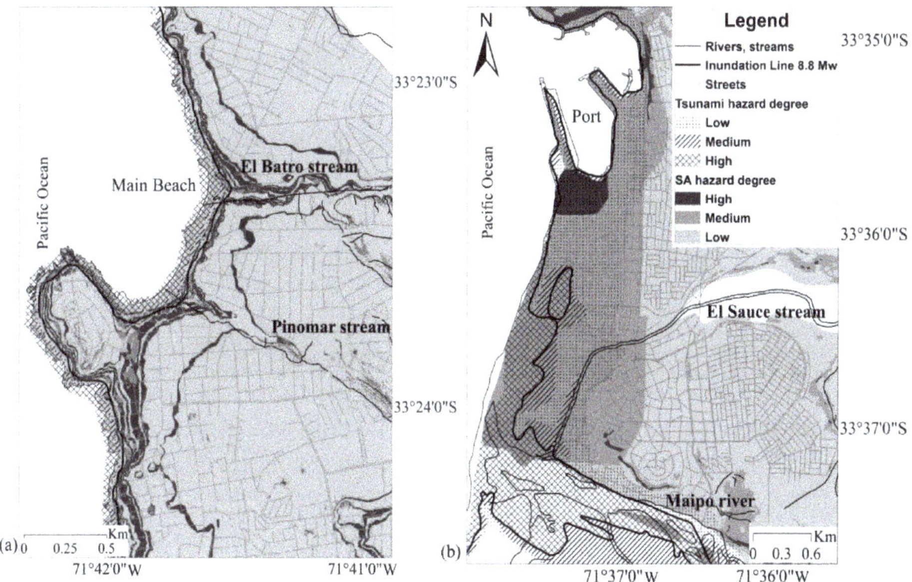

Figure 5. Integrated degree of hazard due to seismic wave amplification (SA) and tsunami inundation at (**a**) El Quisco and (**b**) San Antonio.

Regarding the fluvial flood modeling, simulated inundation areas due to fluvial floods are larger for San Antonio than for El Quisco (Figure 6a,b). Large flooded areas are simulated in the urban sectors bordering the streams and the Maipo river. Maximum velocities of ~5 m/s are reached under the 100-year flood. Urban areas adjacent to El Sauce stream and the Maipo river become flooded with the 2-year flood and the 10-year flood respectively, whereas considerable large flooded areas are obtained for events larger than the 25-year flood. Furthermore, inundations are simulated for flood events with a recurrence interval of 2-years or larger near the San Juan stream and in the crops yields located between this stream and the Maipo River. On the other hand, simulations at El Quisco show that the streams banks start overflowing in some urban sectors for events larger than the 2-year flood. Floods are simulated at the Cordova stream for events larger than the 5-year flood.

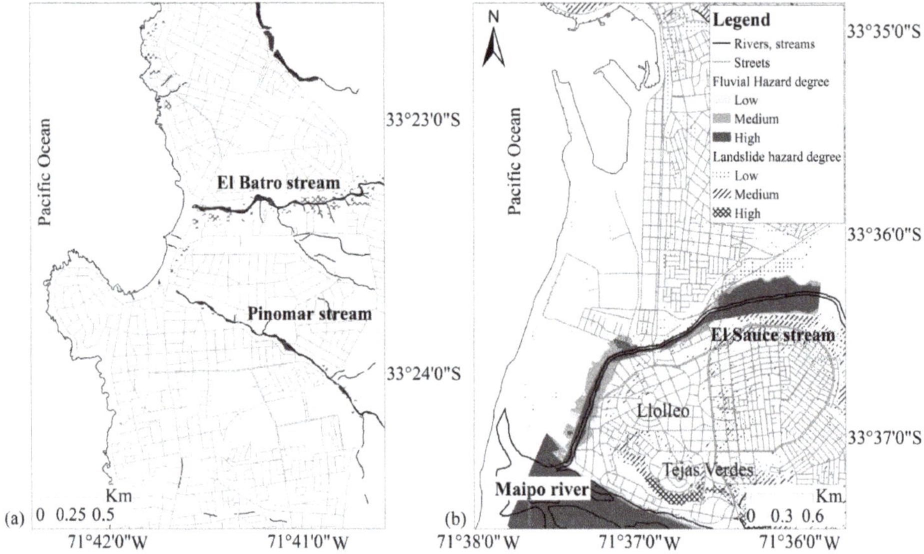

Figure 6. Integrated degree of hazard due to fluvial flood inundations and landslides triggered by 10-, 20-, 25-, and 50-year rainfall at (**a**) El Quisco and (**b**) San Antonio.

San Antonio is more likely to be affected by landslides than El Quisco (Figure 6a,b). Landslide events are prone to occur only in the north part of the main beach of El Quisco, which are expected to be initiated by a 5-year rainfall critical threshold. In the case of San Antonio, the simulations indicate that Llolleo, the Tejas Verdes sectors and the road to San Juan, all located in the southern part of the city, are prone to landslides triggered by a 5-year rainfall threshold. On the other hand, in downtown area landslide events can be initiated by a 50-year rainfall event.

Finally, the analysis of the seismic wave amplification shows that a large part of El Quisco is founded on soils type II which are associated with a low hazard degree, while San Antonio is majorly founded on soils type II and III (i.e., medium degree of hazard), and soils type IV (high degree of hazard) in some locations. Sectors with steep slopes in both cities are prone to be affected by topographic amplification, which increases the hazard degree (Figure 5a,b).

3.2. Vulnerability Evaluation

The housing elements at El Quisco and San Antonio have similar vulnerability to natural hazards (Figure 7a,b). According to the results, housing at El Quisco predominately has a medium-high vulnerability (65%) and medium-low (30%) vulnerability to tsunami and flood inundations respectively. Furthermore, housing in El Quisco has High vulnerability (95%) to landslides and low and medium-Low (50% and 48%, respectively) vulnerability to seismic amplification. The vulnerability of facilities at El Quisco ranges from Low to High considering different elements and hazards. Mostly medium-high to high vulnerability to landslides and medium-low and low vulnerability to seismic amplification is observed. Most vulnerable facilities are those related to health and the educational services, which are crucial as they provide support and shelter in the aftermath of natural disasters. In El Quisco these elements are located in areas prone to fluvial inundations.

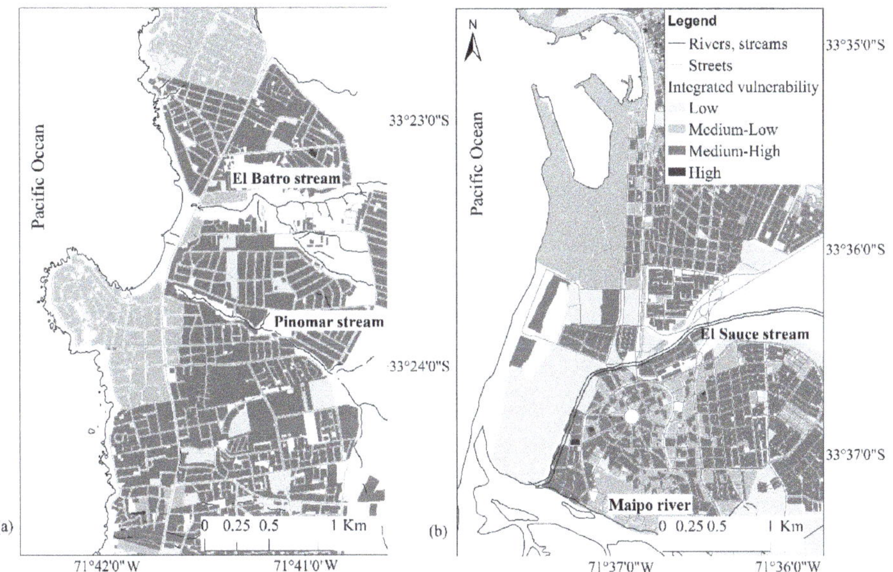

Figure 7. Integrated vulnerability of housing in (**a**) El Quisco and (**b**) San Antonio.

The evaluation of housing at San Antonio indicates that the city presents mostly medium-low (46%) and medium-high (28%) vulnerability to tsunami and flood inundations, high (94%) vulnerability to landslides processes and low and medium-low (62% and 33%, respectively) vulnerability to seismic amplification. Likewise, depending on the evaluated element, facilities have vulnerabilities ranging between high to low.

The vulnerability in both cities ranges from low to medium-high for electricity and telecommunication infrastructure, and from medium-high to high for water supply systems. Major differences are observed for road networks, as El Quisco has mostly unpaved roads and the vulnerability to inundations and to landslides ranges from high to medium-high. On the other hand, roads at San Antonio have a low vulnerability to tsunami and fluvial inundations.

3.3. Risk Evaluation

As indicated in Section 2.2.3, although is very unlikely that the four hazards here considered have the same probability of occurrence, with the purpose of informing and designing conservative mitigation plans, the local authority SUBDERE recommends aggregating the levels of risk of all the natural hazards. Despite that, the levels of risk of every hazard in an independently manner was also assessed (not showed).

Overall, San Antonio is more at risk to natural hazards than El Quisco when considering housing and facilities, but both cities have a similar risk with respect to infrastructure elements. Because of the low exposure of houses at El Quisco, they have a low risk to tsunami and flood inundations and landslides processes. Regarding facilities, only little harbors have a medium-high risk to inundations (Figure 8a). Furthermore, housing and facilities risk to seismic amplification waves range between low to medium-low and between medium-low to medium-high, respectively.

Housing risk to tsunamis at San Antonio ranges between medium-low to medium-high. In particular, a high risk is identified in the Llolleo lagoon sector (North of the Maipo estuary, Figure 8b), where the inundation depth exceeds 2.5 m and houses are highly vulnerable to the threat. The San Antonio housing sector has all the levels of risk to fluvial floods depending on the location. Because of the high vulnerability and the high degree of hazard to inundation of the sectors bordering the streams in San Antonio, they have high level of risk. On the other hand, the housing sector is at medium-low to low risk to landslides, except for the Tejas Verdes and San Juan locations, where a high risk is identified due to their High exposure to the hazard. Regarding facilities, a medium-high risk to tsunami is observed in the city, and only around the Llolleo area a high risk to fluvial flood is detected. A high, medium-low, and low risk to landslide events is identified for the facility sector, whereas a medium-low, medium-high, and high risk to seismic amplification is detected in different zones of the city. Finally, no risk to inundations or landslides is detected for the infrastructure in both cities, as there is no exposure. On the contrary, almost all the infrastructure sector is at high risk to seismic amplification in both cities.

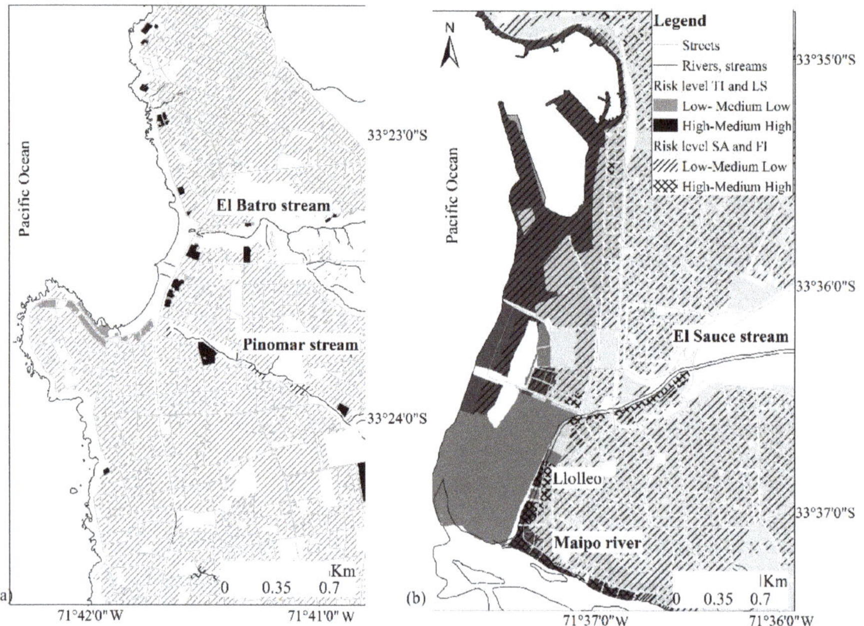

Figure 8. Housing integrated level of risk in (**a**) El Quisco and (**b**) San Antonio. The figures consider the Tsunami inundation (TI), landslide (LS), seismic amplification (SA), and fluvial inundation (FI) risk levels.

3.4. Multi-Risk Analysis

The threats analyzed in this study can potentially occur simultaneously, increasing the risk to natural hazards in the region. For instance, a large earthquake in the coastal sector can trigger a tsunami, or a strong rainfall event might cause floods and landslides simultaneously. The results for exposure to seismic amplification and tsunami hazards are presented in Figure 5a,b. El Quisco has a low degree of exposure to tsunami and seismic wave amplification, excepting some regions like the coastal line (Figure 5a). Figure 5b shows that the port and the coastline of San Antonio is highly exposed to these natural hazards. Despite half of the population is in areas with a low degree of exposure to these hazards, large urban sectors are located in areas with a medium degree of exposure to the combined effect of tsunami and seismic wave amplification. On the other hand, the aggregated analysis of landslides and fluvial flood hazards is presented in Figure 6. High exposure to fluvial flood and landslide is identified along the streams at El Quisco; however, no urban areas are involved (Figure 6a). On the contrary, significant urban areas are affected by one of the two natural hazards in San Antonio (Figure 6b); the sector bordering the Maipo river and streams are highly exposed to landslide and fluvial flood hazards.

Regarding a crossed vulnerability analysis, both cities have a medium-high housing (Figure 7) and facility vulnerability. The assessment of the multi-hazard risk for housing (Figure 8) indicates that risk is mainly low and medium-low in both cities. A larger risk is detected in some sectors of San Antonio, because of the large exposure to floods and landslide previously discussed. Despite the high facility vulnerability, facilities in El Quisco are at low risk, as they are not exposed to floods or landslides (Figure S1, Supplementary Material). Conversely, the facilities in San Antonio are at high risk because of their high exposure to these hazards. Out of the possible risks, the seismic wave amplification risk is the highest in both cities due to the high vulnerability to this hazard.

Overall, El Quisco is better prepared than San Antonio to deal with natural hazards (Figure 8). However, some of the low risk areas correspond to uninhabited places such as location 1 in Figure S2 (Supplementary Material), where future urban development might increase the risk to natural hazards. Finally, note that the risk analysis is carried out at a block (i.e., polygon) scale, while the exposure to the hazard is calculated at the grid-cell level. Thus, every block is given the highest risk computed for each of the grid-cells belonging to that block.

3.5. Mitigation and Evacuation Plans

The results of the multi-risk assessment show that integrated risk management plans to reduce life and properties losses and to increase resilience, are to be undertaken urgently in the study area. Although the risk zoning presented in this paper can be used to design and evaluate engineering measures to alleviate the risk, this is out of the scope of this research, which mostly focuses on characterizing the current risk state. This characterization provides the basis for the subsequent planning and assessment of mitigation measures. Recommendations from the literature [62,84] are adapted to the local conditions to design a risk mitigation plan, which considers the following main guidance:

- Relocation of the population living nearby the Llolleo lagoons (estuary of Maipo river) in San Antonio. Such action will enhance the development of dune systems, creating a buffer zone against tsunamis.
- Local and regional impact assessment and definition of mitigation infrastructure in case that any expansion of the San Antonio port is planned. Furthermore, a safe location of the stacking areas must be ensured, along with a contingency plan to control the damage caused by ships that can be dragged during tsunamis events.
- Restriction to new developments near streams and the Maipo River. In addition, mitigation infrastructure should be considered to protect existing constructions in the area. Special attention should be drawn to south Llolleo, which is highly exposed to landslides events.
- More control over the extraction of aggregates from the Cordova stream in El Quisco to reduce flood and tsunami risks.

In addition to these measures, the development of evacuation plans for both cities to cope with tsunamis becomes essential [85]. Using the risk zoning previously identified along with the topography and the existent road infrastructure, we identify evacuation routes and characterize traveling times to safe zones for both cities (Figure 9). For this purpose, four steps are considered:

1. Construction and updating of the information of road axes, including the road impedance and traveling times in each road.
2. Definition of safe areas and evacuation times under the worst modeled scenario (i.e., an 8.8 Mw earthquake and tsunami with critical evacuation times of 10, 15, and 60 min [86].
3. Definition of evacuation roads based on the urban layout, the relief and the safe zones in the city, which were located in open and flat areas able to receive a large number of people [87].
4. Identification of evacuation roads with slopes lower than 12%, in which stairs must be avoided. Areas with evacuation times over 15 min must be considered as uninhabitable zones in the urban design of the cities.

Overall, El Quisco is better prepared than San Antonio for an evacuation due to Tsunami, as the majority of the most distant locations to the safe zone are not residential (Figure 9). Nevertheless, alternative routes to the main roads and clear signals indicating the evacuation roads must be considered for El Quisco in order to facilitate the possible evacuation of summer floating population.

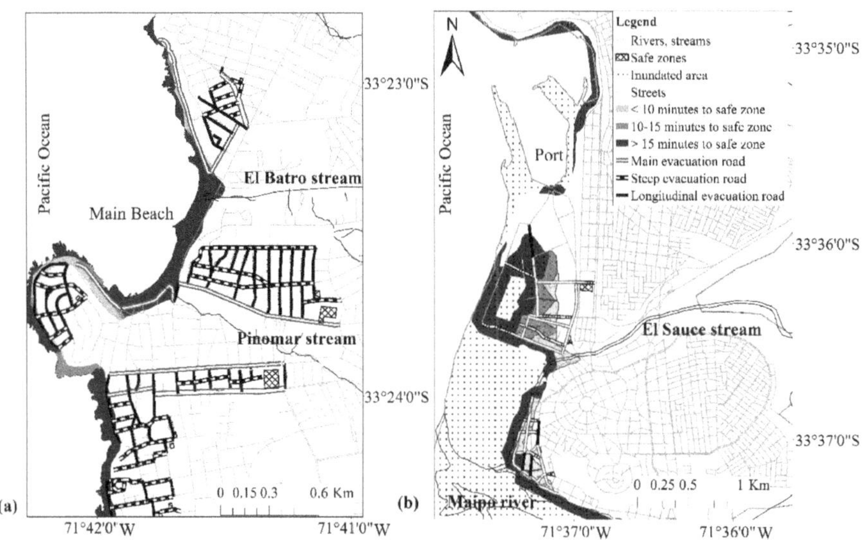

Figure 9. Tsunami evacuation plan at (**a**) El Quisco and (**b**) San Antonio.

4. Conclusions

This study successfully performs a simplified multi-risk evaluation of areas prone to be affected by natural hazard in El Quisco and San Antonio, two coastal cities in Central Chile. The natural hazards considered include tsunamis, fluvial floods, landslides, and seismic amplification, with the aim of supporting urban planning instruments and mitigation plans to increase resilience. Using simple calculations based on available data, a cross-analysis of multiple hazards and the multi-vulnerability estimation of housing, facilities, and infrastructure from a physical and socio-economical perspective was conducted. Adhering to [1,14], a matrix-based multi-risk methodology was designed that considers the local needs incorporated through the experience and opinion of the regional stakeholders. Such approach increases local level engagement in building and implementing disaster risk reduction and resilience thinking plans. Finally, mitigation measures for both cities were recommended based on the results of the assessment. Although the methodology implemented in this study was based on local-level data and stakeholder appreciation, we hope the results can be extended to other coastal cities and regions in Chile and other areas prone to be affected by the hazards here considered. The main conclusions of the study are:

1. Despite the short distance between the cities and similar vulnerability to the evaluated natural hazards, their local characteristics produce very different levels of risk. This highlights the importance of local characteristics when assessing the risk and developing suitable mitigation plans.
2. Disaster risk management plans should consider multi-risk methodologies instead of a single hazard approach. The multi-risk approach demonstrates that exposure to hazards and the risk increase when multiple hazards are considered.
3. Evacuation plans for tsunamis are particularly important for these cities. Plans should consider clear and safe routes that can be used during summer, when floating population increases the vulnerability to natural disaster. Another relevant aspect is the development of urban planning tools that restrict future urban development near streams to reduce flood and landslide risks.

The lack of extensive and detailed data is an important shortcoming when implementing physically based hazard models. Thus, future assessments should consider a probabilistic approach such as the FEMA P-58 frame [88] to account for uncertainties in the calculations and improve the characterization and evaluation of hazards and vulnerability. Furthermore, seismic instrumentation, detailed soil characterization, and bathymetric profiles of streams and the shore should also be incorporated. On the other hand, if the qualitative information used to characterize the vulnerability is improved, the methodology here presented should be updated to consider both the physical and the socio-economical dimensions of vulnerability for every component under assessment.

Despite risk management in Chilean coastal cities is critical, the country lacks a methodology and criteria to assess and manage risk. Next stages in Chilean risk management should focus on (1) the consolidation of a methodology like the one presented here, which considers a probabilistic and cascade effect of the multi-hazard and multi-vulnerability; and on (2) the implementation of risk mitigation plans oriented to exposure and vulnerability reduction, to be prepared for the next disaster. These plans should be constantly evaluated and updated as new natural disasters take place and new data are available.

Supplementary Materials: The following are available online at http://www.mdpi.com/2073-4441/11/3/572/s1, Figure S1: Aggregated natural disaster risk in facilities at (**a**) El Quisco and (**b**) San Antonio. Figure S2: (**a**) Degree of hazard, (**b**) Vulnerability, and (**c**) Risk evaluation in a highly exposed to tsunami events sector at San Antonio.

Author Contributions: Conceptualization and formal analysis, P.B., R.M., R.C., J.G., C.E., C.B., and C.L.; Methodology and Software, M.L.C., M.G., R.R., P.B., and A.T.; Writing—original draft preparation, P.B. and M.L.C.; Writing—review and editing, P.B., R.M., R.C., J.G., C.E., C.B., C.L., R.R., and M.G.; Visualization, P.B.; Project administration, R.M. and A.T.; Funding acquisition, R.M., R.C., J.G., C.E., C.B., and C.L.

Funding: This research was funded by Conicyt/Fondap grant no. 15110017.

Acknowledgments: Support from Conicyt/Fondap 15110020, Cities of San Antonio and El Quisco and MINVU are also acknowledged.

Conflicts of Interest: The authors declare no conflict of interest.

References

1. Field, C.B.; Barros, V.; Stocker, T.F.; Dahe, Q. *Managing the Risks of Extreme Events and Disasters to Advance Climate Change Adaptation: Special Report of the Intergovernmental Panel on Climate Change*; Cambridge University Press: Cambridge, UK, 2012.

2. Neumann, B.; Vafeidis, A.T.; Zimmermann, J.; Nicholls, R.J. Future Coastal Population Growth and Exposure to Sea-Level Rise and Coastal Flooding-a Global Assessment. *PLoS ONE* **2015**, *10*, e0118571. [CrossRef] [PubMed]

3. Stocker, T.F.; Qin, D.; Plattner, G.-K.; Tignor, M.; Allen, S.K.; Boschung, J.; Nauels, A.; Xia, Y.; Bex, V.; Midgley, P.M. *Climate Change 2013: The Physical Science Basis. Contribution of Working Group I to the Fifth Assessment Report of the Intergovernmental Panel on Climate Change (AR5)*; Cambridge Univ Press: New York, NY, USA, 2013.

4. Álvarez, G.; Quiroz, M.; León, J.; Cienfuegos, R. Identification and Classification of Urban Micro-Vulnerabilities in Tsunami Evacuation Routes for the City of Iquique, Chile. *Nat. Hazards Earth Syst. Sci.* **2018**, *18*, 2027–2039. [CrossRef]

5. Crichton, D. The Risk Triangle. In *Natural Disaster Management*; Tudor Rose: London, UK, 1999; pp. 102–103.

6. ISDR. *Living with Risk: A Global Review of Disaster Reduction Initiatives*; United Nations Publications, International Strategy for Disaster Reduction: Geneva, Switzerland, 2004; Volume 1.

7. UNISDR. *Terminology: Basic Terms of Disaster Risk Reduction*; UNISDR: Santiago, Chile, 2009.

8. Camarasa-Belmonte, A.M.; Soriano-García, J. Flood Risk Assessment and Mapping in Peri-Urban Mediterranean Environments Using Hydrogeomorphology. Application to Ephemeral Streams in the Valencia Region (Eastern Spain). *Landsc. Urban Plan.* **2012**, *104*, 189–200. [CrossRef]

9. Sepúlveda, S.A.; Rebolledo, S.; McPhee, J.; Lara, M.; Cartes, M.; Rubio, E.; Silva, D.; Correia, N.; Vásquez, J.P. Catastrophic, Rainfall-Induced Debris Flows in Andean Villages of Tarapacá, Atacama Desert, Northern Chile. *Landslides* **2014**, *11*, 481–491. [CrossRef]

10. Chen, H.X.; Zhang, S.; Peng, M.; Zhang, L.M. A Physically-Based Multi-Hazard Risk Assessment Platform for Regional Rainfall-Induced Slope Failures and Debris Flows. *Eng. Geol.* **2016**, *203*, 15–29. [CrossRef]

11. Villagra, P.; Herrmann, G.; Quintana, C.; Sepúlveda, R.D. Resilience Thinking and Urban Planning in a Coastal Environment at Risks of Tsunamis: The Case Study of Mehuín, Chile. *Rev. Geogr. Norte Gd.* **2016**, *82*, 63–82. [CrossRef]

12. Gallina, V.; Torresan, S.; Critto, A.; Sperotto, A.; Glade, T.; Marcomini, A. A Review of Multi-Risk Methodologies for Natural Hazards: Consequences and Challenges for a Climate Change Impact Assessment. *J. Environ. Manag.* **2016**, *168*, 123–132. [CrossRef] [PubMed]

13. Fleming, K.M.; Zschau, J.; Gasparini, P.; Modaressi, H.; Consortium, M. New Multi-Hazard and MulTi-RIsk Assessment MethodS for Europe (MATRIX): A Research Program towards Mitigating Multiple Hazards and Risks in Europe. In *AGU Fall Meeting Abstracts*; American Geophysical Union: San Francisco, CA, USA, 2011.

14. Carpignano, A.; Golia, E.; Di Mauro, C.; Bouchon, S.; Nordvik, J. A Methodological Approach for the Definition of Multi-risk Maps at Regional Level: First Application. *J. Risk Res.* **2009**, *12*, 513–534. [CrossRef]

15. CNID-CREDEN. *Hacia Un Chile Resiliente Frente a Desastres: Una Oportunidad*; CNID-CREDEN: Santiago, Chile, 2016.

16. Fundacion Chile. *Desafíos Del Agua Para La Región Latinoamericana*; Fundacion Chile: Santiago, Chile, 2017.

17. Camus, P.; Arenas, F.; Lagos, M.; Romero, A. Visión Histórica de La Respuesta a Las Amenazas Naturales En Chile y Oportunidades de Gestión Del Riesgo de Desastre. *Rev. Geogr. Norte Gd.* **2016**, *64*, 9–20. [CrossRef]

18. UNISDR AM. *Diagnóstico de La Situación de La Reducción Del Riesgo de Desastres Naturales En Chile*; UNISDR AM: Geneva, Switzerland, 2010.

19. ONEMI. *Política Nacional Para La Gestión de Riesgo de Desastres*; ONEMI: Santiago, Chile, 2014.

20. USGS. *Magnitude 8.8 – Offshore Maule, Chile, February 27, 2010*; USGS: Reston, VA, USA, 2010.

21. Ministerio del Interior. *Balance de Reconstrucción*; Ministerio del Interior: Santiago, Chile, 2011.

22. Fariña, L.M.; Opaso, C.; Vera-Puz, P. *Impactos Ambientales Del Terremoto y Tsunami En Chile. Las Réplicas Ocultas Del F, 27. Fund*; TERRAM: Santiago, Chile, 2012.

23. DOH. *Plan Maestro de Evacuación y Drenaje de Aguas Lluvias, San Antonio y Cartagena, V Región*; DOH: Santiago, Chile, 2003.

24. Sepúlveda, S.A.; Rebolledo, S.; Vargas, G. Recent Catastrophic Debris Flows in Chile: Geological Hazard, Climatic Relationships and Human Response. *Quat. Int.* **2006**, *158*, 83–95. [CrossRef]

25. Petley, D. Global Patterns of Loss of Life from Landslides. *Geology* **2012**, *40*, 927–930. [CrossRef]

26. MOP. *Plan Especial de Infraestructura MOP de Apoyo Al Turismo Sustentable a 2030*; MOP: Santiago, Chile, 2017.

27. DGA. *Balance Hídrico de Chile*; DGA: Santiago, Chile, 1987.

28. Gobernación Provincial de San Antonio. *Plan de Seguridad Provincia de San Antonio*; Gobernación Provincial de San Antonio: San Antonio, Chile, 2010.

29. IMSA. *Plan Regulador Comuna de San Antonio*; IMSA: San Antonio, Chile, 2006.

30. INE. *Censo de Población y Vivienda 2002*; INE: Santiago, Chile, 2002.

31. Butz, W.P.; Lutz, W.; Sendzimir, J. Education and Differential Vulnerability to Natural Disasters. *Ecol. Soc.* **2014**.

32. Muttarak, R.; Lutz, W. Is Education a Key to Reducing Vulnerability to Natural Disasters and Hence Unavoidable Climate Change? *Ecol. Soc.* **2014**, *19*, 42. [CrossRef]

33. Lagos López, M. *Tsunamis de Origen Cercano a Las Costas de Chile*; Pontificia Universidad Católica de Chile: Santiago, Chile, 2000.

34. Fritz, H.M.; Petroff, C.M.; Catalán, P.A.; Cienfuegos, R.; Winckler, P.; Kalligeris, N.; Weiss, R.; Barrientos, S.E.; Meneses, G.; Valderas-Bermejo, C. Field Survey of the 27 February 2010 Chile Tsunami. *Pure Appl. Geophys.* **2011**, *168*, 1989–2010. [CrossRef]

35. IMSA. *Efectos de La Ola Sísmica o Tsunami En La Laguna Llolleo*; IMSA: San Antonio, Chile, 2010.

36. Habersack, H.; Schober, B.; Hauer, C. Floodplain Evaluation Matrix (FEM): An Interdisciplinary Method for Evaluating River Floodplains in the Context of Integrated Flood Risk Management. *Nat. Hazards* **2015**, *75*, 5–32. [CrossRef]

37. Kantamaneni, K.; Gallagher, A.; Du, X. Assessing and Mapping Regional Coastal Vulnerability for Port Environments and Coastal Cities. *J. Coast. Conserv.* **2018**, *23*, 59–70. [CrossRef]

38. Fitton, J.M.; Hansom, J.D.; Rennie, A.F. A National Coastal Erosion Susceptibility Model for Scotland. *Ocean Coast. Manag.* **2016**, *132*, 80–89. [CrossRef]

39. Fitton, J.M.; Hansom, J.D.; Rennie, A.F. A Method for Modelling Coastal Erosion Risk: The Example of Scotland. *Nat. Hazards* **2018**, *91*, 931–961. [CrossRef]

40. Kourgialas, N.N.; Karatzas, G.P. A Hydro-Sedimentary Modeling System for Flash Flood Propagation and Hazard Estimation under Different Agricultural Practices. *Nat. Hazards Earth Syst. Sci.* **2014**, *14*, 625–634. [CrossRef]

41. Te Chow, V.; Maidment, D.R.; Mays, L.W. *Hidrología Aplicada*; McGraw-Hill: New York, NY, USA, 1994.

42. Snyder, F.F. Synthetic Unit-graphs. *Eos Trans. Am. Geophys. Union* **1938**, *19*, 447–454. [CrossRef]

43. Rossman, L.A.; Huber, W. *Storm Water Management Model Reference Manual Volume I—Hydrology (Revised)*; US Environ. Prot. Agency: Cincinnati, OH, USA, 2016.

44. Verni, F.; King Farias, H. *Estimación de Crecidas En Cuencas No Controladas*; Sociedad Chilena Ingenieria Hidraulica: Santiago, Chile, 1977; pp. 357–374.

45. DGA. *Manual de Cálculo de Crecidas y Caudales Mínimos En Cuencas Sin Información Fluviométrica*; DGA: Santiago, Chile, 1995.

46. Espey, W.; Altman, D.G.; Graves, C. *Nomographs for Ten-Minute Unit Hydrographs for Small Urban Watersheds*; Issue 32 of Technical Memorandum; American Society of Civil Engineers Urban Water Resources Council: Reston, VA, USA, 1977.

47. Viessman, W.; Lewis, G.L. *Introduction to Hydrology*, 5th ed.; Prentice Hall/Pearson Education: Upper Saddle River, NJ, USA, 2003.

48. NASA. *ASTER Global Digital Elevation Model [Data Set]*; NASA: Washington, DC, USA, 2009.

49. IMEQ. *Plan Municipal de Cultura Comuna de El Quisco 2014–2016*; IMEQ: El Quisco, Chile, 2014.

50. USACE. *HEC-GeoRAS GIS Tools for Support of HEC-RAS Using ArcGIS© User's Manual Version 4.3*; USACE: Washington, DC, USA, 2011.

51. USACE. *HEC RAS, River Analysis System User's Manual, v. 4.0*; USACE: Washington, DC, USA, 2008.

52. CIREN. *Estudio Agrológico V Región*; CIREN: Santiago, Chile, 1997.

53. OAS. *Primer on Natural Hazard Management in Integrated Regional Development Planning*; Organization of American States, Department of Regional Development and Environment ExecutiveSecretariat for Economic and Social Affairs: Washington, DC, USA, 1991.

54. Highland, L.; Bobrowsky, P.T. *The Landslide Handbook: A Guide to Understanding Landslides*; US Geological Survey Reston: Reston, VA, USA, 2008.

55. Sepúlveda, S.A.; Padilla, C. Rain-Induced Debris and Mudflow Triggering Factors Assessment in the Santiago Cordilleran Foothills, Central Chile. *Nat. Hazards* **2008**, *47*, 201–215. [CrossRef]

56. Guzzetti, F.; Peruccacci, S.; Rossi, M.; Stark, C.P. The Rainfall Intensity—Duration Control of Shallow Landslides and Debris Flows: An Update. *Landslides* **2008**, *5*, 3–17. [CrossRef]

57. Montgomery, D.R.; Dietrich, W.E. A Physically Based Model for the Topographic Control on Shallow Landsliding. *Water Resour. Res.* **1994**, *30*, 1153–1171. [CrossRef]

58. Guimaraes, R.F.; Montgomery, D.R.; Greenberg, H.M.; Fernandes, N.F.; Gomes, R.A.T.; de Carvalho Júnior, O.A. Parameterization of Soil Properties for a Model of Topographic Controls on Shallow Landsliding: Application to Rio de Janeiro. *Eng. Geol.* **2003**, *69*, 99–108. [CrossRef]

59. Gonzalez, C. *Estudio de Áreas de Riesgo Geomorfológico de La Zona Urbana y de Expansión de La Comuna de San Antonio*; Universidad de Chile: Santiago, Chile, 2005.

60. Brito, J.L. *San Antonio: Nuevas Crónicas Para Su Historia y Geografía*; Sales. Impr.: San Antonio, TX, USA, 2019.

61. Kawagoe, S.; Kazama, S.; Sarukkalige, P.R. Probabilistic Modelling of Rainfall Induced Landslide Hazard Assessment. *Hydrol. Earth Syst. Sci.* **2010**, *14*, 1047–1061. [CrossRef]

62. Heintz, J.; Mahoney, M. *Guidelines For Design Of Structures For Vertical Evacuation From Tsunamis*; FEMA: Washington, DC, USA, 2008.

63. Okada, Y. Surface Deformation Due to Shear and Tensile Faults in a Half-Space. *Bull. Seismol. Soc. Am.* **1985**, *75*, 1135–1154.

64. DIHA. *Propagación Regional de Tsunamis Basados En Eventos Históricos*; DIHA: Santiago, Chile, 2011.

65. PRDW-AV. *Estudio de La Propagación Regional de Tsunamis Basados En El Evento de 1730*; PRDW-AV: Santiago, Chile, 2011.

66. Mader, C.L. *Numerical Modeling of Water Waves*; CRC Press: Boca Ratón, FL, USA, 2004.

67. Marche, F. Derivation of a New Two-Dimensional Viscous Shallow Water Model with Varying Topography, Bottom Friction and Capillary Effects. *Eur. J. Mech.* **2007**, *26*, 49–63. [CrossRef]

68. Guerra, M.; Cienfuegos, R.; Escauriaza, C.; Marche, F.; Galaz, J. Modeling Rapid Flood Propagation over Natural Terrains Using a Well-Balanced Scheme. *J. Hydraul. Eng.* **2014**, *140*, 4014026. [CrossRef]

69. Cienfuegos, R.; Barthélemy, E.; Bonneton, P. A Fourth-order Compact Finite Volume Scheme for Fully Nonlinear and Weakly Dispersive Boussinesq-type Equations. Part II: Boundary Conditions and Validation. *Int. J. Numer. Methods Fluids* **2007**, *53*, 1423–1455. [CrossRef]

70. Kotani, M. Tsunami Run-up Simulation and Damage Estimation Using GIS. *Proc. Coast. Eng. JSCE* **1998**, *45*, 356–360.

71. Walsh, T.J.; Titov, V.V.; Venturato, A.J.; Mofjeld, H.O.; Gonzalez, F.I. Tsunami Hazard Map of the Elliott Bay Area, Seattle, Washington: Modeled Tsunami Inundation from a Seattle Fault Earthquake. *Wash. Div. Geol. Earth Resour. Open File Rep.* **2003**, *14*.

72. Johnson, L.R.; Silva, W. The Effects of Unconsolidated Sediments upon the Ground Motion during Local Earthquakes. *Bull. Seismol. Soc. Am.* **1981**, *71*, 127–142.

73. Pilz, M.; Parolai, S.; Picozzi, M.; Zschau, J. Evaluation of Proxies for Seismic Site Conditions in Large Urban Areas: The Example of Santiago de Chile. *Phys. Chem. Earth Parts A/B/C* **2011**, *36*, 1259–1266. [CrossRef]

74. Code, P. *Eurocode 8: Design of Structures for Earthquake Resistance-Part 1: General Rules, Seismic Actions and Rules for Buildings*; European Committee for Standardization: Brussels, Belgium, 2005.

75. Ciurean, R.L.; Schroter, D.; Glade, T. Conceptual Frameworks of Vulnerability Assessments for Natural Disasters Reduction. In *Approaches to Disaster Management-Examining the Implications of Hazards, Emergencies and Disasters*; InTech: London, UK, 2013.

76. Guillard-Gonçalves, C.; Zêzere, J. Combining Social Vulnerability and Physical Vulnerability to Analyse Landslide Risk at the Municipal Scale. *Geosciences* **2018**, *8*, 294. [CrossRef]

77. UNESCO-RAPCA. *Introduction to the UNESCO-RAPCA Project*; ITC: Enschede, The Netherlands, 2000.

78. Susman, P.; O'Keefe, P.; Wisner, B. Global Disasters, a Radical Interpretation. *Interpret. Calam.* **1983**, 263–283.

79. Hewitt, K. *Regions of Risk: A Geographical Introduction to Disasters*; Routledge: London, UK, 2014.

80. Thywissen, K. *Components of Risk: A Comparative Glossary*; UNU-EHS: Bonn, Germany, 2006.

81. SUBDERE. *Guía Análisis de Riesgos Naturales Para El Ordenamiento Territorial*; SUBDERE: Santiago, Chile, 2011.

82. FADEU-UC. *Estudio de Riesgo de Sismos y Maremoto Para Comunas Costeras de Las Regiones de O'Higgins y Del Maule*; FADEU-UC: Providencia, Chile, 2012.

83. Marzocchi, W.; Garcia-Aristizabal, A.; Gasparini, P.; Mastellone, M.L.; Di Ruocco, A. Basic Principles of Multi-Risk Assessment: A Case Study in Italy. *Nat. Hazards* **2012**, *62*, 551–573. [CrossRef]

84. ICPR. *Non Structural Flood Plain Management: Measures and Their Effectiveness*; International Commission for the Protection of the Rhine (ICPR): Koblenz, Germany, 2002.

85. Johnstone, W.M.; Lence, B.J. Use of Flood, Loss, and Evacuation Models to Assess Exposure and Improve a Community Tsunami Response Plan: Vancouver Island. *Nat. Hazards Rev.* **2011**, *13*, 162–171. [CrossRef]

86. Vargas, G.; Farías, M.; Carretier, S.; Tassara, A.; Baize, S.; Melnick, D. Coastal Uplift and Tsunami Effects Associated to the 2010 Mw8. 8 Maule Earthquake in Central Chile. *Andean Geol.* **2011**, *38*, 219–238.

87. UNESCO. *Preparación Para Casos de Tsunami. Guía Informativa Para Los Planificadores Especializados En Medidas de Contingencia Ante Catástrofes*; UNESCO: Paris, France, 2008.

88. FEMA. *Seismic Performance Assessment of Buildings (Volume 1—Methodology)*; FEMA P-58; Federal Emergency Management Agency: Washington, DC, USA, 2012.

Article

Impact of Urban Growth and Changes in Land Use on River Flood Hazard in Villahermosa, Tabasco (Mexico)

Omar S. Areu-Rangel [1], Luis Cea [2], Rosanna Bonasia [3,*] and Victor J. Espinosa-Echavarria [1]

[1] Departamento de Ciencias de la Tierra, Tecnológico Nacional de México/Instituto Tecnológico de Pachuca, Pachuca de Soto 42080, Hidalgo, Mexico; oareu@outlook.com (O.S.A.-R.); victor_j@live.com.mx (V.J.E.-E.)

[2] Department of Civil Engineering, Universidade da Coruña Environmental and Water Engineering Group, 15001 A Coruña, Spain; luis.cea@udc.es

[3] CONACYT—Instituto Politécnico Nacional, ESIA, UZ, Miguel Bernard, S/N, Edificio de Posgrado, Mexico City 07738, Mexico

* Correspondence: rosannabonasia017@gmail.com

Received: 29 December 2018; Accepted: 31 January 2019; Published: 12 February 2019

Abstract: The city of Villahermosa, a logistical center in the State of Tabasco's economy, is affected by recurrent river floods. In this study, we analyzed the impact of two factors that are the most probable causes of this increase in flood hazard: changes in land use in the hydrological catchments upstream of the city, and the uncontrolled urbanization of the floodplains adjacent to the main river channels. Flood discharges for different return periods were evaluated, considering land uses of the catchments, both as they were in 1992 and as they are today. These flood discharges were then used in a 2D shallow water model to estimate the increase of water depths in the city from 1992 to the present day. To evaluate the influence of urban expansion on inundation levels, three future urbanization scenarios were proposed on the basis of the urban growth rate forecast for 2050. Results confirm that the change in land use in the hydrological catchments is the main factor that explains the increase in inundation events observed over recent years. This study also provides useful insights for future city planning that might help to minimize the flood impact on Villahermosa.

Keywords: flood hazard; land use; urban growth; Villahermosa

1. Introduction

River floods due to excessive rainfall–runoff are responsible for recurrent damage to basins where soil surface characteristics have been modified as a consequence of deforestation and urban growth. A decrease in the connectivity of river networks and the loss of fluvial floodplains, in combination with the conversion of agriculture land to less permeable urban surfaces, can significantly reduce the basin's response time during storm events, which in turn increases peak discharges and the risk of flooding.

The State of Tabasco (Mexico) recurrently experiences inundation events throughout its territory. It is located within the delta of two main rivers: the Grijalva and the Usumacinta. These rivers account for approximately 30% of the total runoff in Mexico, and converge in the same system along their course towards the Gulf of Mexico, about 40 km downstream of the city of Villahermosa. A lack of environmental planning has drastically modified the characteristics of the terrain of this state over recent decades. The elimination of vegetable cover has led to a loss of soils and a reduction in infiltration capacity, causing higher volumes of surface runoff, silting, and hillslope erosion. In the whole territory of Mexico, from 1970, there have been substantial changes in land use, which have drastically increased the number of floods [1]. The rapid transformation of the territory has led to significant loss of vegetation [2] that, together with an increase in the frequency of extreme weather phenomena, has increased the flood risk throughout the country. Zúñiga and Magaña [3] analyzed the frequency of floods in Mexico by comparing observed and modeled frequency values for the period 1970–2010, finding that most flood events had occurred as a result of the rapid process of deforestation that affected the basins of the main rivers of the country. Results of this study show that the regions experiencing substantial increases in flood frequency over recent years are located mainly along the coasts, although the authors also stressed that the spatial distribution of floods depends to a great extent on the vulnerability of the regions in which rainfall occurs. In the specific case of the State of Tabasco, Aparicio et al. [4] pointed out that the change of land use from forest to agriculture leads to the transport of large quantities of sediments due to the increased erosion of basins, and they found that this can be considered to be one of the major causes of disasters in the region. The grass cover of the basins increases the roughness of the soil and reduces the runoff coefficient [5], which determines the magnitude of the discharges downstream. Grassland can significantly reduce runoff, compared to bare and agriculture land [6].

Over the last 40 years, the construction of housing complexes and containment bridges within Villahermosa have not only modified the landscape, but have also caused a phenomenon of irregular settlements on the margins of the same embankments. This process is expected to continue in the years to come: for the next 30 years, the city's growth rate is expected to be greater than 0.50%. In addition, the exposure of the population to river flooding has dramatically increased in recent years due the presence of hurricanes in the coastal areas of the Pacific and Atlantic oceans, which produce heavy rainfall, leading to the transportation and deposition of large quantities of silts in the Tabasco plain [7]. From 1940 to 1990, the State of Tabasco lost 97% of its forest resources due to excessive exploitation of forests, causing wind and water erosion to affect almost 50% of the territory. In October 2007, Tabasco was hit by an intense flooding event. Extraordinary rains in the Grijalva basin generated intense runoff throughout the Tabasco plain, flooding about 70% of the state. The capital city, Villahermosa, was especially badly hit, suffering economic losses of more than three billion dollars. From 2008 to 2011, the city was continuously affected by the flooding of the rivers that traverse it [8].

In recent decades, the analysis of floods in urban areas has improved considerably with the introduction of accurate numerical models [9–11] and advanced methodologies, which aim in particular at the combination and integration of land use change forecasting models with hydrological models, with the objective of being able to determine increasingly precise peak discharge rates as a function of the predicted urbanization [12,13]. In particular, Wang et al. [13] presented the application of a cellular automata based model (Celular Automata Dual-DraInagE Simulation, CADDIES) to a small study area to analyze flood inundation, and demonstrate the importance of identifying and using key urban features, obtained from terrain data, to better reproduce high resolution flood processes. The effect of urbanization and changes in land use on floods has been studied from different perspectives, focusing both on river feeding catchments and on potentially floodable areas [14–17], with the conclusion that an appropriate study of land use and consequent management are crucial in flood attenuation.

This study analyzed two of the main causes of the increase in flood frequency and magnitude in Villahermosa: changes in land use in the hydrological basins located upstream of the city, and the expansion of the urban area in the city.

In the present study, we first analyzed the effect of the changes in land use of the catchments that drain to Villahermosa. Flood discharges for different return periods were calculated using both the oldest information on the basin's land use (up until 1992) and the most recent (2013). The scenarios obtained were numerically simulated with the 2D flood inundation model Iber [18]. Secondly, the effect of the increase in the area occupied by buildings in the city was analyzed by comparing the inundation depths for three possible future urbanization scenarios.

The remainder of this paper is organized as follows. Section 2 presents the estimation of flood discharges for different return periods, considering both the land uses of the basins as they were in 1992 and the present ones. The inundation model used to evaluate the flood depths in the city for past and present scenarios is described in Section 3. The results of the numerical simulations are presented and discussed in Section 4. The influence of changes in land use (from 1992 to the present day) on inundation levels in the city was quantified and analyzed. Then, three possible future urbanization scenarios were proposed, and their effects on inundation hazard were analyzed.

2. Estimation of Flood Discharges

The city of Villahermosa is found in the municipality of Centro and is located within the hydrological zone of Grijalva-Usumacinta [1], which has a territorial extension of 102,456 km^2. This hydrological zone is divided into 83 basins, eight of which drain into the city of Villahermosa. Four rivers are responsible for the flood events in the urban area of the city: the Carrizal, the Viejo Mezcalapa, the Pichucalco, and the de la Sierra rivers. Figure 1 shows the catchments of these rivers, while their hydrological characteristics are listed in Table 1.

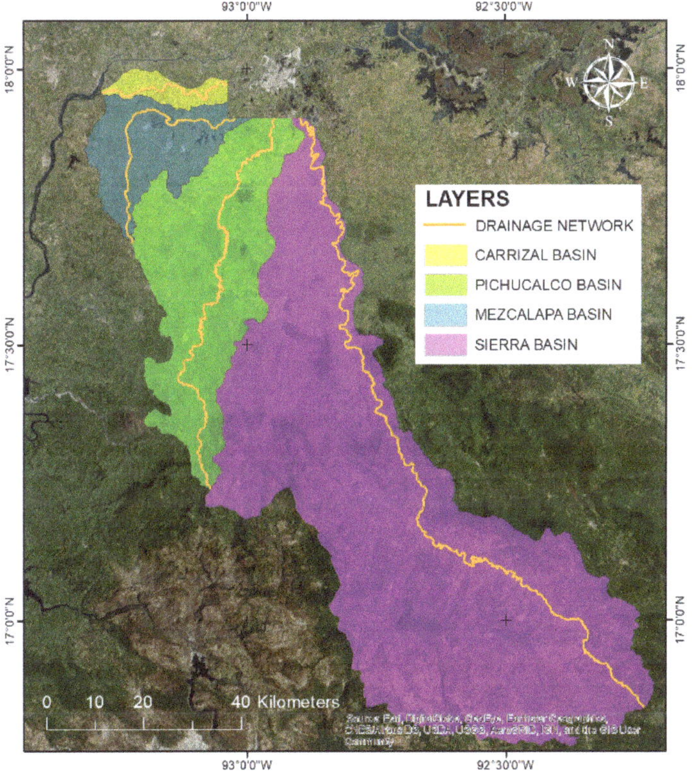

Figure 1. Map of the basins that drain into Villahermosa.

Table 1. Hydrological characteristics of the main rivers that drain into Villahermosa.

	Carrizal	V. Mezcalapa	Pichucalco	de la Sierra
Basin area (km^2)	133.71	566.13	1314.23	1073.57
Length of the main channel (m)	51,524	98,544	155,134	145,102
Minimum elevation of the main channel (m)	8	7	5	4
Average elevation of the main channel (m)	12	33	1071	1188
Maximum elevation of the main channel (m)	17	60	2137	2373
Average slope of the main channel (%)	0.017	0.054	1.37	1.63

Flood discharges of each river for different return periods were estimated with Chow's method [19], including two adjustment factors to account for the spatial and temporal variability of extreme rainfall in large catchments. The flood discharge for a given return period is calculated as:

$$Q_d = 2.78 \cdot A \cdot X \cdot Z \cdot K \cdot KA \tag{1}$$

where A is the basin area (km^2), X is the runoff factor (cm/h), Z is the peak reduction factor, K is a correction coefficient to account for the temporal variability of extreme rainfall, and KA is an areal reduction factor that relates extreme area-averaged rainfall with extreme point rainfall. Factor Z relates to the morphological parameters of each basin, and is calculated according to a relationship between the concentration time and the retrace time with the support of a nomogram [19]. For the Carrizal and Mezcalapa basins, this was estimated to 1, and for the Pichucalco and De La Sierra basins to 0.95. The runoff factor X is calculated as:

$$X = \frac{P_e}{T_c} \tag{2}$$

where P_e is the excess precipitation depth (given in cm) for a given storm duration and return period, and T_c is the catchment concentration time (given in hours). The concentration time was calculated as the average of the values obtained with Kirpich's formula [20].

The coefficients K and KA are given as:

$$K = 1 + \frac{T_c^{1.25}}{T_c^{1.25} + 14} \tag{3}$$

$$KA = 1 - \left(\frac{\log_{10}(A)}{15} \right) \tag{4}$$

where the concentration time (T_c) is given in hours and the area of the catchment (A) in km^2.

Following the recommendations of the Soil Conservation Service, the excess precipitation depth was computed from the total precipitation by means of the Curve Number (N), which is based mainly on land use:

$$P_e = \frac{(P - \frac{508}{N} + 5.08)^2}{P + \frac{2032}{N} - 20.32} \tag{5}$$

where P is the total precipitation depth associated with a given return period, given in cm. The precipitation depths (P) were obtained from the rainfall database of the Secretary of Communications and Transportation (SCT, Mexico).

Information on the land uses of the catchments under study over recent years was provided by the National Institute of Statistics and Geography (INEGI) [1]. Geospatial Information provided by INEGI [1] shows the distribution of agricultural land use, natural land, and induced vegetation in the country. It also indicates livestock and forestry uses and other uses that occur in the territory related to vegetation cover. The use of agricultural land is represented according to the availability of water for different types of crops during their agricultural cycle. The vegetation is represented in accordance with the provisions of the Guidelines for the use of the Catalog of Types of Natural and Induced Vegetation of Mexico for statistical and geographic purposes. INEGI's information comprises five series that provide the spatial distribution of land use until 1992 (Series I), from 1993 to 1996 (Series II), from 2000 to 2004 (Series III), from 2007 to 2010 (Series IV) and from 2011 to 2013 (Series V). From the analysis of these data, it was observed that, from 1992 to 2013, the terrains located upstream of Villahermosa mainly changed from pasture cover to agriculture cover. Figure 2 shows thematic maps of land use that have been reconstructed from all the available land use data. Table 2 shows the correspondence of land use types with their curve numbers, as well as the medium curve numbers for Series I and V.

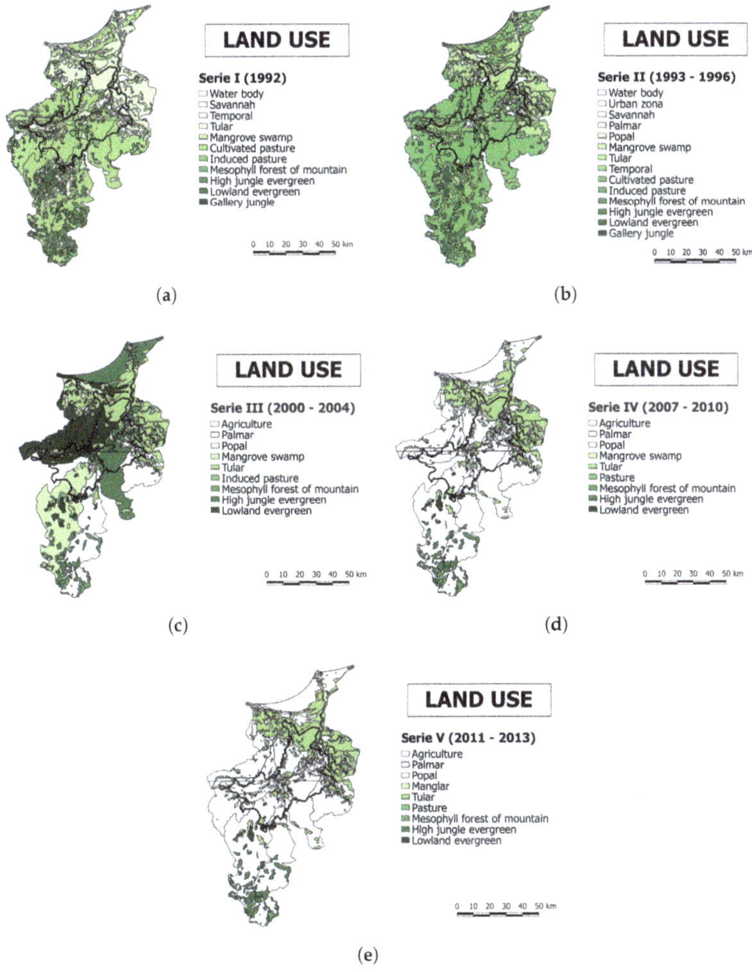

Figure 2. Thematic maps of land use distribution in the basins located upstream of Villahermosa: (**a**) land use of Series I (until 1992); (**b**) land use of Series II (1993–1996); (**c**) land use of Series III (2000–2004); (**d**) land use of Series II (2007–2010); and (**e**) land use of Series II (2011–2013) [1].

Table 2. Correspondence of land use types with their curve number, for SI and SV.

Land Use	Curve Number (N)
Series I	
Normal forest with medium transpiration	76
Forest with low transpiration	84
Pastureland	80
Average	80
Series V	
Forest with low transpiration	91
Agriculture in flat areas	90
Poor pastureland	89
Little permeable urban area	92
Average	91

To analyze the impact that changes in land use have on the inundation levels, the flood discharges for the return periods of 10, 25, 50 and 100 years were calculated using the curve numbers (N) corresponding to the land uses of Series I (1992) and Series V (2011–2013). Table 3 summarizes the parameters and coefficients used in Equation (1). The effective precipitation depths and flood discharges calculated for each return period are shown in Table 4.

Table 3. Parameters used to calculate flood discharges from Equation (1).

	Carrizal	V. Mezcalapa	Pichucalco	de la Sierra
Average N—Series I	80	80	80	80
Average N—Series V	91	91	91	91
Tc (h)	38.7	41.3	16.8	14.9
K	1.87	1.88	1.71	1.68
KA	0.86	0.82	0.79	0.76

Table 4. Excess precipitation depths and flood discharges estimated for every basin and each return period. Flood discharges are estimated for Series I (SI) and Series V (SV) land use in the catchments.

	Return Period (years)	SI Pe (mm)	SI Q_d (m³/s)	SV Pe (mm)	SV Q_d (m³/s)
Carrizal	T = 5 years	12.6	227	15.8	284.9
	T = 10 years	13.3	240	16.6	298.6
	T = 25 years	22.7	409	26.4	474.5
	T = 50 years	23.4	421	27.1	486.7
	T = 100 years	30.2	544	34.1	612.7
V. Mezcalapa	T = 5 years	12.9	924	16.1	1155.5
	T = 10 years	13.6	977	16.9	1210.8
	T = 25 years	23.2	1661	26.8	1922.8
	T = 50 years	23.8	1710	27.5	1972.5
	T = 100 years	30.8	2209	34.6	2482.4
Pichucalco	T = 5 years	3.1	1070	5.1	1793.7
	T = 10 years	3.3	1152	5.4	1895.3
	T = 25 years	6.5	2285	9.1	3225.0
	T = 50 years	6.7	2369	9.4	3318.9
	T = 100 years	9.2	3245	12.1	4287.3
de la Sierra	T = 5 years	2.2	2623	4.1	4763.4
	T = 10 years	2.4	2838	4.3	5044.2
	T = 25 years	5	5475	7.5	8745.8
	T = 50 years	5.2	6101	7.7	9008.3
	T = 100 years	7.2	8492	10.0	11,720.5

3. Flood Inundation Model

The inundation scenarios in Villahermosa were modeled with the software Iber [18]. This numerical model solves the 2D shallow water equations to compute the spatial distribution of the water depth and the two horizontal components of the depth-averaged velocity. The shallow water equations were solved with an explicit unstructured finite volume solver. The performance of the software Iber could not be validated in the study area because there are no available historical data to do so. However, the model has been extensively validated and applied in previous studies related to river inundation, tidal currents in estuaries, and rainfall–runoff modeling [21–27], showing its ability to represent 2D free surface shallow flows and river inundation processes. The 2D shallow water equations solved by the software Iber are a high-fidelity physically-based model, and they were assumed to be the best representation of the inundation process for the purposes of this work. The only parameter to calibrate in the equations is the Manning coefficient. Plausible values of this parameter can be established from

River Engineering Manuals, considering the land uses in the study area. Those values are expected to give a good approximation of the inundation extent [28,29].

Figure 3 shows the location map of the area under study. In the figure, the city of Villahermosa plus water networks and water bodies are presented. The computation domain, which is delimited by a red line, extends over an area of 290 km^2. It encompasses the entire city and the rural areas surrounding it that have been most affected by historical inundation events.

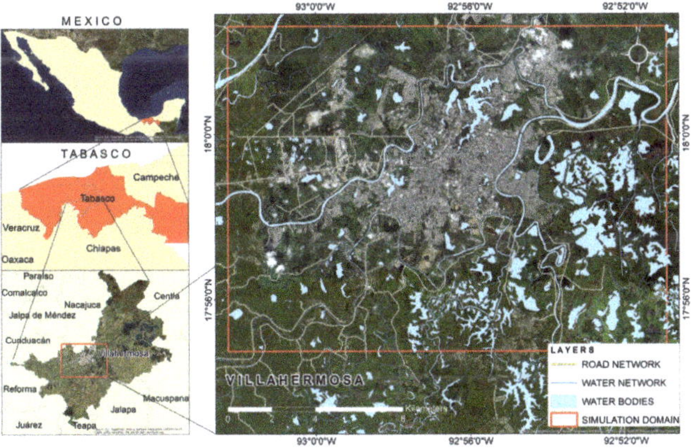

Figure 3. Location map of the area under study. Limits of the city, as well as water networks, are shown. The computation domain is limited by a red square.

The study area was discretized with an unstructured mesh to obtain different levels of spatial resolution in the different zones of the spatial domain. After some preliminary tests to evaluate the effect of the mesh size on the results, an unstructured triangular mesh of 281,289 elements was generated, with element sizes of 50 m in the urban areas, 100 m in the agricultural areas, and 15 m in the main river channels.

Three land uses were considered within the simulation domain to characterize surface roughness: urban, agricultural and river main channel. Given the dense and complex urbanization pattern of the city, it was not possible to define in the numerical model an explicit representation of buildings. As an alternative, the effect of buildings on drag resistance was accounted for using an augmented Manning coefficient to characterize the surface roughness in the urban districts. This kind of approach has been used in previous studies, such as by Liang et al. [30], Neelz and Pender [31] and Huang et al. [32]. Here, we used the formulation proposed by Huang et al. [32] to obtain the augmented Manning roughness based on the blockage percentage of buildings. From airborne photography, the buildings' blockage percentage ratio was estimated to be 75–80%, which, according to the formulation of Huang et al. [32], gives a Manning roughness of 0.20 in the urban area. In the agricultural areas and in the river's main channel, the Manning coefficients used were set to 0.12 and 0.035, respectively.

4. Results

4.1. Land Use Changes in the Catchments

Figure 4 shows water depth differences computed for the flood discharges obtained with the basin's land use of Series I and Series V, for different return periods. From a general qualitative observation of the maps, it is clear that, for lower return periods (10 and 25 years), the change in land use from mainly pasture soils (Series I) to agricultural soils (Series V) produces a significant increase in the flood depths, with maximum values between 1.0 and 1.6 m. These values are reduced to increases between 0.6 and 1.2 m as the return period increases (50 and 100 years).

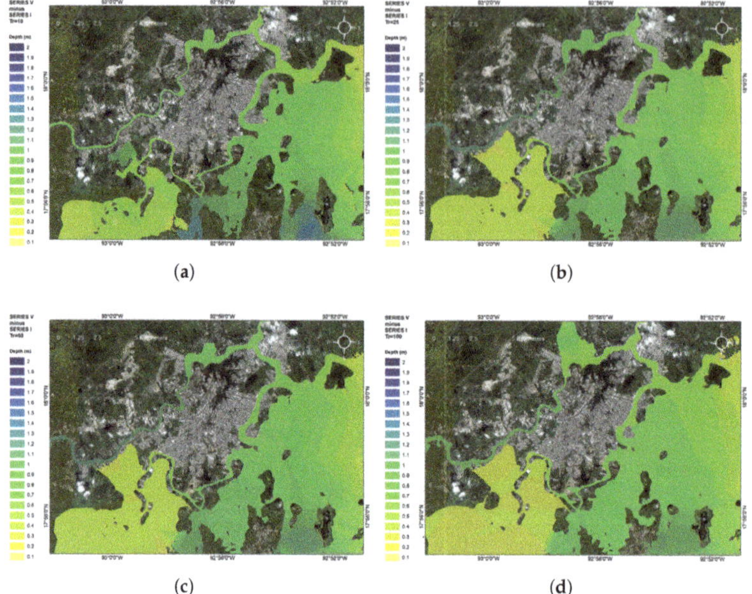

Figure 4. Maps of water depth differences between flood scenarios constructed with the land use data of Series I and Series V. Differences are calculated for the return periods: (**a**) 10 years; (**b**) 25 years; (**c**) 50 years; and (**d**) 100 years.

In recent years, national news media have reported several floods hitting the city, as a consequence of tropical depressions and hurricanes. Heavy rainfall caused flood levels of over 1.5 m, especially in the neighborhoods of Gaviotas (located in the south and southeast of the city) and Tamulte (southwest of the city), resulting in damage to houses, hospitals and government offices. Maps in Figure 4 show that these neighborhoods, along with smaller neighborhoods in the south of the city, are the most affected by the increase in inundation levels due to changes in land use in the catchments upstream. This, in addition to the increase in frequency of extreme weather phenomena recorded over recent years in the area under study, increases the likelihood of serious damage due to flooding. Consequently, our results confirm the predisposition of these districts to be hit by high flood levels.

To quantify the average difference in water depth over the whole study area, the following flood depth index (DI) was computed:

$$DI = \frac{1}{N} \sum_{i=1,N} \frac{h_i(SeriesV)}{h_i(SeriesI)} \tag{6}$$

where h_i is the flood depth at the control point i computed from Series I and Series V, and N is the number of control points, which are shown in Figure 5. To compute DI, the water depth was sampled at 1104 control points distributed over the flooded areas only, at a distance of 300 m from each other. Unflooded areas were not considered for the DI calculation. The depth index DI was computed for each return period considered in the analysis.

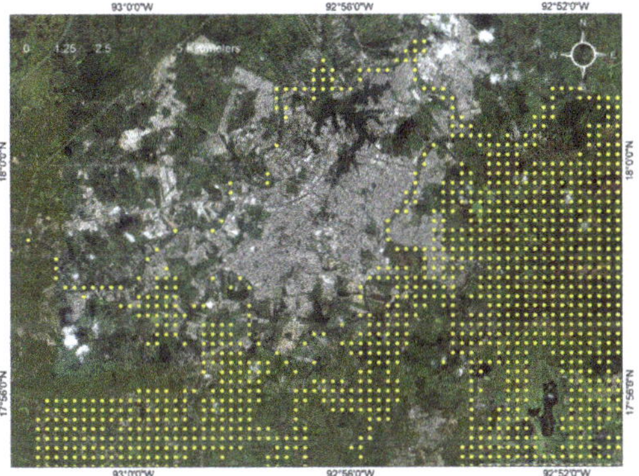

Figure 5. Location of the control points used to evaluate the flood Depth Index (DI).

Table 5 shows the values of DI obtained for each return period. The values of DI indicate that the influence of changes in land use on the inundation levels is very notable. A measure of the effects of changes in land use on the flood level is also shown in Figure 6 where flood areas with water depths greater than 1 m are compared between Series I and Series V, for each return period.

Table 5. Flood Depth Index values obtained as the difference between flood levels calculated for Series I and Series V of land use, for each return period.

T_r	DI
5	2.22
10	1.96
25	1.60
50	1.34
100	1.07

Figure 6. Extension of flood areas with depths greater than 1 m for Series I and Series V. Units in hectares.

4.2. Future Urbanization Scenarios

During the 1970s, the population of Villahermosa had approximately 100,000 inhabitants, settled in an area of 16 km^2. Nowadays, the population is about 350,000 in an area of 50 km^2. According to the population projections for the period 2010–2030 provided by the National Information System and Housing Indicators of the Secretary of Agrarian, Territorial and Urban Development of Mexico, in 2030, the population of Villahermosa will have risen at a rate of 0.50% per year. Extrapolating this population growth rate to 2050, it was estimated that the population will reach 496,381, settled in an area of 69.15 km^2. This corresponds to an increase in the urban area of about 15 km^2.

To analyze the impact on flood inundation of this urban expansion, three urbanization scenarios have been proposed, taking into account the areas that are most prone to future urbanization. The plain areas of the city, as well as the land covered by agricultural soil, were considered to be the districts with the highest probability of being urbanized in the coming years. Figure 7 presents the three proposed urbanization scenarios, which will be referred to as Scenarios A–C. In Scenario A, the new urban districts are located to the southwest–northeast of the city, at the entrance of the Carrizal River and at the exit of the Grijalva River. Scenario B assumes a future expansion of the urban area towards the east of Villahermosa, at the exit of the Grijalva River. Finally, in Scenario C, it is assumed that the city will grow towards the south. For each of these scenarios, inundation depths were computed for the flood discharge corresponding to the return period of 100 years and the land uses of Series V (see Table 4).

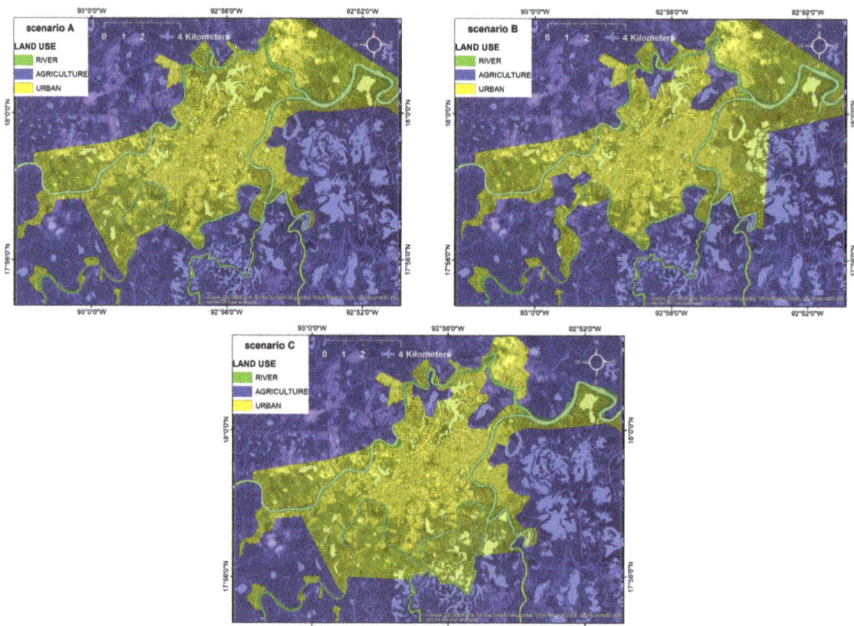

Figure 7. Proposed urbanization scenarios in the city of Villahermosa for 2050.

The computational mesh and numerical parameters used in these simulations, as well as the Manning's coefficients assigned to each land use, are the same as those described in Section 3. Thus, the only difference among Scenarios A–C is the extension of the urban districts and, thus, the spatial distribution of the Manning's coefficients.

Figure 8 shows the increase in inundation depths in Scenario A with respect to the present situation. Under this urbanization scenario, a future flood would produce an increase of water levels of up to 0.3–0.35 m. Figure 9 shows the difference in water depths between Scenario B and the present. In this case, the increase in water depths would be of 0.55 m at the east of the city, one of the most affected areas today. Finally, Figure 10 shows that, under Scenario C, the increase in inundation levels would reach 0.7 m in the rural area to the south of the city, and 0.45 m in the urban area.

Table 6 shows the calculated Depth Index defined by Equation (6). Compared to the present urbanization of Villahermosa, the relative increase in water depth in the whole city is of 1.5%, 3.3% and 4% for Scenarios A–C, respectively.

Table 6. Depth Index (DI) computed for the three future urban scenarios.

Scenario	DI
A	1.015
B	1.033
C	1.040

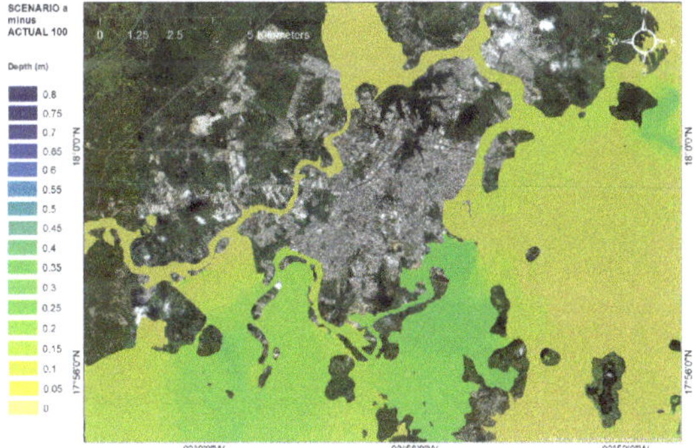

Figure 8. Differences in the water depth between Scenario A (2050) and the current situation.

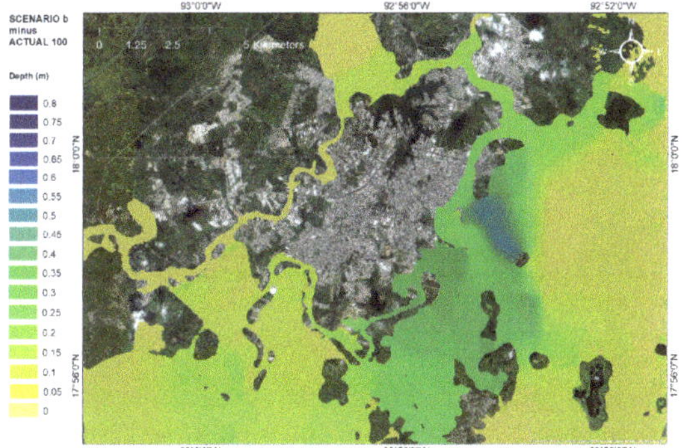

Figure 9. Differences in the water depth between Scenario B (2050) and the current situation.

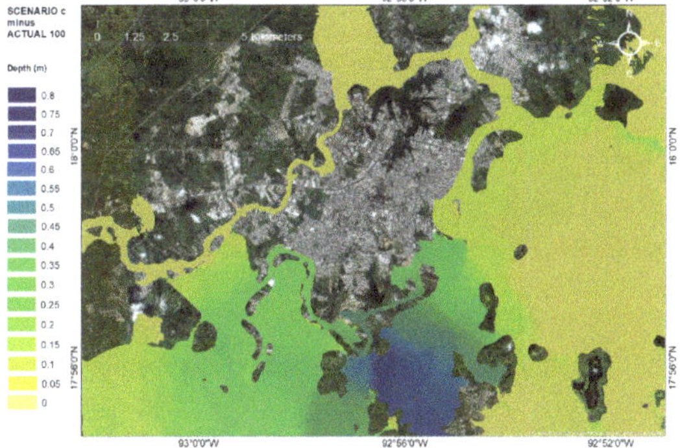

Figure 10. Differences in the water depth between Scenario C (2050) and the current situation.

Table 7 shows, for each urbanization scenario, the extension of the flood areas where the water depth exceeds a given threshold of 1, 2, 3 and 4 m. The extension of the flood is similar for urbanization Scenarios A–C (Figure 11 and Table 7) with a slightly greater increase found in the case of Scenario C. Although the expansion of the urban area does not seem to have a considerable effect on the increase of flood levels, especially when compared to the effect of changes in land use in the basins (Table 7 and Figure 6), urban expansion should not be overlooked. In fact, it is important to bear in mind that, since the 1980s, flooding events have increased in the city of Villahermosa, causing unprecedented damage. This was mostly due to uncontrolled demographic growth since the end of the 1970s, which led to the legal and illegal expansion of the urban area within the districts most prone to flooding [4]. The areas currently at flood risk, located on the edge of the rivers that cross the city, run the risk of being even more vulnerable in the event of any further expansion of the urban area.

Table 7. Flood areas in hectares.

CASES	Total Flood	Depth > 1 m	Depth > 2 m	Depth > 3 m	Depth > 4 m
SI Tr5	9911	6821	3647	2066	1044
SI Tr10	10,122	7238	4360	2254	1280
SI Tr25	12,081	9333	6817	3997	2176
SI Tr50	12,189	9459	6932	4146	2298
SI Tr100	13,362	10,863	7976	5613	2968
SV Tr5	11,319	8433	5960	3002	1669
SV Tr10	11,449	8606	6175	3215	1735
SV Tr25	13,305	10,582	7851	5557	2967
SV Tr50	13,432	10,745	7950	5674	3039
SV Tr100	14,775	11,906	8855	6693	4255
Scenario A	14,794	12,091	9127	6836	4383
Scenario B	15,058	12,139	9191	6922	4462
Scenario C	14,977	12,246	9218	6912	4409

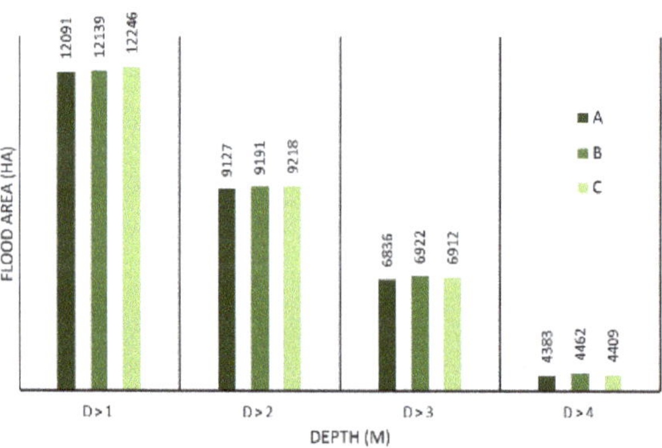

Figure 11. Extension of flood areas with flood depths greater than 1, 2, 3 and 4 m, for Scenarios A–C. Units in hectares.

5. Conclusions

Results show that, for relatively frequent inundation events (return periods up to 50 years), changes in the basin's land use cause an increase of maximum water depths in the city of up to 1.5 m. For less frequent events (return periods of 50 and 100 years) the maximum increase of water depth is between 0.7 and 1 m. The areas most affected by the rise of flood levels are those located from the southwest to the east of the city (e.g., the urban areas of Tamulte, Miguel Hidalgo, Anacleto Canabal, Ixtacomitan and Gaviotas). Such results are in accordance with recent inundations caused by tropical weather events that have caused maximum water depths of over 1.5 m in some urban districts.

The estimations obtained in this study show that during storm events the increase in flood depths due to changes in land use in the basin can range from 7% to 22%, depending on the frequency of the storm event.

On the basis of existing estimations of the city's growth rate, three possible urban expansion scenarios were proposed for 2050. Results show that, even in the most favorable scenario, flooding levels could increase by up to 0.3 m. In the worst scenario, inundation levels could rise by up to 0.7 m in areas already affected by floods every year. As stated by Zhang et al. [33], urban/construction land will always increase the problem of flooding. In the case of Villahermosa, which is a center for business and administration for Mexico's oil industry, as well as the connection between Mexico City and the largest cities in the southeast of the country, urban land increase will be a growing problem in years to come. It is important to note here that the present study does not consider climate change for the 2050 scenarios. According to Jenkins et al. [34], climate change and the rapid urbanization will be the two main factors responsible for increased flood rates in the coming decades. For this reason, the application of an approach that determines the impact of climate change on the future flood scenarios in Villahermosa, similar to the one proposed by Liu et al. [35], is a crucial aspect to be considered in further developments of this study.

The present study provides an indication of how much the flood levels are expected to increase if the expansion of the urban area continues in an uncontrolled manner. The model developed can be considered as a useful tool for future territorial planning and could be used to evaluate alternative measures to reduce river inundation hazard in Villahermosa. Those could include the design of new embankments and retaining walls within the city, or a more sustainable management of land uses in the basins located upstream of the city, to increase the infiltration capacity of the soils, and reduce the flood river discharges.

Author Contributions: Conceptualization, R.B.; Data curation, O.S.A.-R. and L.C.; Formal analysis, O.S.A.-R., L.C. and R.B.; Investigation, O.S.A.-R., L.C. and R.B.; Methodology, O.S.A.-R., L.C., R.B. and V.J.E.-E.; Project administration, R.B.; Software, L.C.; Supervision, R.B.; Validation, O.S.A.-R. and L.C.; Visualization, O.S.A.-R. and V.J.E.-E.; Writing—original draft, R.B.; and Writing—review and editing, O.S.A.-R., L.C. and R.B.

Funding: This research received no external funding.

Conflicts of Interest: The authors declare no conflict of interest.

References

1. Instituto Nacional de Estadística, Geografía e Informática (INEGI) Uso del suelo y vegetación Series 1 a 5. 2011. Available online: http://www.beta.inegi.org.mx (accessed on 15 November 2017).
2. Mas, J.F.; Velázquez, A.; Díaz-Gallegos, J.R.; Mayorga-Saucedo, R.; Alcántara, C.; Bocco, G.; Castro, R.; Pérez-Vega, A.; et al. Assessing land use/cover changes: A nationwide multidatespatial database for Mexico. *Int. J. Appl. Earth Obs. Geoinf.* **2004**, *5*, 249–261. [CrossRef]
3. Zúñiga, E.; Magaña, V. Vulnerability and risk to intense rainfall in Mexico: The effect of land use cover change. *Investig. Goegr.* **2018**. [CrossRef]
4. Aparicio, J.; Martínez-Austria, P.F.; Güitrón, A.; Ramírez A.I. Floods in Tabasco, Mexico: A diagnosis and proposal for courses of action. *J. Flood Risk Manag.* **2009**, *2*, 132–138. [CrossRef]
5. Xia, J.; Qiao, Y.F.; Song, X.F.; Ye, A.Z.; Zhang, X.C. Analysis about Effect Rules of Underlying Surface Change to the Relationship between Rainfall and Runoff in the Chabagou Catchment. *Resour. Sci.* **2007**, *29*, 71–76.
6. Feng, H.; Wu, S.F.; Wu, P.T.; Li, M. Study on Scouring Experiment of Regulating Runoff in Grassland Slope. *J. Soil Water Conserv.* **2005**, *19*. (In Chinese)
7. Nuñez, A.J. Plan Estatal de Desarrollo 2013–2018. Available online: https://tabasco.gob.mx//sites/all/files/sites/tabasco.gob.mx/files/pled-2013-2018_web.pdf (accessed on 2 February 2019).
8. Perevochtchikova, M.; Lezema, J.D. *Causas de un Desastre: Inundaciones del 2007 en Tabasco, México*; Centro de Estudios Demográficos, Urbanos y ambientales, El Colegio de México: Mexico City, Mexico, 2010.
9. Bates, P.D.; Horritt, M.S.; Fewtrell, T.J. A simple inertial formulation of the shallow water equations for efficient two-dimensional flood inundation modelling. *J. Hydrol.* **2010**, *387*, 33–45. [CrossRef]

10. Smith, L.S.; Liang, Q. Towards a generalized GPU/CPU shallow-flow modelling tool. *Comput. Fluyds* **2013**, *88*, 334–343. [CrossRef]

11. Ghimire, B.; Chen, A.S.; Guidolin, M.; Keedwell, E.C.; Djordjeviç, S.; Savić, D.A. Simulation of a fast 2D urban pluvial flood model using a cellular automata approach. *J. Hydroinf.* **2012**, *15*, 676–686. [CrossRef]

12. McColl, C.; Aggett G. Land-use forecasting and hydrologic model integration for improved land-use decision support. *J. Environ. Manag.* **2007**, *84*, 494–512. [CrossRef] [PubMed]

13. Wang, Y.; Chen, A.S.; Fu, G.; Djordjeviç, S.; Zhang, C.; Savić, D.A. An integrated framework for high resolution urban flood modelling considering multiple information sources and urban features. *Environ. Model. Softw.* **2018**, *107*, 85–95. [CrossRef]

14. Norman, L.M.; Huth, H.; Levick, L.; Burns, I.S.; Guertin, D.P.; Lara-Valencia, F.; Semmens, D. Flood hazard awareness and hydrologic modelling at Ambo Nogales, United States-Mexico border. *J. Flood Risk Manag.* **2010**, *3*, 151–165. [CrossRef]

15. Booth, D.B.; Jackson, C.R. Urbanization of aquatic system-degradation thresholds, stormwater detention, and the limit of mitigation. *J. Am. Water Resour. Assoc.* **1997**, *22*, 1077–1090. [CrossRef]

16. Goonetilleke, A.; Thomas, E.; Ginn, S.; Gilbert, D. Understanding the role of land use in urban stormwater quality management. *J. Environ. Manag.* **2005**, *74*, 31–42. [CrossRef] [PubMed]

17. Hollis, G.E. The effect of urbanization on floods of different recurrence interval. *Water Resour. Res.* **1975**, *11*, 431–435. [CrossRef]

18. Bladé, E.; Cea, L.; Corestein, G.; Escolano, E.; Puertas, J.; Vázquez-Cendón, J.; Dolz, J.; Coll, A. IBER: Herramienta de simulación numérica de flujo en ríos. *Rev. Int. Métodos Numer. para Cálc. y Diseño en Ing.* **2014**, *30*, 1–10. [CrossRef]

19. Chow, V.T.; Maidment, D.R.; Mays, L.W. *Applied Hydrology*; McGraw-Hill: New York, NY, USA, 1988.

20. Kirpich, Z.P. Time of concentration in small agricultural watersheds. *Civ. Eng.* **1940**, *1*, 362.

21. Fraga, I.; Cea, L.; Puertas, J. Effect of rainfall uncertainty on the performance of physically-based rainfall-runoff models. *Hydrol. Process.* **2018**. [CrossRef]

22. Bladé, E.; Cea, L.; Corestein, G. Modelización numérica de inundaciones fluviales. *Ing. del agua* **2014**, *18*, 68. [CrossRef]

23. Bonasia, R.; Areu-Rangel, O.S.; Tolentino, D.; Mendoza-Sanchez, I.; González-Cao, J.; Klapp, J. Flooding hazard assessment at Tulancingo (Hidalgo, Mexico). *J. Flood Risk Manag.* **2018**, *11*, S1116–S1124. [CrossRef]

24. Cea, L.; French, J.R. Bathymetric error estimation for calibration and validation of estuarine hydrodynamic models. *Estuar. Coast. Shelf Sci.* **2012**, *100*, 3317–3339. [CrossRef]

25. Cea, L.; Bladé, E.; Coristein, G.; Fraga, I.; Espinal, M.; Puertas, J. Comparative analysis of several sediment transport formulations applied to dam-break flows over erodible beds. In Proceedings of the EGU General Assembly 2014, Vienna, Austria, 27 April–2 May 2014.

26. Cea, L.; Bladé, E. A simple and efficient unstructured finite volume scheme for solving the shallow water equations in overland flow applications. *Water Resour. Res.* **2015**, *51*, 5464–5486. [CrossRef]

27. Sopelana, J.; Cea, L.; Ruano, S. Determinación de la inundación en tramos de ríos afectados por marea basada en la simulación continúa de nivel. *Ing. del agua* **2017**, *21*, 231. [CrossRef]

28. Bermúdez, M.; Neal, J.C.; Bates, P.D.; Coxon, G.; Freer, J.E.; Cea, L.; Puertas, J. Quantifying local rainfall dynamics and uncertain boundary conditions into a nested regional-local flood modeling system. *Water Resour. Res.* **2017**, *53*, 2770–2785. [CrossRef]

29. Horritt, M.S.; Bates, P.D. Evaluation of 1D and 2D numerical models for predicting river flood inundation. *J. Hydrol.* **2002**, *268*, 87–99. [CrossRef]

30. Liang, D.F.; Falconer, R.A.; Lin, B.L. Coupling surface and subsurface flows in a depth averaged flow wave model. *J. Hydrol.* **2007**, *337*, 147–158. [CrossRef]

31. Neelz, S.; Pender, G. Parameterisation of square-grid hydrodynamic models of inundation in the urban area. In Proceedings of the Congress-International Association of Hydraulic Research, Venice, Italy, 1–6 July 2007; Volume 1, pp. 41–50.

32. Huang, C.; Hsu, M.; Teng, W.; Wang Y. The impact of building coverage in the metropolitan area on the flow calculation. *Water* **2014**, *6*, 2449–2466. [CrossRef]

33. Zhang, Y.; Zhao, Y.; Wang, Q.; Wang, J.; Li, H.; Zhai, J.; Zhu, Y.; Li, J. Impact of Land Use on Frequency of Floods in Yongding River Basin, China. *Water* **2016**, *8*, 401. [CrossRef]

34. Jenkins, K.; Surminski, L.; Hall, J.; Crick, F. Assessing surface water flood risk and management strategies under future climate change: Insights from an Agent-Based model. *Sci. Total Environ.* **2017**, *595*, 159–186. [CrossRef]

35. Liu, H.; Wang, Y.; Zhang, C.; Chen, A.S.; Fu, G. Assessing real options in urban surface water flood risk management under climate change. *Nat. Hazards* **2018**, *94*, 1–18. [CrossRef]

Article

Nonstationary Flood Frequency Analysis Using Univariate and Bivariate Time-Varying Models Based on GAMLSS

Ting Zhang [1], Yixuan Wang [1,*], Bing Wang [2], Senming Tan [3] and Ping Feng [1]

[1] State Key Laboratory of Hydraulic Engineering Simulation and Safety, Tianjin University, Tianjin 300072, China; zhangting_hydro@tju.edu.cn (T.Z.); fengping@tju.edu.cn (P.F.)

[2] Tianjin Zhongshui Science and Technology Consulting Co., Ltd., Tianjin 300170, China; iceking1985@163.com

[3] Pearl River Comprehensive Technology Center (Information Center) of Pearl River Water Resources Commission of the Ministry of Water Resources, Guangzhou 510611, China; tansenming@yahoo.com

* Correspondence: wjxlch@tju.edu.cn

Received: 16 May 2018; Accepted: 14 June 2018; Published: 21 June 2018

Abstract: With the changing environment, a number of researches have revealed that the assumption of stationarity of flood sequences is questionable. In this paper, we established univariate and bivariate models to investigate nonstationary flood frequency with distribution parameters changing over time. Flood peak Q and one-day flood volume W_1 of the Wangkuai Reservoir catchment were used as basic data. In the univariate model, the log-normal distribution performed best and tended to describe the nonstationarity in both flood peak and volume sequences reasonably well. In the bivariate model, the optimal log-normal distributions were taken as marginal distributions, and copula functions were addressed to construct the dependence structure of Q and W_1. The results showed that the Gumbel-Hougaard copula offered the best joint distribution. The most likely events had an undulating behavior similar to the univariate models, and the combination values of flood peak and volume under the same OR-joint and AND-joint exceedance probability both displayed a decreasing trend. Before 1970, the most likely combination values considering the variation of distribution parameters over time were larger than fixed parameters (stationary), while it became the opposite after 1980. The results highlight the necessity of nonstationary flood frequency analysis.

Keywords: nonstationarity; univariate model; GAMLSS; bivariate model; copulas

---

1. Introduction

Flood frequency analysis is the premise and foundation of water conservancy project planning and construction. The current flood frequency analysis methods usually assume that the flood series satisfies consistency, that is, that the distribution form or the statistical law of the flood sequence is fixed [1]. However, with climate change and intensification of human activities, especially the construction of large-scale water conservancy and water conservation engineering and the urbanization process, the runoff yield and concentration mechanism, and the temporal and spatial distribution of flooding have been changed [2]. This results in the inconsistency of flood series and the unreliability of the frequency obtained from current frequency analysis methods [3]. Therefore, it is of great significance to study the nonstationary flood frequency analysis methods [4].

Existing nonstationary flood frequency analysis methods in literature include mixture distribution methods, conditional probability distribution methods, and time-varying moment methods. The main idea of the nonstationary flood frequency analysis methods based on mixture distribution is that the individuals of the extreme series are not from the same population [5]. That is, the series formed by different hydrological processes does not follow the same distribution, thus it was assumed to consist of several sub-distributions. The nonstationary flood frequency analysis methods based on conditional probability distribution divide the flood into several periods based on the differences in flood formation mechanism, analyze the occurrence probability of annual maximum value in the different periods, and then obtain the probability density function of the extreme series [6].

Different from the mixture distribution models and conditional probability distribution models, models with time-varying moment consider the change of climate and land surface to have resulted in a change in the physical processes and mechanism of flood generation, such that the parameters of the distribution followed by the flood sequence are functions of time rather than constant. Much attention has been paid to time-varying moment models [7]. Vasiliades et al. [8] applied a time-varying moment model based on Generalized Extreme Value (GEV) distribution to analyze nonstationary frequency, through assuming the parameters of GVE distribution to be the functions of time or other factors and conducting a goodness-of-fit test to the model, thus verifying the significance of the nonstationarity of hydrological sequences.

This study addressed models with time as covariate for nonstationary flood frequency analysis based on GAMLSS (Generalized Additive Models for Location, Scale, and Shape) theory. GAMLSS was first proposed by Rigby and Stasinopoulos [9,10]. This model overcomes the limitations of the GLM and GAM models, greatly expands the range of the distribution types, and provides a variety of ways to produce different distributions, including a series of continuous and discrete distributions with high skewness and/or high peak. In addition, the systematic components provide more plentiful content. For example, it can introduce more complex parametric (linear or non-linear), semi parametric, non-parametric, or random-effect terms to establish models between the distribution parameters (mean, variance, etc.) and the explanatory variables. The GAMLSS model has been widely applied in the military, economics, medicine, and other fields [11–13]. Hydrologists have also done many researches using GAMLSS in recent years. Serinaldi and Kilsby [14] used the GAMLSS model to analyze the monthly rainfall data of 6 stations in England and Wales. They found that the model could better describe the characteristics of rainfall series, and had better performance for fitting the relation between extreme rainfall events, rainfall and atmospheric circulation index, and sea surface temperature. López and Francés [15] proposed two methods based on the GAMLSS model to investigate the nonstationary frequency analysis of the annual maximum flood records of 20 Spanish inland rivers. The results illustrate that the nonstationarity of flood series caused by the effects of climate change and human activities can not to be ignored, and GAMLSS provides a convenient and flexible model framework for considering the influence of climate factors and human activities in the analysis of non-stationary flood frequency.

In flood frequency analysis, univariate probability distribution functions are usually used to estimate the occurrence probability or magnitude of the flood peak or volume in a certain region. However, flood events involve more than one characteristic variable such as flood peak discharge, flood volume, flood water level, etc. In order to estimate the probability of flooding, one needs to know not only the high and extreme values of each variable, but also the likelihood of their occurring simultaneously [16]. The main issue of the univariate models is their difficulties in capturing the underlying joint probability among multiple physical processes, and this will lead to underestimation of the associated occurring probability [17]. For instance, if only the rainfall is considered to estimate the flood risk for a catchment, the resultant estimation would be significantly lower than its true risk, when there is a statistically significant dependence between the rainfall on the catchment and the downstream high water levels. To this end, bivariate models are used to address this issue [18].

Copula functions, for which the marginal distribution of each variable is uniform, were adopted in the bivariate model with time as covariate in this study. They are popular in high-dimensional statistical applications because they allow one to easily model and estimate the distribution of random vectors by estimating marginals and copulae separately. They can describe the linear, non-linear, symmetric, and asymmetric relations between variables, and are simple, flexible, and adaptable in application.

Therefore, copulas are effective mathematical tools to construct the multivariate joint distribution and correlation between variables. In recent years, they have been widely used in multivariate hydrological frequency analysis. In drought characteristics analysis, Mirabbasi et al. [19] used a copula function to establish the joint probability distribution between drought duration and drought degree. In rainfall frequency analysis, Zhang and Singh [20] adopted copula functions to construct the bivariate joint distribution between rainfall intensity and depth, rainfall intensity and duration, and rainfall depth and duration, respectively. The results were compared with a Gumbel mixture model and a two-dimensional normal transformation distribution model. Fu G. and Butler D. [21] used the copula method to separate the dependence structure of rainfall variables from their marginal distributions, and analyzed the different impacts of dependence structure and marginal distributions on system performance.

This paper takes Wangkuai Reservoir, which is undergoing substantial change in climate, land use/land cover, and increased number of soil-water conservation projects in Daqing River Basin, to construct both univariate and bivariate time-varying moment models for flood frequency analysis based on GAMLSS theory. The inflow flood peak and flood volume time series (1956–2004) of Wangkuai Reservoir were selected as basic data to discuss the nonstationary univariate and peak-volume bivariate joint flood frequency analysis methods. Flood quantiles and the combined values of flood peak and flood volume under certain exceedance probabilities have been worked out. This study aims to provide new ideas and approaches for nonstationary flood frequency analysis method under a changing environment.

2. Study Region and Data

Wangkuai Reservoir is located in the upstream of Sha River, Daqinghe Catchment (Figure 1). Its construction started in June 1958 and finished in September 1960. The control area of the reservoir is 3770 km^2, and the storage capacity is 13.89×10^8 m^3. The currently used design floods were calculated with flood data series under the assumption of stationarity. The watershed receives an average precipitation of 626.4 mm annually, mostly in the summer (70–80%). The annual mean temperature is 7.4 °C.

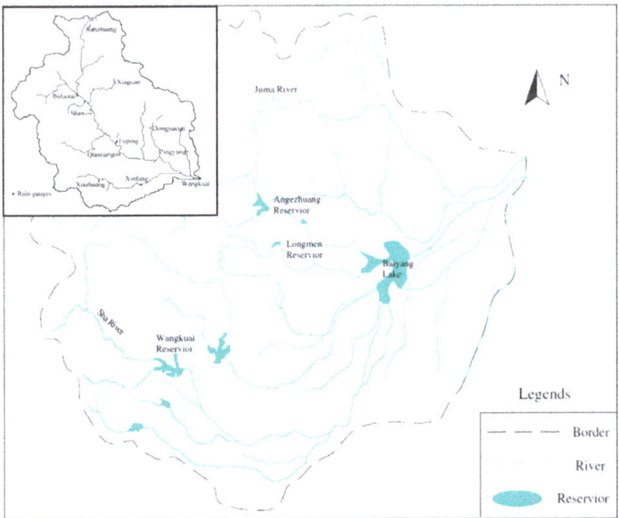

Figure 1. The study area: Wangkuai Reservoir in the Daqinghe river basin with the Wangkuai Reservoir watershed in the upper-left corner.

Since 1980, a series of water conservation measures have been carried out in the Wangkuai Reservoir catchment, such as closing land for reforestation and returning farmland to forest. Meanwhile, a number of small hydraulic structures have been constructed. These factors have increased the vegetation coverage rate and significantly changed the land surface, which has affected the flood process in this watershed and thus resulted in nonstationarity of flood series, as revealed by many studies [22–24].

Flooding runoff data have been monitored for a period of 49 years from 1956 to 2004, and collected on hourly basis. The data were provided by Hydrology and Water Resources Survey Bureau of Hebei Province. Maximum flood peak Q and maximum one-day flood volume W_1 of each year are selected in this work.

3. Methods

3.1. Generalized Additive Models in Location, Scale, and Shape (GAMLSS)

In this study, we adopted a general class of regression models which was called the Generalized Additive Models in Location, Scale, and Shape (GAMLSS) to analyze the nonstationary flood frequency. In GAMLSS models, one assumes that the vector of actual observation values $y^T = (y_1, y_2, \cdots, y_n)$ follows a probability (density) distribution function $f(y_i | \theta_i)$, where $\theta_i = (\theta_{1i}, \theta_{2i}, \theta_{3i}, \theta_{4i}) = (\mu_i, \sigma_i, v_i, \tau_i)$ is a parametric vector. The first two parameters of the model are usually defined as position and scale parameters, which are the mean vector and the mean variance vector of random variables. If there are other parameters in the distribution, such as the skewness vector v_i and the kurtosis vector τ_i of random variable series, they are all designated shape parameters. In this paper, we only consider the first two parameters. A GAMLSS model can be expressed as a known monotonic link function which demonstrates the explanatory variables and random effects:

$$f(y_i | \theta_i) = X_i \beta_i \tag{1}$$

where θ_i are n-length vectors, and $\theta_i = (\theta_{1i}, \theta_{2i}, \cdots, \theta_{ni})^T$, $\beta_i = (\beta_{1i}, \beta_{2i}, \cdots, \beta_{I_i i})^T$ is a parameter vector of length I_i, and X_i is an explanatory matrix of order $n \times I_i$.

There are two basic algorithms for parameter estimation in the GAMLSS models. The first is the CG algorithm [25], and the other is the RS algorithm [9,10]. The latter algorithm is more suited for fitting mean and dispersion additive models, thus, it was used herein for parameter estimation.

In order to estimate the parameters in GAMLSS models, a penalized likelihood function L is usually introduced:

$$L = \sum_{i=1}^{n} \log f(y_i | \theta_i) \tag{2}$$

Generally, the objective function is that the logarithmic likelihood function takes its maximum value. Then the regression parameter vector β_i can be estimated by RS algorithm.

Analyzing the normality and independence of the residuals of the models can verify the quality of model fitting when there are no validated models. The model should be adequate with their mean nearly zero, variance nearly one, coefficient of skewness and kurtosis close to zero and three, respectively, and the Filliben coefficient [26] greater than the critical value given a certain sample size.

3.2. Univariate Time Varying Model Based on GAMLSS Theory

Four two-parameter distributions can reflect the actual hydrological process, and have thus been adopted within the framework of GAMLSS. The parametric model is given as:

$$f(y_i | \mu) = X_1 \beta_1, \; f(y_i | \sigma) = X_2 \beta_2 \tag{3}$$

The four two-parameter distributions for nonstationary flood frequency analysis are expressed as following:

(1) Gumbel distribution

$$f_Y(y|\mu,\sigma) = \frac{1}{\sigma}\exp\left[\left(\frac{y-\mu}{\sigma}\right) - \exp\left(\frac{y-\mu}{\sigma}\right)\right], \quad -\infty < y < \infty, \; -\infty < \mu < \infty, \; \sigma > 0 \tag{4}$$

(2) Weibull distribution

$$f_Y(y|\mu,\sigma) = \frac{\sigma y^{\sigma-1}}{\mu^\sigma}\exp\left[-\left(\frac{y}{\mu}\right)^\sigma\right], \quad y > 0, \; \mu > 0, \; \sigma > 0 \tag{5}$$

(3) Gamma distribution

$$f_Y(y|\mu,\sigma) = \frac{1}{(\sigma^2\mu)^{1/\sigma^2}} \frac{y^{1/\sigma^2-1}e^{-y/(\sigma^2\mu)}}{\Gamma(1/\sigma^2)}, \quad y > 0, \; \mu > 0, \; \sigma > 0 \tag{6}$$

(4) Log-normal distribution

$$f_Y(y|\mu,\sigma) = \frac{1}{\sqrt{2\pi\sigma^2}}\frac{1}{y}\exp\left\{-\frac{[\log(y)-\mu]^2}{2\sigma^2}\right\}, \quad y > 0, \; \mu > 0, \; \sigma > 0. \tag{7}$$

3.3. Bivariate-Joint Time Varying Model Based on Copulas

For modeling the dependence structure of two or more random variables, copula functions are efficient mathematical tools. They were proposed first by Sklar [27], and have been widely used over the last decades. Considering a pair of random variables X and Y, with marginal distribution functions $u = F_X(x) = P(X \le x)$ and $v = F_Y(y) = P(Y \le y)$, there will be a copula function C to describe the associated relationship, which can be expressed as:

$$F_{X,Y}(x,y) = C[F_X(x), F_Y(y)] = C(u,v) \tag{8}$$

where $F_{X,Y}(x,y)$ is a joint cumulative distribution function (cdf) with margins u and v, all $(u,v) \in (0,1)^2$ [28].

One kind of frequently-used copula is the Archimedean, which has three types, written as:

(1) Gumbel-Hougaard copula

$$C_\theta(u,v) = \exp\left\{-\left[(-\ln u)^\theta + (-\ln v)^\theta\right]^{\frac{1}{\theta}}\right\}, \quad \theta \in [-1,\infty); \tag{9}$$

(2) Clayton copula

$$C_\theta(u,v) = \left(u^{-\theta} + v^{-\theta} - 1\right)^{\frac{1}{\theta}}, \quad \theta \in (0,\infty); \tag{10}$$

(3) Frank copula

$$C_\theta(u,v) = -\frac{1}{\theta}\ln\left\{1 + \frac{[\exp(-\theta u)-1][\exp(-\theta v)-1]}{\exp(-\theta)-1}\right\}, \quad \theta \in R. \tag{11}$$

The nonstationary models in this study were constructed through copulas, composed of two marginal distributions and a copula parameter θ. The marginal distributions were determined by the best nonstationary univariate models mentioned in Section 3.1, and only the copula parameter

θ needed to be solved herein. In literature the copula parameter θ was estimated by the inference function of margins method (IFM), which is given as:

$$L(u, v; \theta) = \sum \ln C[F_X(x), F_Y(y); \theta]. \tag{12}$$

Let $\partial L / \partial \theta = 0$, then θ can be calculated.

Two steps were carried out for model selection. Firstly, the Kolmogorov-Smirnov (K-S) test method was adopted to conduct the test of fit for copulae. Secondly, the models which passed K-S test were selected optimally according to the goodness-of-fit (GoF), which was represented by ordinary least squares (OLS) and Akaike information criterion (AIC) [29]. The K-S test statistic D and estimates of OLS and AIC were calculated by the following formulae:

$$D = \max\left\{ \left| C_k - \frac{i}{n} \right|, \left| C_k - \frac{i-1}{n} \right| \right\}, \tag{13}$$

where C_k is the copula value of the measured sample series, i is the number of samples which meet the requirements that $x \leq x_i$, $y \leq y_i$, and n is the length of the sample series.

$$OLS = \sqrt{\frac{1}{n}\sum_{i=1}^{n}(Pe_i - P_i)^2}, \tag{14}$$

$$Pe_i = \frac{i}{n+1}, \tag{15}$$

where Pe_i and P_i are the empirical frequency and theoretic frequency of measured sample series, respectively.

$$AIC = n\ln\left(\frac{1}{n}\sum_{i=1}^{n}(Pe_iP_i)^2\right) + 2m \tag{16}$$

where m is the number of model parameters. When the value of AIC is smaller, the model fitting is better.

The concept of return period in stationary frequency analysis is prone to misconceptions and misuses. New methods have been adopted to solve this problem, but have not worked well enough [30]. Since the return period is based on the probability, for the present study we explore the effect of nonstationarity on flood data focusing on the exceedance probability. As the return period is based on the probability, in the present work we studied the effect of nonstationarity on flood data based on exceedance probability. In the bivariate model, the OR-joint exceedance probability describes that at least one of the hydrologic variables X and Y exceed the values x and y respectively, that is, $P^{\cup} = P(X > x \cup Y > y)$. The AND-joint exceedance probability describes that X and Y both exceed the values x and y, that is, $P^{\cap} = P(X > x \cap Y > y)$. The $P^{\cup}$ and $P^{\cap}$ are written as:

$$P^{\cup} = 1 - C[F_X(x), F_Y(y)] \tag{17}$$

$$P^{\cap} = 1 - F_X(x) - F_Y(y) + C[F_X(x), F_Y(y)] \tag{18}$$

For a given data set, all the copula $C(u, v)$ at the same probability level have the same exceedance probability. However, at least one combination of a given probability is more likely than others, namely the most-likely events. Therefore, the most-likely events can be selected as the point with the largest joint probability on the level curve, which was given by Gräler et al. [31]:

$$(u, v) = \underset{C_{UV}(u,v)=k}{\operatorname{argmax}} f_{XY}\left(F_X^{-1}(u), F_Y^{-1}(v)\right) \tag{19}$$
$$x = F_X^{-1}(u), \ y = F_Y^{-1}(v)$$

where k is a given value of copula, and x and y can be calculated by the inverse cdf of the marginal distributions.

4. Results

In this study, following the identification of the nonstationarity for the Wangkuai Reservoir inflow flood sequences, two nonstationary models based on GAMLSS theory were established, in which the flood peak Q and flood volume W_1 were considered as the independent response variables, and time t was adopted as the explanatory variable.

4.1. Identification of Nonstationarity for Flood Sequences

Firstly, we adopted the pettitt test and Mann-Kendall test to detect the change point and trend of the annual maximum flood peak Q and one-day flood volume W_1 time series. Through pettitt test, the possible change points for the Q sequence are 1979 and 1996, and for W_1 1979 and 1982. Since the test probability P of 1979 is the largest, and it is among the change points of both Q and W_1 sequences, the most possible change point is 1979. This agrees with the results obtained by other researchers using different methods [32].

Before the trend analysis, the autocorrelation analysis of the flood sequences should be carried out. Since the autocorrelation coefficients of both the Q and W_1 sequences are less than 0.1, it is considered that the autocorrelation of these sequences is not significant, so the trend test can be conducted directly. Without considering the change point, the non-parametric Mann-Kendall test was used to analyze the trend of the flood sequences. The statics U_n of flood sequences Q and W_1 are both less than -1.96, which shows that the flood sequences have passed the test at a significance of 5% and present a downward trend.

In order to consider the influence of the change point on the trend test, the Q and W_1 sequences were divided into two subsequences and tested by non-parametric Mann-Kendall test method, respectively. Figure 2 shows the test results of the subsequences: although the subsequence of Q before 1979 shows a slight upward trend, and the subsequences of Q and W_1 after 1979 show a downward trend, none of them passed the test of significance. Meanwhile, no trend has been found for the sequence of W_1 before 1979.

In the case of full flood sequences presenting a significant downward trend, while the subsequences with 1979 as the dividing point showing no significant trend, the latter is more suitable for describing the inconsistency of the flood sequences.

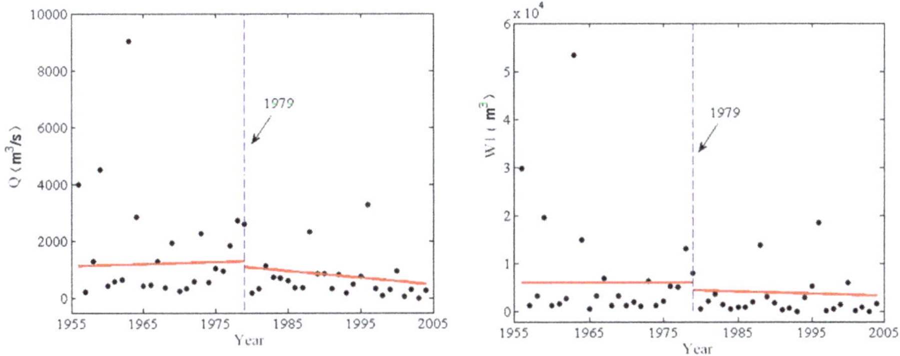

Figure 2. Trend test results of the flood subsequences with consideration of change point.

4.2. Nonstationary Model of Univariate Flood Frequency Analysis with Time as Covariate

4.2.1. Model Fitting Evaluation

A univariate nonstationary model with different distribution functions based on GAMLSS with time as covariate has been constructed based on the inflow flood sequences of Wangkuai Reservoir. In order to avoid an over-fitting problem due to excessive freedom degree, this study only considered the linear relationship between the distribution parameters and time t. The parameters include the distribution parameter θ_1 (mean value of flood sequence) and θ_2 (variance of flood sequence). The optimal fitting distribution, the optimal covariates of the distribution parameters, and the functional relationship between the distribution parameters and the optimal covariates were determined by the AIC criterion. Table 1 shows that, for flood peak (Q) and flood volume (W_1) sequences, Weibull distribution and Gamma distribution had similar fitting effect, while log-normal distribution performed best with a minimum value of AIC. The functional relationships are written as Equation (20) for Q and Equation (21) for W_1. For all the distribution models, the optimal covariates of the distribution parameters have been proved to satisfy the significant level 0.05 via χ^2 test. Besides, for both Q and W_1 time series, the distribution parameter θ_1 presents a linear dependent relationship with time t, whereas θ_2 is constant. Therefore, it can be concluded that the variation of the inflow flood sequences of Wangkuai Reservoir is mainly reflected by the mean value rather than the variance.

$$\theta_1 = 77.769 - 0.036t, \ \theta_2 = 1.030 \ (for \ Q) \tag{20}$$

$$\theta_1 = 84.876 - 0.039t, \ \theta_2 = 1.193 \ (for \ W_1) \tag{21}$$

Based on the above analysis, log-normal distribution was selected as the optimal distribution of both flood peak and volume. The corresponding residuals of the optimal distributions and the Filliben coefficients are shown in Table 2. For a sample size of 49, when the Filliben coefficients are both greater than 0.975, they satisfy the significant level 0.05 via T test. Thus, the model residuals of both Q and W_1 sequences are acceptable and in normal distribution.

Table 1. The functional relationships of the explanatory variables and distribution parameters.

Distributions	Q			W_1		
	AIC	θ_1	θ_2	AIC	θ_1	θ_2
Gumbel	895.80	t	-	1074.23	t	-
Weibull	786.60	t	-	928.57	t	-
Gamma	785.99	t	-	930.55	t	-
Log-Normal	783.92	t	-	919.26	t	-

Table 2. The residuals of the optimal distributions and corresponding Filliben coefficients.

Flood Series	Optimal Distribution	Mean	Variance	Coefficient of Skewness	Coefficient of Kurtosis	Filliben Coefficient
Q	Log-Normal	0	1.021	-0.357	3.408	0.9895
W_1	Log-Normal	0	1.021	0.249	2.402	0.9925

4.2.2. Model Results Analysis

The summary of the associated results of the univariate nonstationary model is shown in Figure 3. Most of the observed data points are distributed in the grayscale range between 5% to 95% quantiles, indicating that the model results are able to capture the nonstationarity of the flood data. With time t as covariate, flood peak and volume time series show a downward trend over time. Especially when the quantile is larger (such as 95% quantile), the decrease trend is more significant. This proves that variation of flood sequences has occurred under the changing environment, and the traditional "stationarity" hypothesis has become questionable. Thus, use of the traditional hydrological frequency analysis method may result in inaccurate results. The change of the flood sequences are mainly caused by climate change, land use change, and construction of Water conservancy projects [33–35]. Besides, the downward trend is significant before 1980, and tends to be gentle after 1980. This may be due to the change of land use. As revealed by Li et al. [34], water area, agricultural land, and grassland area decreased, while the forest area increased significantly in the control basin of Wangkuai Reservoir during 1970–1980, but from 1980 to 2000, the land use types kept almost invariant.

Figures 4 and 5 show the worm plot and normal QQ of the peak discharge Q and one-day volume W_1. As can be seen, no significant departure from normality has been highlighted. Therefore, the results of residuals supported the inference that log-normal distribution provides a good fit to both Q and W_1.

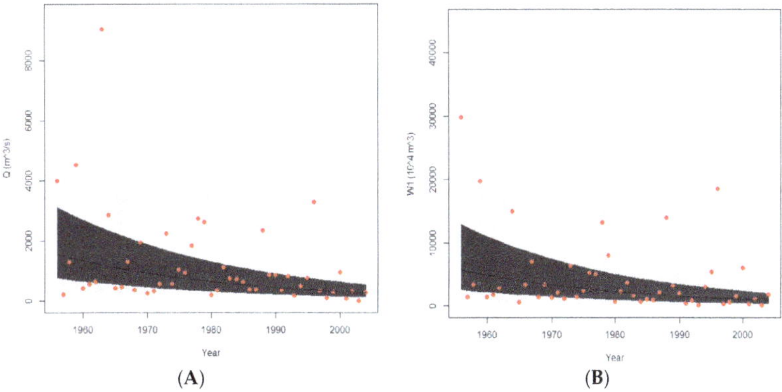

(**A**)   (**B**)

Figure 3. Summary of results of the univariate nonstationary model with Generalized Additive Models in Location, Scale, and Shape (GAMLSS) implementation. Red points represent the observed time series of peak discharge Q (**A**) and flood volume W_1 (**B**). The solid black line represents 0.5 quantile; the dark grey region is the area between 0.25 and 0.75 quantiles; the light grey region is the area between 0.05 and 0.95.

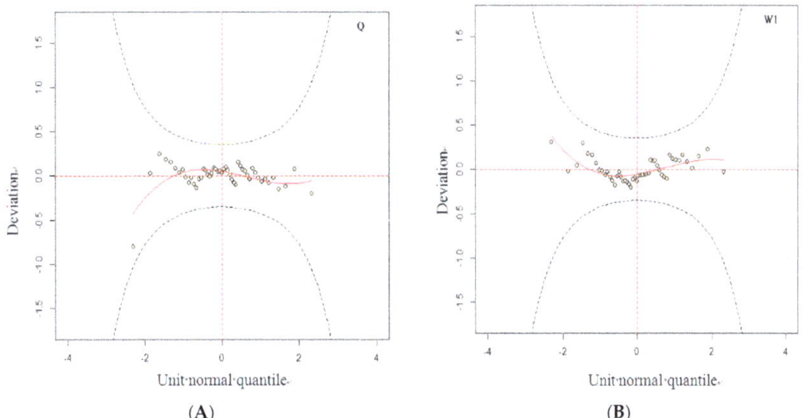

(**A**)   (**B**)

Figure 4. Worm plot of the residuals of non-stationary time-varying model with GAMLSS implementation (Q in **A**, W_1 in **B**). For a satisfactory fit of worm plot, all the observations should fall inside the two elliptic curves.

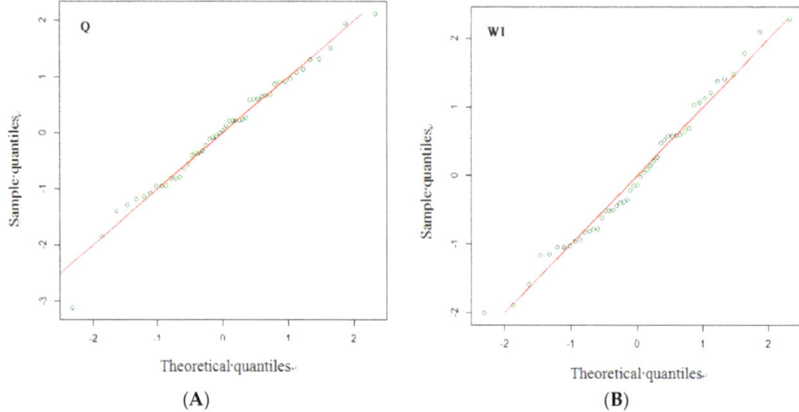

Figure 5. QQ plot of the residuals of non-stationary time-varying model with GAMLSS implementation (Q in **A**, W_1 in **B**).

4.3. Nonstationary Model of Bivariate-Joint Flood Frequency Analysis with Time as Covariate

4.3.1. Parameters Calculation, Fitting Test, and Result Optimization of Joint Probability Distribution

With the above two log-normal distributions (Equations (19) and (20)) as marginal distributions, the joint distribution model of flood peak and volume based on copulas have been constructed according to Equations (9)–(11). The parameter of Copula functions θ, K-S test statistic D, and values of OLS and AIC were estimated by Equations (12)–(16). A significance level of $\alpha = 0.05$ was used for the K-S test, and the corresponding standard quantile was approximately 0.1943 for $n = 49$. This means that when D-value is less than 0.1943, it passes the test. The results of the calculation, test, and optimization are presented in Table 3.

Table 3. The estimation of copula parameters and the associated fitting test and model selection.

Copulas	Copula Parameter (θ)	Indices of Fitting Test and Model Selection		
		D (K-S)	OLS	AIC
Gumbel-Hougaard Copula	2.9224	0.1648	0.0404	−312.3751
Clayton Copula	1.6444	0.1940	0.0470	−297.7266
Frank Copula	9.1480	0.2537	0.0469	−297.7374

It can be observed in Table 3 that Gumbel-Hougaard copula function and Clayton copula function have passed the K-S test, while the Frank Copula has not passed the test. Among them, the Gumbel-Hougaard copula function has the minimum values of OLS and AIC, indicating that the Gumbel-Hougaard copula function is more suitable to describe the extreme sequence of hydrological variables, such as the joint distribution of flood peak and volume. Therefore, the G-H Copula with parameter $\theta = 2.9224$ was considered as the optimal copula for the bivariate-joint distribution of the Q and W_1 time series of Wangkuai Reservoir.

4.3.2. Most Likely Combination of the Flood Peak and Volume

The "most likely" design events derived from Equation (11) have their maximum value of likelihood function on each probability-isoline, and all combinations of Q and W_1 were illustrated in Figure 6. They have the similar undulating behavior as the univariate models shown in Figure 3.

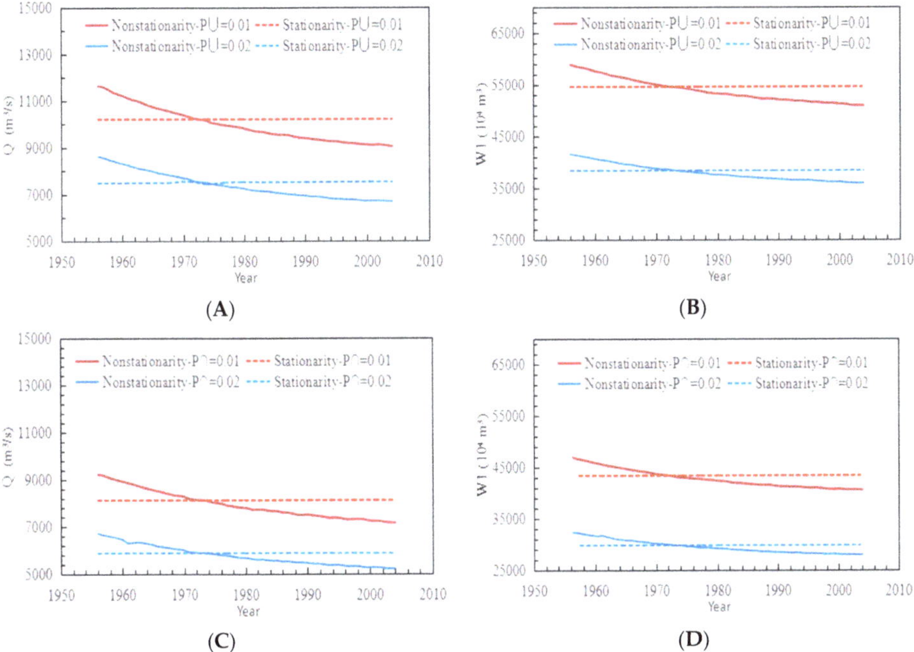

Figure 6. Results of the most likely design events (combinations of flood peak and volume). Case illustration for $P^{\cup} = 0.02$, 0.01 (**A,B**) and $P^{\cap} = 0.02$, 0.01 (**C,D**).

Figure 6 shows that, under changing environmental conditions, with time t as the covariate of the marginal distribution, the combination values of the flood peak and volume under the same OR-joint and AND-joint exceedance probability both display a decreasing trend. Meanwhile, the combination values of the nonstationary model intersect with the traditional stationary results between 1970–1980. That is, before 1970, the most likely combination values considering the variation of distribution parameters over time were larger than fixed parameters (stationary), however, they were smaller after 1980.

5. Discussion and Conclusions

The changing environment makes the assumption of stationarity of flood sequences questionable. In this context, this paper constructed both univariate and bivariate nonstationary models with time as covariate based on GAMLSS theory for flood frequency analysis. The inflow flood peak Q and flood volume W_1 series of Wangkuai Reservoir were used as basic data.

Within the framework of nonstationary flood frequency, this paper adopted four two-parameter distributions (Gumbel, Weibull, Gamma, and Log-Normal) as alternative distributions, which have the characteristics of power distribution and exponential distribution simultaneously. In the univariate nonstationary model with time as covariate, log-normal distribution performed best according to AIC criterion. The flood peak and volume time series presented a decreasing trend over time. Especially when the quantile is high (such as 95% quantile), the downward trend is more significant. Besides, the decreasing trend is significant before 1980, and tends to be gentle after 1980. This proves that variation of flood sequences has occurred under the changing environment.

Based on the optimal univariate models, copula functions were addressed to construct the dependence structure of Q and W_1, with the two optimal log-normal distributions as marginal distributions. The results showed that only the Gumel-Hougaard copula can provide the best joint distribution. The most likely events have similar undulating behavior to the univariate models, and the combination values of the flood peak and volume under the same OR-joint and AND-joint exceedance probability both display a decreasing trend. Meanwhile, the combination values of the nonstationary model intersect with the traditional stationary results between 1970–1980. That is, before 1970, the most likely combinations considering the variation of distribution parameters over time were larger than fixed parameters (stationary), whereas it became the opposite after 1980.

Author Contributions: Conceptualization, P.F.; Methodology, T.Z.; Software, S.T.; Validation, Y.W.; Formal Analysis, T.Z.; Investigation, B.W.; Resources, B.W.; Data Curation, Y.W.; Writing Original Draft Preparation, T.Z.; Writing Review & Editing, T.Z.; Visualization, S.T.; Supervision, P.F.; Project Administration, P.F.; Funding Acquisition, T.Z.

Funding: This research was funded by National Natural Science Foundation of China grant number 51609165, China Postdoctoral Science Foundation grant number 2017T100159, and the Foundation of State Key Laboratory of Hydraulic Engineering Simulation and Safety (Tianjin University) grant number HESS-1607.

Acknowledgments: We are grateful to Hydrology and Water Resource Survey Bureau of Hebei Province for providing the hydrometeorological data. The authors also acknowledge the reviewers for providing valuable advices to improve our papers.

Conflicts of Interest: The authors declare no conflict of interest. The funding sponsors had no role in the design of the study; in the collection, analyses, or interpretation of data; in the writing of the manuscript, and in the decision to publish the results.

References

1. Zheng, F.; Westra, S.; Leonard, M.; Sisson, S.A. Modeling dependence between extreme rainfall and storm surge to estimate coastal flooding risk. *Water Resour. Res.* **2014**, *50*, 205–2071. [CrossRef]
2. Milly, P.C.D.; Betancourt, J.; Falkenmark, M.; Hirsch, R.M. Stationarity is dead: Whither water management. *Science* **2008**, *319*, 573–574. [CrossRef] [PubMed]
3. Ishak, E.H.; Rahman, A.; Westra, S.; Sharma, A.; Kuczera, G. Evaluating the non-stationarity of Australian annual maximum flood. *J. Hydrol.* **2013**, *494*, 134–145. [CrossRef]
4. Salas, J.D.; Obeysekera, J. Revisiting the concepts of return period and risk for nonstationary hydrologic extreme events. *J. Hydrol. Eng.* **2014**, *19*, 554–568. [CrossRef]
5. Waylen, P.; Woo, M.K. Prediction of annual floods generated by mixed processes. *Water Resour. Res.* **1982**, *18*, 1283–1286. [CrossRef]
6. Singh, V.P.; Wang, S.X.; Zhang, L. Frequency analysis of nonidentically distributed hydrologic flood data. *J. Hydrol.* **2005**, *307*, 175–195. [CrossRef]
7. Cunderlik, J.M.; Burn, D.H. Non-stationary pooled flood frequency analysis. *J. Hydrol.* **2003**, *276*, 210–223. [CrossRef]
8. Vasiliades, L.; Galiatsatou, P.; Loukas, A. Nonstationary frequency analysis of annual maximum rainfall using climate covariates. *Water Resour. Manag.* **2015**, *29*, 339–358. [CrossRef]
9. Rigby, R.A.; Stasinopoulos, D.M. Generalized additive models for location, scale and shape. *J. R. Stat. Soc. Ser. C (Appl. Stat.)* **2005**, *54*, 507–554. [CrossRef]
10. Stasinopoulos, D.M.; Rigby, R.A. Generalized additive models for location scale and shape (GAMLSS) in R. *J. Stat. Softw.* **2007**, *23*, 1–46. [CrossRef]
11. Boutselis, P.; Ringrose, T.J. GAMLSS and networks in combat simulation metamodelling: A case study. *Expert Syst. Appl.* **2013**, *40*, 6087–6093. [CrossRef]
12. Luo, J.W.; Chen, L.N.; Liu, H. Distribution characteristics of stock market liquidity. *Phys. A Stat. Mech. Appl.* **2013**, *382*, 6004–6014. [CrossRef]
13. Wahl, S.; Fenske, N.; Zeilinger, S.; Suhre, K.; Gieger, C.; Waldenberger, M.; Grallert, H.; Schmid, M. On the potential of models for location and scale for genome-wide DNA methylation data. *BMC Bioinform.* **2014**, *15*, 1471–2105. [CrossRef] [PubMed]
14. Serinaldi, F.; Kilsby, C.G. A modular class of multisite monthly rainfall generators for water resource management and impact studies. *J. Hydrol.* **2012**, *464–465*, 528–540. [CrossRef]

15. López, J.; Francés, F. Non-stationary flood frequency analysis in continental Spanish rivers, using climate and reservoir indices as external covariates. *Hydrol. Earth Syst. Sci.* **2013**, *17*, 3189–3203. [CrossRef]
16. Hawkes, P.J. Joint probability analysis for estimation of extremes. *J. Hydraul. Res.* **2008**, *46*, 246–256. [CrossRef]
17. Fiorentino, M.; Gioia, A.; Iacobellis, V.; Manfreda, S. Regional analysis of runoff thresholds behaviour in Southern Italy based on theoretically derived distributions. *Adv. Geosci.* **2011**, *26*, 139–144. [CrossRef]
18. Zheng, F.; Westra, S.; Sisson, S.A. Quantifying the dependence between extreme rainfall and storm surge in the coastal zone. *J. Hydrol.* **2013**, *505*, 172–187. [CrossRef]
19. Mirabbasi, R.; Fakheri-Fard, A.; Dinpashoh, Y. Bivariate drought frequency analysis using the copula method. *Theor. Appl. Climatol.* **2012**, *108*, 191–206. [CrossRef]
20. Zhang, L.; Singh, V.P. Bivariate rainfall frequency distributions using Archimedean copulas. *J. Hydrol.* **2007**, *332*, 93–109. [CrossRef]
21. Fu, G.; Butler, D. Copula-based frequency analysis of overflow and flooding in urban drainage systems. *J. Hydrol.* **2014**, *510*, 49–58. [CrossRef]
22. Li, J.; Liu, X.; Chen, F. Evaluation of Nonstationarity in Annual Maximum Flood Series and the Associations with Large-scale Climate Patterns and Human Activities. *Water Resour. Manag.* **2015**, *29*, 1653–1668. [CrossRef]
23. Li, J.; Tan, S. Nonstationary Flood Frequency Analysis for Annual Flood Peak Series, Adopting Climate Indices and Check Dam Index as Covariates. *Water Resour. Manag.* **2015**, *29*, 5533–5550. [CrossRef]
24. Zeng, H.; Feng, P.; Li, X. Reservoir flood routing considering the non-stationarity of flood series in north China. *Water Resour. Manag.* **2014**, *28*, 4273–4287. [CrossRef]
25. Cole, T.J.; Green, P.J. Smoothing reference centile curves: The lms method and penalized likelihood. *Stat. Med.* **2010**, *11*, 1305–1319. [CrossRef]
26. Filliben, J.J. The Probability Plot Correlation Coefficient Test for Normality. *Technometrics* **1975**, *17*, 111–117. [CrossRef]
27. Sklar, A. *Fonctions de Répartition À N Dimensions Et Leurs Marges*; Institut de Statistique Université de Paris: Paris, France, 1959; pp. 229–231.
28. Nelsen, R.B. *An Introduction to Copulas*; Springer: New York, NY, USA, 2006.
29. Akaike, H. A new look at the statistical model identification. *IEEE Trans. Autom. Control* **1974**, *19*, 716–723. [CrossRef]
30. Olsen, J.R.; Stedinger, J.R.; Matalas, N.C.; Stakhiv, E.Z. Climate variability and flood frequency estimation for the Upper Mississippi and Lower Missouri Rivers. *J. Am. Water Resour. Assoc.* **1999**, *35*, 1509–1524. [CrossRef]
31. Gräler, B.; van den Berg, M.J.; Vandenberghe, S.; Petroselli, A.; Grimaldi, S.; De Baets, B.; Verhoest, N.E.C. Multivariate return periods in hydrology: A critical and practical review focusing on synthetic design hydrograph estimation. *Hydrol. Earth Syst. Sci.* **2013**, *17*, 1281–1296. [CrossRef]
32. Li, X.; Zeng, H.; Feng, P. Flood frequency analysis considering variation in flood time series. *J. Hydroelectr. Eng.* **2014**, *33*, 11–19.
33. Wang, Y.; Li, J.; Feng, P.; Chen, F. Effects of large-scale climate patterns and human activities on hydrological drought: a case study in the luanhe river basin, China. *Nat. Hazards* **2015**, *76*, 1687–1710. [CrossRef]
34. Li, J.; Li, G.; Zhou, S.; Chen, F. Quantifying the Effects of Land Surface Change on Annual Runoff Considering Precipitation Variability by SWAT. *Water Resour. Manag.* **2016**, *30*, 1071–1084. [CrossRef]
35. Gu, X.; Zhang, Q.; Chen, X.; Jiang, T. Nonstationary flood frequency analysis considering the combined effects of climate change and human activities in the East River Basin. *Trop. Geogr.* **2014**, *34*, 746–757.

Article

Use of Artificial Intelligence to Improve Resilience and Preparedness Against Adverse Flood Events

Sara Saravi [1,*], Roy Kalawsky [1], Demetrios Joannou [1], Mónica Rivas Casado [2], Guangtao Fu [3] and Fanlin Meng [3]

[1] Wolfson School of Mechanical, Electrical & Manufacturing Engineering, Advanced VR Research Centre, Loughborough University, Loughborough LE11 3TU, UK; r.s.kalawsky@lboro.ac.uk (R.K.); d.joannou@lboro.ac.uk (D.J.)

[2] School of Water, Energy and Environment, Cranfield University, Cranfield, Bedfordshire MK43 0AL, UK; m.rivas-casado@cranfield.ac.uk

[3] Centre for Water Systems, College of Engineering, Mathematics and Physical Sciences, University of Exeter, Exeter, Devon EX4 4QF, UK; g.fu@exeter.ac.uk (G.F.); m.fanlin@exeter.ac.uk (F.M.)

* Correspondence: s.saravi@lboro.ac.uk; Tel.: +44-(0)-1509-222-938

Received: 12 February 2019; Accepted: 6 May 2019; Published: 9 May 2019

Abstract: The main focus of this paper is the novel use of Artificial Intelligence (AI) in natural disaster, more specifically flooding, to improve flood resilience and preparedness. Different types of flood have varying consequences and are followed by a specific pattern. For example, a flash flood can be a result of snow or ice melt and can occur in specific geographic places and certain season. The motivation behind this research has been raised from the Building Resilience into Risk Management (BRIM) project, looking at resilience in water systems. This research uses the application of the state-of-the-art techniques i.e., AI, more specifically Machin Learning (ML) approaches on big data, collected from previous flood events to learn from the past to extract patterns and information and understand flood behaviours in order to improve resilience, prevent damage, and save lives. In this paper, various ML models have been developed and evaluated for classifying floods, i.e., flash flood, lakeshore flood, etc. using current information i.e., weather forecast in different locations. The analytical results show that the Random Forest technique provides the highest accuracy of classification, followed by J48 decision tree and Lazy methods. The classification results can lead to better decision-making on what measures can be taken for prevention and preparedness and thus improve flood resilience.

Keywords: Artificial Intelligence; machine learning; flood; preparedness; resilience; flood resilience

1. Introduction

Climate change is expected to increase the frequency and intensity of extreme events, including flooding. Across the world, flooding has an enormous economic impact and cost millions of lives. The number of large scale natural disasters have significantly increased in the past few years; this results in considerable impact to human lives, environment and buildings, and substantial damage to societies. During these disasters, vast quantities of data are collected on the characteristics of the event via governmental bodies, society (e.g., citizen science), emergency responders, loss adjusters and social media, amongst others. However, there is a lack of research on how this data can be used to inform how different stakeholders are/can be directly or indirectly affected by large scale natural disasters pre-, during and post-event disaster management decisions. There is a growing popularity and need for the use of Artificial Intelligence (AI) techniques [1] that bring large-scale natural disaster data into real practice and provide suitable tools for natural disaster forecasting, impact assessment, and societal resilience. This in turn will inform on resource allocation, which can lead to better preparedness and prevention for a natural disaster, save lives, minimize economic impact, provide better emergency respond, and make communities stronger and more resilient.

The majority of the work done in the area of AI in flooding has been on the use of social media [2] (e.g., Facebook, Twitter or Instagram) where status update, comments and photo sharing have been used for data mining to improve flood modelling and risk management [3–5]. The author of [6] has used Artificial Neural Networks (ANN) in flash flood prediction using data from soil moisture and rainfall volume. Further research [7] has focused on the use of the Bursty Keyword technique combined with the Group Burst algorithm to retrieve co-occurring keywords and derive valuable information for flood emergency response. AI has also been used on images provided by citizens affected by flooding for emergency responders to have situational awareness. In [8], the authors explored the use of algorithms based on ground photography shared within social networks. Use of specific algorithms for satellite images or aerial imagery [9] to detect flood extent was also explored. Within this context, the resolution of the imagery collected is of key relevance to detect features of interest due to the complexity of the imagery acquired in urban areas [10]. Some studies have focused on the analysis of high resolution, real-time data processing to derive flood information [11,12].

Overall, the majority of disaster-monitoring methods are based on change detection algorithms, where the affected area is identified through a complex elaboration on images from pre- and post-event. Change detection can be applied to the amplitude or intensity, filtered or elaborated versions of the amplitude [2,13,14]. For example, in [15], a technique based on change detection applied to quantities related to the fractal parameters of the observed surface was developed to address change detection. In [16], information extracted from images taken and shared on social media by people in flooded regions was combined with the embedded metadata within them to detect flood patterns. In this study, a convolutional inception network was applied on pre-trained weights on ImageNet to extract rich visual information from the social media imagery. A word embedding was used for the metadata to represent the textual information continuously and feed it to a bidirectional Recurrent Neural Networks (RNN). The word embedding was initialized using Glove vectors, and finally, the image and text features were concatenated to find out probability of the sample, including related information about flooding. Similarly, in [17], an AI system was designed to retrieve social media images containing direct evidence of flooding events and derive visual properties of images and the related metadata via a multimodal approach. For that purpose, an image pre-processing including cropping and test-set pre-filtering based on image colour or textual metadata and ranking for fusion was implemented. In [18,19], Convolutional Neural Networks and Relation Networks were used for end-to-end learning for disaster image retrieval and flood detection from satellite images.

2. Methodology

Flood management strategies and emergency response depend upon the type of area affected (e.g., agricultural or urban) as well as on the flood type (e.g., fluvial, pluvial or coastal). Resilience measures are generally deployed by governmental agencies to reduce the impact of flooding. The use of AI to derive flood information for specific events is well documented in the scientific literature. However, little is known about how AI could inform future global patterns of flood impact and associated resilience needs.

The main focus of this paper is on the use of AI and more explicitly Machine Learning (ML) applied to natural disasters involving flooding to estimate the flood type from the weather forecast, location, days event lasted, begin/end location, begin/end latitude and longitude, injuries direct/indirect, death direct/indirect and property and crop damage.

The proposed method uses historical information collected from 1994 to 2018, to learn the patterns and changes in various parameters' behaviours in flood events and make remarks for the future events. This paper focuses only on providing an insight on how floods behave differently in terms of damage. Using the historic data, the models developed adapt to all the changes over time by learning from past information and can provide high accuracy of classification. The proposed technique is highly adjustable to use for estimating any other desired parameters, providing a detailed set of historic data. This technique combined with other proposed techniques from literature, such as satellite imagery, social media information, etc. can provide a very powerful tool for having insight to flood events and help with preparedness, reduce impact, and better decision making.

The flood pathways and key variables are first described, and data sourcing and ML techniques used in this study are then explained, and finally the model evaluation metrics are provided.

2.1. Flood Pathways and Key Variables

An important step in the process is to create influence maps as visual aids to illustrate how related variables interact and affect each other. Figure 1 indicates an overall causal loop diagram for a full flooding scenario. This map includes all the stakeholders and their interaction i.e., natural climate change, man-made facilities, businesses, public and governmental sectors, and social media.

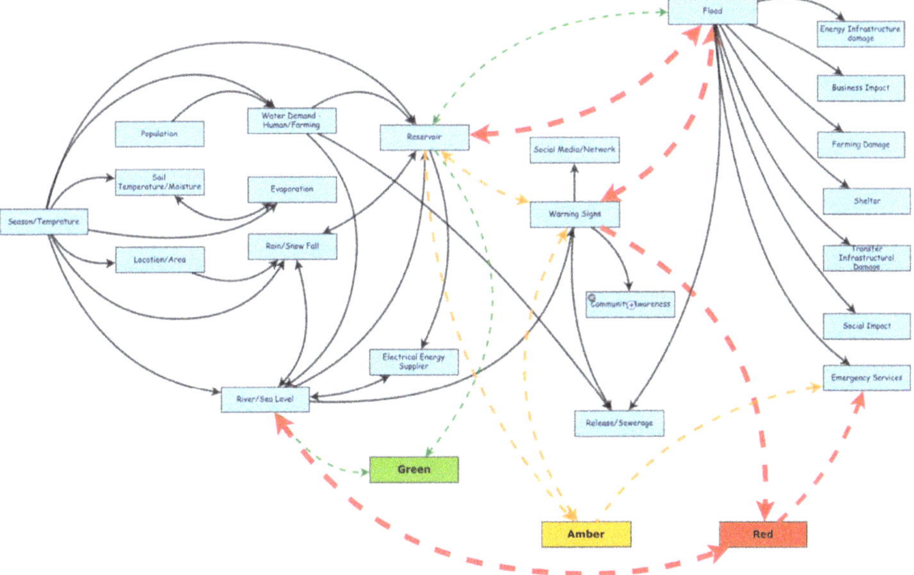

Figure 1. An overall causal loop of flood pathways. Green refers to normal conditions, amber refers to caution for a probability of flooding, and red refers to a very high risk or event of flooding.

The season, temperature, location of the area (highland/inland, coastal/urban), and rain/snowfall can affect the levels of the sea or river and reservoirs. Usage of water by energy suppliers, human/farming water demand can change the balance of the reservoirs and river water levels and indicate a warning sign for flooding. The use of social media and public awareness can help tackle the risks of a flooding event. When the flooding occurs, many sectors are affected i.e., road/rail way damage, gas/water pipe damage, power cut, farming damage, etc. Emergency response and access to food and local amenities are restricted. Grocery prices spike due to lack of supply and businesses are affected by physical building damage or lack of human resources. In this loop, public awareness, emergency responses (local/public), early release of sewerage system, and shelters can help save lives.

There are three states in the diagram in Figure 1: Green refers to normal conditions, amber refers to caution for a probability of flooding, and red refers to a very high risk or event of flooding.

2.2. Data Collation and Preparation

One of the most important requirements for this research was a detailed historic and inclusive data set, which was acquired from Federal Emergency Management Agency (FEMA) [20], National Oceanic and Atmospheric Administration (NOAA) [21] and National Climatic Data Centre (NCDC) [21]. The data used in this study covers the period of 1950 until 2018. However, the data of flooding events is recorded from the year 1994 onwards and is inclusive of all event types, i.e., heavy snow, thunderstorm, fog, hail, flood, high wind, etc. Table 1 summarises the different attributes used to build models within the ML based framework.

Table 1. Description of the attributes used to build models within the ML based framework to inform flood resilience and resistance actions (Source: NCDC-NOAA [21] and FEMA [20]).

Attributes	Description
Begin Date	Year, Month, Day
End Date	Year, Month, Day
Month Name	Jan, Feb, March, ...
State	CA, AZ, TX ...
County/Zone Name	Utah, Leon, Beaver, ...
Injuries Direct/Indirect	Number of direct/indirect injuries
Deaths Direct/Indirect	Number of direct/indirect deaths
Weather Forecast /Flood Cause	Heavy snow/Rain, Thunderstorm, Fog/Thunderstorm, hail ...
Days Lasted	Total number of days event lasted
Begin/End Location	The location where the event started (Name)
Begin/End Lat/Long	Begin/End Latitude and Longitude
Property Damage	$
Crop Damage	$
Class	**Description**
Event type	Flash food, Lakeshore flood, ...

The data sets collated were inspected for outliers and extreme values, missing data and redundant information via a bespoke MATLAB application known as Flood Data Aggregation Tool (FDAT) developed for this purpose by the authors. FDAT removes all existing outliers and missing data and re-orders the data based on specific categories chosen for the implementation of the ML techniques and it converts the alphanumeric and alphabetic data to numeric data using one-hot encoding.

The processed dataset is then divided into training and testing data sets. The training data set is used to develop the model whereas the testing data set is used to quantify the accuracy of the model built. A larger portion of data is separated for training and the remaining is used for testing and validation to ensure accuracy of the classification model built and software performance. Figure 2 shows an overview of the overall analytical process employed in this study. The raw data collected is fed to FDAT tool for data cleaning, normalisation, aggregation, and other pre-processing steps. The output data is divided into testing and training data and passed through the ML/data mining tool, the patterns are extracted, and the model is built, followed by analysis to verify its quality.

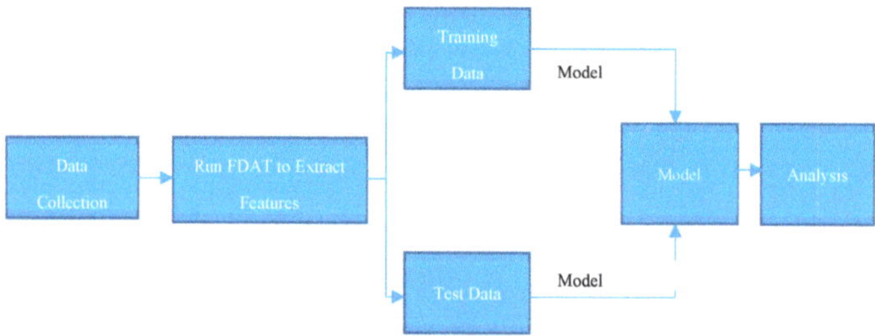

Figure 2. Work flow summarising the analytical steps followed. "Data" includes both data collation and extraction.

Figure 3 illustrates a sample of input data prepared for training, which is an output of the ADAT application. All the attributes have been described prior to data definitions. The detailed attributes can be found in Table 1.

```
1    @relation Flood
2
3    @attribute Begin_Date date yyyy-mm-dd
4    @attribute End_Date date yyyy-mm-dd
5    @attribute Days_Last_Total numeric
6    @attribute Month_Name {April,June,August,January,February,May,July,November,December,September,October,March}
7    @attribute Event_Type {'THUNDERSTORM WINDS/FLOODING','HAIL FLOODING','THUNDERSTORM WINDS/FLASH FLOOD','THUNDERSTORM WINDS/ FLOOD',Flood,'Flash Flood','Coastal Flood','Lakeshore Flood'}
8    @attribute State {'SOUTH CAROLINA',NEVADA,TEXAS,INDIANA,'NEW YORK',VIRGINIA,OHIO,'WEST VIRGINIA',MAINE,MASSACHUSETTS,KENTUCKY,CONNECTICUT,MARYLAND,'NEW JERSEY','NORTH CAROLINA',VERMONT,GEORGIA,WYOMING,'DISTRICT OF COLUM
9    @attribute Injuries_Direct numeric
10   @attribute Injuries_Indirect numeric
11   @attribute Death_Direct numeric
12   @attribute Death_Indirect numeric
13   @attribute Begin_Location {Dillon,0,'Extreme Western Nevad','Santa Monica',Countywide,COUNTYWIDE,LYNCHBURG,TRIADELPHIA,BUFFALO,ALL,'WEST SENECA',GALAX,HELEN,ALTAVISTA,AMHERST,PUNXSUTAWNEY,BIRMINGHAM,WESTERN,FLEISCHMANNS
14   @attribute End_Location string
15   @attribute Begin_Lat numeric
16   @attribute Begin_Long numeric
17   @attribute End_Lat numeric
18   @attribute End_Long numeric
19
20   @data
21   1994-04-15,1994-04-15,0,April,'THUNDERSTORM WINDS/FLOODING','SOUTH CAROLINA',0,0,0,0,Dillon,0,0,0,0,0
22   1995-06-01,1995-06-01,0,June,'HAIL FLOODING',NEVADA,0,0,0,0,0,0,0,0,0,0
23   1995-04-04,1995-04-04,0,April,'THUNDERSTORM WINDS/ FLOOD',TEXAS,0,0,0,0,'Santa Monica',0,0,0,0,0
24   1995-04-04,1995-04-04,0,April,'THUNDERSTORM WINDS/ FLOOD',TEXAS,0,0,0,0,Countywide,0,0,0,0,0
25   1995-08-05,1995-08-05,0,August,'THUNDERSTORM WINDS/FLASH FLOOD',NEVADA,0,0,0,0,'Extreme Western Nevad','West Central Nevada',0,0,0,0
26   1996-04-01,1996-04-01,0,April,'Flash Flood',OHIO,0,0,0,0,COUNTYWIDE,COUNTYWIDE,0,0,0,0
27   1996-05-01,1996-05-31,30,May,Flood,ILLINOIS,0,0,0,0,0,0,0,0,0,0
28   1996-05-01,1996-05-01,0,May,Flood,'NEW YORK',0,0,0,0,0,0,0,0,0,0
29   1996-05-01,1996-05-01,0,May,Flood,'NEW YORK',0,0,0,0,0,0,0,0,0,0
30   1996-05-01,1996-05-01,0,May,Flood,'NEW YORK',0,0,0,0,0,0,0,0,0,0
31   1996-05-01,1996-05-31,30,May,Flood,MISSOURI,0,0,0,0,0,0,0,0,0,0
32   1996-05-01,1996-05-01,0,May,Flood,VERMONT,0,0,0,0,0,0,0,0,0,0
33   1996-05-01,1996-05-31,30,May,Flood,ILLINOIS,0,0,0,0,0,0,0,0,0,0
34   1996-05-01,1996-05-31,30,May,Flood,ILLINOIS,0,0,0,0,0,0,0,0,0,0
35   1996-05-01,1996-05-20,19,May,Flood,INDIANA,0,0,0,0,0,0,0,0,0,0
36   1996-05-01,1996-05-26,25,May,Flood,INDIANA,0,0,0,0,0,0,0,0,0,0
37   1996-05-01,1996-05-26,25,May,Flood,ILLINOIS,0,0,0,0,0,0,0,0,0,0
38   1996-05-01,1996-05-31,30,May,Flood,ILLINOIS,0,0,0,0,0,0,0,0,0,0
39   1996-08-01,1996-08-01,0,August,'Flash Flood',LOUISIANA,0,0,0,0,HAUGHTON,HAUGHTON,0,0,0,0
40   1996-08-01,1996-08-01,0,August,'Flash Flood',LOUISIANA,0,0,0,0,COUSHATTA,COUSHATTA,0,0,0,0
41   1996-08-01,1996-08-01,0,August,'Flash Flood',COLORADO,0,0,0,0,'CANON CITY','CANON CITY',0,0,0,0
42   1996-08-01,1996-08-01,0,August,Flood,ARIZONA,1,0,0,0,WILLIAMS,WILLIAMS,0,0,0,0
43   1996-08-01,1996-08-01,0,August,'Flash Flood','PUERTO RICO',0,0,0,0,'GURABO CAGUAS','GURABO  CAGUAS',0,0,0,0
44   1996-12-01,1996-12-01,0,December,'Flash Flood','WEST VIRGINIA',0,0,0,0,COUNTYWIDE,COUNTYWIDE,0,0,0,0
45   1996-12-01,1996-12-02,1,December,Flood,KENTUCKY,0,0,0,0,0,0,0,0,0,0
46   1996-12-01,1996-12-02,1,December,'Flash Flood','NEW YORK',0,0,0,0,'WEBB MILLS','WEBB MILLS',0,0,0,0
47   1996-12-01,1996-12-02,1,December,'Flash Flood','NEW YORK',0,0,0,0,COUNTYWIDE,COUNTYWIDE,0,0,0,0
48   1996-12-01,1996-12-02,1,December,'Flash Flood',PENNSYLVANIA,0,0,0,0,COUNTYWIDE,COUNTYWIDE,0,0,0,0
49   1996-12-01,1996-12-02,1,December,Flood,KENTUCKY,0,0,0,0,0,0,0,0,0,0
50   1996-12-01,1996-12-01,0,December,'Flash Flood',ALABAMA,0,0,0,0,PRATTVILLE,PRATTVILLE,0,0,0,0
51   1996-12-01,1996-12-01,0,December,'Flash Flood',ALABAMA,0,0,0,0,MILLBROOK,MILLBROOK,0,0,0,0
52   1996-12-01,1996-12-02,1,December,Flood,KENTUCKY,0,0,0,0,0,0,0,0,0,0
53   1996-02-02,1996-02-02,0,February,'Flash Flood',ALABAMA,0,0,0,0,COUNTYWIDE,COUNTYWIDE,0,0,0,0
54   1996-05-02,1996-05-08,6,May,Flood,'NEW YORK',0,0,0,0,0,0,0,0,0,0
55   1996-05-02,1996-05-08,6,May,Flood,'NEW YORK',0,0,0,0,0,0,0,0,0,0
56   1996-05-02,1996-05-08,6,May,Flood,VERMONT,0,0,0,0,0,0,0,0,0,0
57   1996-05-02,1996-05-08,6,May,Flood,VERMONT,0,0,0,0,0,0,0,0,0,0
58   1996-05-02,1996-05-08,6,May,Flood,VERMONT,0,0,0,0,0,0,0,0,0,0
59   1996-07-02,1996-07-02,0,July,'Flash Flood','WEST VIRGINIA',0,0,0,0,HAMLIN,HAMLIN,0,0,0,0
60   1996-07-02,1996-07-02,0,July,'Flash Flood','WEST VIRGINIA',0,0,0,0,HEPZIBAH,SPELTER,0,0,0,0
61   1996-08-02,1996-08-02,0,August,'Flash Flood',VIRGINIA,0,0,0,0,'N PORTION','N PORTION',0,0,0,0
62   1996-08-02,1996-08-02,0,August,'Flash Flood','WEST VIRGINIA',0,0,0,0,FAIRMONT,FAIRMONT,0,0,0,0
63   1996-08-02,1996-08-02,0,August,'Flash Flood','WEST VIRGINIA',0,0,0,0,'S PORTION','S PORTION',0,0,0,0
64   1996-09-02,1996-09-02,0,September,'Flash Flood',ALABAMA,0,0,0,0,TUSCALOOSA,TUSCALOOSA,0,0,0,0
65   1996-10-02,1996-10-02,0,October,'Flash Flood',KENTUCKY,0,0,0,0,'GREASY CREEK','GREASY CREEK',0,0,0,0
66   1996-10-02,1996-10-02,0,October,'Flash Flood',FLORIDA,0,0,0,0,'LAKE CITY','LAKE CITY',0,0,0,0
67   1996-10-02,1996-10-02,0,October,'Flash Flood',VIRGINIA,0,0,0,0,VANSANT,PRATER,0,0,0,0
68   1996-10-02,1996-10-02,0,October,'Flash Flood',FLORIDA,0,0,0,0,'LIVE OAK','LIVE OAK',0,0,0,0
69   1996-12-02,1996-12-02,0,December,'Flash Flood','NEW YORK',0,0,0,0,COUNTYWIDE,COUNTYWIDE,0,0,0,0
70   1996-12-02,1996-12-03,1,December,Flood,CONNECTICUT,0,0,0,0,0,0,0,0,0,0
71   1996-12-02,1996-12-03,1,December,Flood,'NEW YORK',0,0,0,0,0,0,0,0,0,0
72   1996-12-02,1996-12-02,0,December,Flood,VERMONT,0,0,0,0,0,0,0,0,0,0
73   1996-12-02,1996-12-02,0,December,'Flash Flood','NEW YORK',0,0,0,0,WHITEHALL,WHITEHALL,0,0,0,0
74   1996-12-02,1996-12-02,0,December,'Flash Flood','NEW YORK',0,0,0,0,COUNTYWIDE,COUNTYWIDE,0,0,0,0
75   1996-12-02,1996-12-03,1,December,Flood,'NEW YORK',0,0,0,0,0,0,0,0,0,0
76   1996-12-02,1996-12-02,0,December,'Flash Flood','NEW YORK',0,0,0,0,COUNTYWIDE,COUNTYWIDE,0,0,0,0
```

Figure 3. A sample of input data prepared for training.

2.3. Machine Learning—Model Development

AI is human intelligence demonstrated by machines and ML is an approach to achieve AI. In this study, the focus will be on supervised ML to learn from historic data, find clustered data, and build classification model for future events. This type of ML works particularly best when used in combination with historic data (results included). For this purpose, a number of data mining tools such as: Weka [22], MATLAB [23] and Orange [24] have been deployed. The reason for using two softwares (Weka and Orange) for this purpose is to test more ML techniques with various training and testing dataset sizes. The data is divided into two parts. The first will be used for training and generating the model, and the second will be used for testing and verification.

Several models were developed using different ML techniques to be able to measure and compare their performance and accuracy and choose the best. These techniques included Random Forest (RF), Lazy, J48 tree, Artificial Neural Network (ANN), Naïve Bayes (NB), and Logistic Regression (LR). The class for the model in all cases was set as "event type" (Table 1), which included flash floods, coastal floods, lakeshore floods and other kinds of floods. The independent attributes in all models were weather forecast, location, injuries direct, injuries indirect, death direct, death indirect, property damage ($) and crop damage ($) (Table 1). Two of these ML methods used i.e., RF and NB, are tested in both softwares (Weka and Orange) to ensure the accuracy of results.

2.3.1. Random Forest (RF)

RF [25] is a collaborative learning technique. It is a combination of the Bagging algorithm and the random subspace method and deploys decision trees as the basis for classifier. Each tree is made from a bootstrap sample from the original dataset. The key point is that the trees are not exposed to trimming, allowing them to partly overfit to their own sample of the data. To extend the classifiers at every branch in the tree, the decision of which feature to divide further is limited to a random sub-data from the full data set. The random sub-data is chosen again for each branching point.

2.3.2. Lazy

Lazy [25] learning is a ML approach where learning is delayed until testing time. The calculations within a learning system can be divided as happening at two separate times: training and testing (consultation). Testing time is the time between when an object is introduced to a system for an action to be taken and the time when the action is accomplished. Training time is before testing time during which the system takes actions from training data in preparation for testing time. Lazy learning refers to any ML process that postpones the majority of computation to testing time. Lazy learning can improve estimation precision by allowing a system to concentrate on deriving the best possible decision for the exact points of the instance space for which estimations are to be made. However, lazy learning must store the entire training set for use in classification. In contrast, eager learning need only store a model, which may be more compact than the original data.

2.3.3. J48 Decision Tree

A decision tree is an analytical machine-learning model that estimates the target value of a new test sample data based on several characteristic values of the training data. The nodes within a decision tree represent the attributes, the branches between the nodes represent the probable values that the attributes in training data may have, and the terminal nodes represent the final classification value of the attribute to be estimated.

J48 is an open source Java implementation of the C4.5 algorithm in the Weka data mining tool. In order to classify a new item, the J48 Decision tree [26] first has to generate a decision tree based on the training data attributes. Therefore, when it encounters a training set, it categorises the attribute that separates different samples most clearly. This feature allows most about the data instances to be classified and contains the highest information gain.

Amongst the possible features, if there is any value for which there is no uncertainty, which the data instances falling within its category have the same value for the target variable, then that branch terminates and will be assigned to the target value obtained.

2.3.4. Artificial Neural Networks (ANN)

An ANN [25] is a data processing system that is inspired by the way neurons in biological brain systems process information, which facilitates a computer to learn from the information provided. The crucial component of this system consists of a large number of greatly interrelated processing features (neurones) working uniformly to solve problems. An ANN system is developed without any precise logic. Basically, an ANN system adapts and changes its configuration based on the pattern within the information that flows through the network during the learning phase, and very similar to human beings, it learns by example. An ANN is primarily trained with a large amount of data. Training involves feeding input data and stating what the output would be. ANN use numerous principles, including gradient-based training, fuzzy logic, genetic algorithms, and Bayesian methods.

ANNs are designed to identify patterns in the given information. Particularly classification task which is to classify data into pre-defined classes, clustering task which is to classify data into distinctive undefined groups), and estimation task which is to use past events to estimate future ones.

One of the challenges of using ANNs is the time it takes to train the networks, which can be computationally expensive for more complex tasks. Another challenge is that the ANNs are like a black box, in which the user can feed in information and receive a built model. The user can modify the model, but they do not have access to the exact decision-making process.

2.3.5. Naïve Bayes (NB)

NB [25] is a simple learning algorithm that uses Bayes' rule along with a theory that the features are provisionally independent given the class. Although this independence theory is usually affected in practice, NB usually delivers competitive classification precision. NB is commonly used in practice because of its computational efficiency and many other desirable features such as low variance, incremental learning, direct prediction of posterior probabilities, robustness in the face of noise, and robustness in the face of missing values.

NB provides a system to use the information from training data to estimate the future probability of each class y given an object x. These estimations can be used for classification or other decision support applications.

2.3.6. Logistic Regression (LR)

LR [25] is a mathematical model for estimation of the probability of an episode happening based on the given input data. LR provides a tool for applying the linear regression methods to classification problems. LR is used when the target variable is categorical. Linear regression estimates the data by defining a straight-line equation to model or estimate data points. LR does not look at the relationship between the two variables as a straight line. Instead, LR uses the natural logarithm function to find the relationship between the variables and uses test data to find the coefficients. The function can then estimate the future results using these coefficients in the logistic equation. LR uses the concept of odds ratios to calculate the probability. This is defined as the ratio of the odds of an event happening to its not happening.

2.4. Model Evaluation Metrics

The system has been trained with several different combinations; however, the final system uses one based on the selected attributes, which was an output of the classifier attribute evaluation from an ML tool. All ML models developed were validated using evaluation criteria, i.e., confusion matrix [25], Mean Absolute Error (MAE) [25] and Root Mean Squared Error (RMSE) [25]. These metrics are used for summarising and assessing the quality of the ML model.

A confusion matrix summarises the classifier performance with regards to the test data. It is a two-dimensional matrix, indexed in one dimension by the actual class of an object and in the other by the class that the classifier allocates, and the cells represent: true positives (TP), false positives (FP), true negatives (TN) and false negatives (FN) identified in a classification. Multiple measures of accuracy are derived from the confusion matrix i.e., specificity (SP), sensitivity (SS), positive estimated value (PPV) and negative estimated value (NPV). These are calculated as follows:

$$SP = TN/(TN + FP) \tag{1}$$

$$SS = TP/(TP + FN) \tag{2}$$

$$PPV = TP/(TP + FP) \tag{3}$$

$$NPV = TN/(TN + FN) \tag{4}$$

The MAE is the mean of the absolute value of the error per instance over all samples in the test data. Each estimation error is the difference between the true value and the estimated value for the sample. MAE is calculated as follows:

$$MAE = \frac{\sum_{i=1}^{n} |y_{est,i} - y_i|}{n} \tag{5}$$

where y_i is the true target value for test sample i, $y_{est,i}$ is the estimated target value for test sample i, and n is the number of test samples.

The RMSE of a model with respect to a test data is the square root of the mean of the squared estimation errors over all samples in the test data. The estimation error is the difference between the true value and the estimated value for a sample. RMSE is calculated as follows:

$$RMSE = \sqrt{\frac{\sum_{i=1}^{n}\left(y_{est,i} - y_i\right)^2}{n}} \tag{6}$$

where y_i is the true target value for test sample i, $y_{est,i}$ is the estimated target value for test sample i, and n is the number of test samples.

3. Model Training and Testing Results

The original data consisted of 126,315 samples. After removing the outliers and filtering using ADAT application, 69,558 instances were narrowed down to be used for learning. The data was then divided into two parts: a larger section (data from 1994 to 2017) for training purposes and the smaller section (data from 2018) for testing purposes. A scattered plot of the training data for event type can be seen in Figure 4.

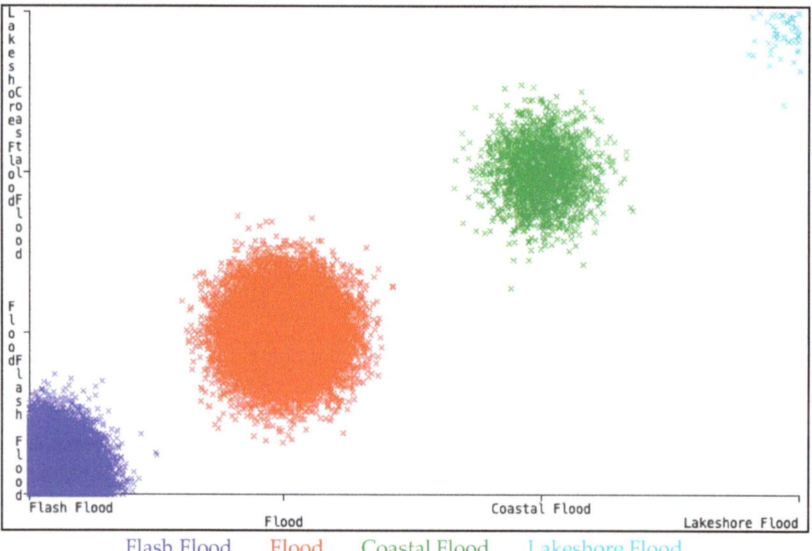

Figure 4. Visual distribution of flood types in training data.

The test data for Orange software consists of 164 instances with target feature of "Event-Type", of which 44 instances are coastal flood, 58 instances are flash flood, 53 are flood and 9 are lakeshore flood. The test data for Weka software consists of 3478 instances of which 100 are coastal flood, 2104 are flash flood, 1266 are flood and 8 are lakeshore flood. The testing dataset size can vary depending on the user desire and performance of the ML software and hardware capabilities. Four different types of ML techniques in Orange and four techniques in Weka are tested and evaluated in order to be able to choose the best performing technique. The techniques tested are RF, Lazy, J48 tree, ANN, NB, and LR. An overview of the model training and testing process is illustrated in Figure 5, which has been implemented in Orange. First, the training data is passed through different classification techniques (i.e., NN, LR, RF and NB) to build the classification models, then the models are tested using the test data. Finally, the evaluation results are produced and can be analysed and/or visualized (i.e., confusion matrix and scatter plot).

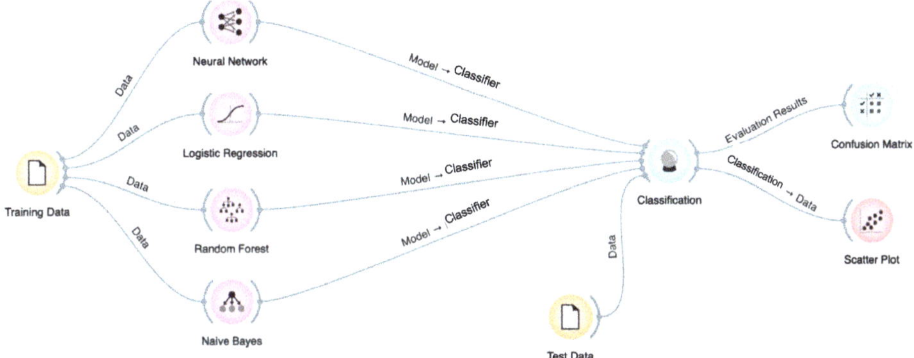

Figure 5. Visual overview of the model training and testing process in Orange.

The results of the models and their performance are discussed below.

Based on the confusion matrix, the RF model using Orange software classified 7 out of 9 instances as Lakeshore Flood, 49 out of 53 as Flood, 32 out of 58 as Flash Flood and 44 out of 44 as Coastal Flood correctly. The correctly classified instances in total was 132 (80.49%). According to the proportion of the classifications on the test data, the RF was ahead of all other techniques. Figure 6 shows the evaluation results and confusion matrix for the RF model based on the supplied test set. Based on the confusion matrix, the RF model using Weka software classified 1850 out of 2104 as Flash Flood, 820 out of 1266 as Flood, 100 out of 100 as Coastal Flood and six out of eight instances as Lakeshore Flood correctly. The correctly classified instances in total are 2776 (79.83%). The MAE is 0.13 and RMSE is 0.27. The RF technique provides best results as compared to the techniques tested in Weka. Figure 7 indicates a visual classifier error for the RF model. The diagram shows the distribution of correctly classified instances in coloured clusters, where the bigger clusters (shown in crosses) are the correctly classified instances and the smaller clusters (shown with small squares) are the misclassified instances.

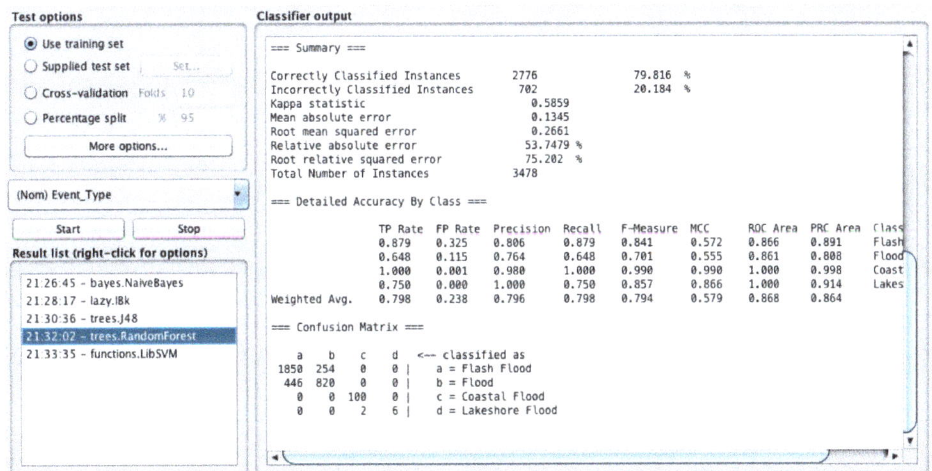

Test options

- ⦿ Use training set
- ◯ Supplied test set   Set...
- ◯ Cross-validation   Folds   10
- ◯ Percentage split   %   95

More options...

(Nom) Event_Type

Start   Stop

Result list (right-click for options)

21:26:45 – bayes.NaiveBayes
21:28:17 – lazy.IBk
21:30:36 – trees.J48
21:32:02 – trees.RandomForest
21:33:35 – functions.LibSVM

Classifier output

```
=== Summary ===

Correctly Classified Instances        2776               79.816 %
Incorrectly Classified Instances       702               20.184 %
Kappa statistic                          0.5859
Mean absolute error                      0.1345
Root mean squared error                  0.2661
Relative absolute error                 53.7479 %
Root relative squared error             75.202  %
Total Number of Instances             3478

=== Detailed Accuracy By Class ===

               TP Rate  FP Rate  Precision  Recall  F-Measure  MCC    ROC Area  PRC Area  Class
               0.879    0.325    0.806      0.879   0.841      0.572  0.866     0.891     Flash
               0.648    0.115    0.764      0.648   0.701      0.555  0.861     0.808     Flood
               1.000    0.001    0.980      1.000   0.990      0.990  1.000     0.998     Coast
               0.750    0.000    1.000      0.750   0.857      0.866  1.000     0.914     Lakes
Weighted Avg.  0.798    0.238    0.796      0.798   0.794      0.579  0.868     0.864

=== Confusion Matrix ===

    a    b    c    d   <-- classified as
 1850  254    0    0 |   a = Flash Flood
  446  820    0    0 |   b = Flood
    0    0  100    0 |   c = Coastal Flood
    0    0    2    6 |   d = Lakeshore Flood
```

Figure 6. Evaluation Results for Random Forest model in Weka.

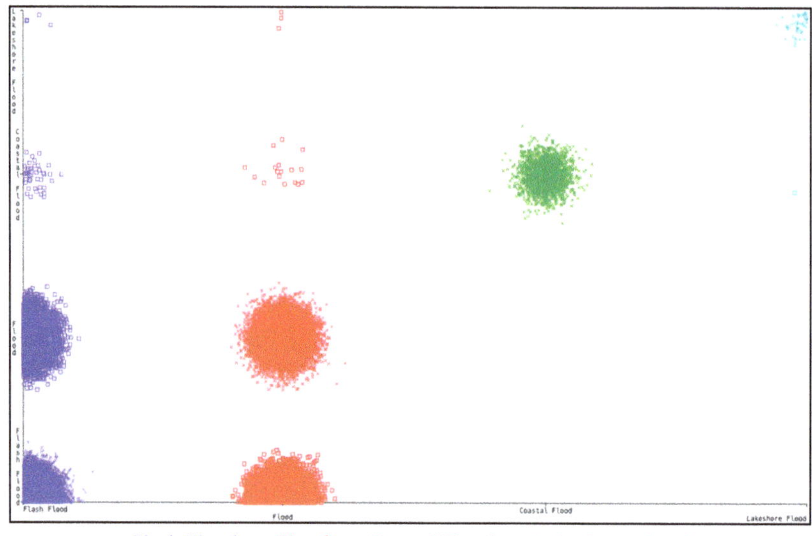

Figure 7. Visual classifier error for Random Forest model in Weka. The crosses indicate correctly classified instances and squares refer to misclassified instances.

The confusion matrix result (Figure 8) indicates that the Lazy model (Weka) correctly classified 1858 out of 2104 as Flash Flood, 809 out of 1266 as Flood, 99 out of 100 as Coastal Flood and six out of eight instances as Lakeshore Flood. The correctly classified instances in total was 2772 (79.70%). The MAE was 0.13 and RMSE was 0.27. The Lazy technique provides better results than Naïve Bayes.

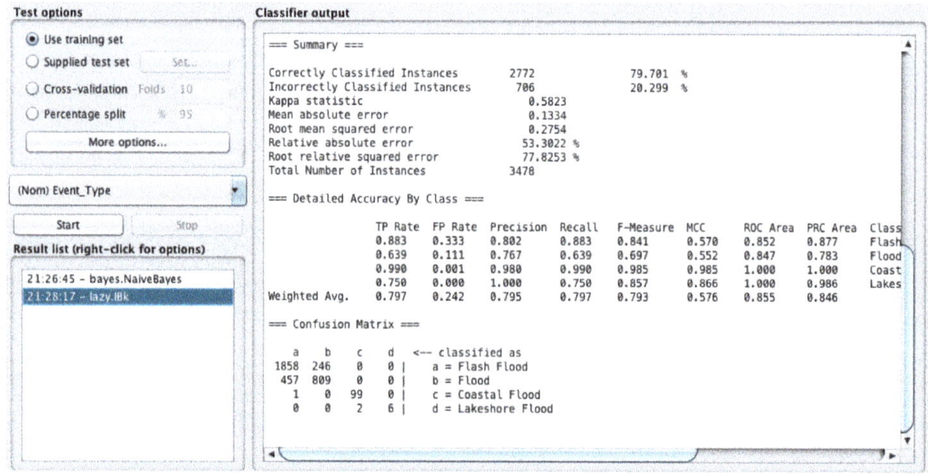

Figure 8. Evaluation Results for Lazy model in Weka.

The confusion matrix (Figure 9) shows that the J48 model (Weka) classified 1882 out of 2104 as Flash Flood, 791 out of 1266 as Flood, 100 out of 100 as Coastal Flood and 0 out of 8 instances as Lakeshore Flood correctly. The correctly classified instances in total is 2773 (79.73% rate of success). The MAE is 0.14 and RMSE is 0.27; the J48 technique provides very similar results to Lazy.

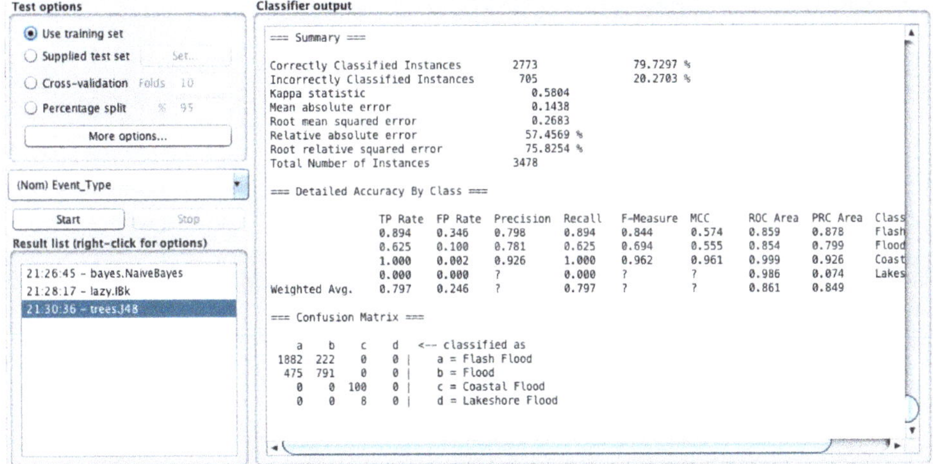

Figure 9. Evaluation Results for J48 model in Weka.

Based on the confusion matrix (Figure 10) result for the ANN model (Orange), successful classifications are 7 out of 9 as Lakeshore Flood, 49 out of 53 instances as Flood, 27 out of 58 as Flash Flood and 44 out of 44 as Coastal Flood correctly. The total of correctly classified instances is 127 (77.44% rate of success).

Confusion Matrix

Confusion Matrix

Neural Network
Naive Bayes
Logistic Regression
SVM
Random Forest

Show: Number of instances

Predicted

Actual		Coastal Flood	Flash Flood	Flood	Lakeshore Flood	Σ
	Coastal Flood	44	0	0	0	44
	Flash Flood	0	27	31	0	58
	Flood	0	4	49	0	53
	Lakeshore Flood	2	0	0	7	9
	Σ	46	31	80	7	164

Output

☑ Predictions ☑ Probabilities

☑ Send Automatically   Select Correct   Select Misclassified   Clear Selection

Figure 10. Evaluation Results for Neural Network model in Orange.

The NB model using Orange software correctly classified nine out of nine instances of Lakeshore Flood, 45 out of 53 as Flood, 27 out of 58 as Flash Flood and 44 out of 44 as Coastal Flood according to the resulting confusion matrix (Figure 10). The total of correctly classified instances was 125 (76.22%). The number of confused instances as Flash Flood and Flood are of a slightly higher proportion compared to the ANN based on 164 instances considered for validation. Correspondingly the confusion matrix showed that the NB model built using Weka software (Figure 11) classified 1614 out of 2104 as Flash Flood, 853 out of 1266 as Flood, 89 out of 100 as Coastal Flood and seven out of eight instances as Lakeshore Flood correctly. The correctly classified instances in total was 2563 (73.69%). The MAE was 0.18 and RMSE was 0.29. This outcome indicates that the NB has a very high classification accuracy when trained on a small data set or larger data set as both software have produced similar results when trained and tested on both large and small data sets.

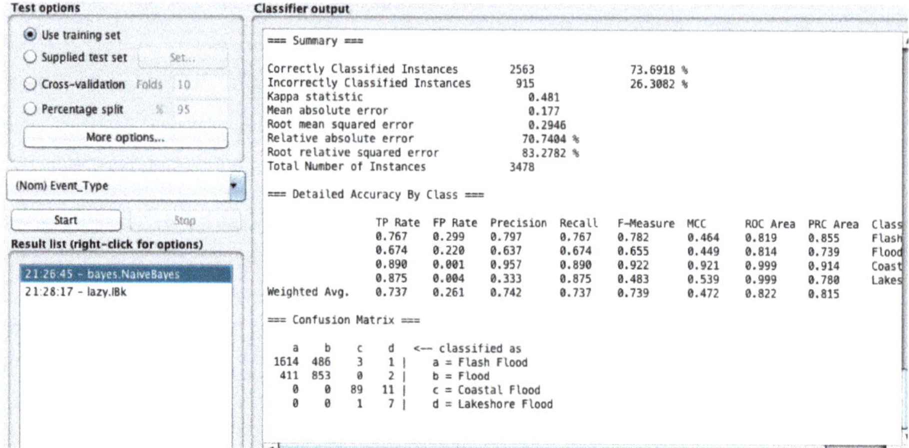

Figure 11. Evaluation Results for Naïve Bayes model in Weka.

The LR model (Orange) is completely disregarded as it has provided as small as 27.44% correctly classified instances (Figure 12).

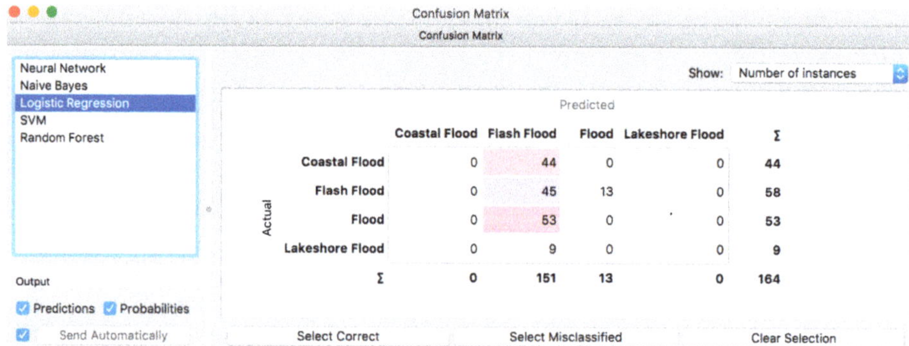

Figure 12. Evaluation Results for Logistic Regression model in Orange.

Table 2 shows the predicative models' performance using the MAE and RMSE evaluation metrics.

Table 2. Summary of evaluation metrics of all classification models' performance in Orange and Weka.

Model	RMSE	MAE	% of correctly classified instances
		Orange software	
RF	2.06	0.19	80.49
NB	2.52	0.24	76.22
ANN	2.45	0.23	77.44
LR	5.52	0.73	27.44
		Weka software	
Model	RMSE	MAE	% of correctly classified instances
RF	0.27	0.13	79.83
NB	0.29	0.17	73.69
Lazy	0.27	0.13	79.70
J48	0.27	0.14	79.73

4. Discussion

4.1. Model Performance

This paper takes advantage of historic data (23 years) collected from flood events to classify the type of the flood that is likely to happen in future. The data was filtered to remove outliers, correct the missing values, order the data, and more, using ADAT application. 69,558 instances (from years 1994 to 2017) were filtered as an output of the ADAT, which has been used in training the machine and the remaining 3642 instances (from 2018) were used for testing. 164 instances were used for test data in Orange software and 3478 instances were used as test data for Weka software with the class of "Event-Type". Five types of ML techniques were tested and evaluated in order to be able to choose the best performing technique. The techniques tested are RF, Lazy, J48 tree, ANN, NB, and LR. Based on the evaluation metrics resulting from the models, it can be concluded that the best performing technique used in Orange software (based on the 164 test cases) proves to be the RF, with RMSE result of 2.06, MAE of 0.19 and correctly classified instance rate of 80.49%, followed by ANN with RMSE result of 2.45, MAE of 0.23 and correctly classified instance rate of 77.44%, NB with RMSE result of 2.52, MAE of 0.24 and correctly classified instance rate of 76.22% and LR with RMSE result of 5.52, MAE of 0.73 and correctly classified instance rate of 27.44%. The best performing technique used in Weka software (based on 3478 test cases) proves to be the RF, with RMSE result of 0.27, MAE of 0.1345 and correctly classified instance rate of 79.83%; followed by J48 with RMSE result of 0.27, MAE of 0.14 and correctly classified instance rate of 79.73%, Lazy with RMSE result of 0.27, MAE of 0.13 and correctly classified instance rate of 79.70% and NB with RMSE result of 0.29, MAE of 0.17 and correctly classified instance rate of 73.69%.

The comparison of the evaluation metrics from the models built using both software tools with different test data sets indicates that the RF performs best amongst all other techniques followed by ANN.

Note that the generated model can be used to provide an insight into the number of flooding events. For example, on using future estimation of weather forecast for the next 10 years from an environment agency, the built model can provide an understanding of the patterns and number of flooding events, and type of flood to be expected for that period.

4.2. Awareness, Preparedness and Resilience

The results of most flooding scenarios indicate that the event is a significant threat to people's lives. Each of the emergency response elements shown in the influence map (Figure 1) such as emergency service, health service, community awareness and media coverage serve as a centre for the organisation of assistance supplies, which involves people who are trained to perform rescue tasks, of which one is for flood incidents. Flooding may result in the loss of a logistic centre and delay in responding to rescue operations. In this scenario, other elements must be used for coordination i.e., there may need to be a collaborative effort amongst response agencies from neighbouring districts and regions to help bring about normality to the affected areas. Sharing of resources and equipment to deal with a flood may be required if the local agencies are operating at their absolute capacity. In the UK, there is existing protocol that agencies adhere to for interagency collaboration and also inter-regional collaboration in times of crisis.

Flooding of infrastructure such as roads and bridges dramatically affect the ability of road users to get from a to b. This not only means that people will be unable to access these routes during a flood, but also that response agencies will be restricted in accessing certain positions, which is clearly problematic if urgent responses are needed. Even once the water withdraws, the residual deposits blocks the usage of the roadways and special equipment must be used to clear them. Flooding of businesses and schools could cause a disruption, preventing employees and students, respectively, from attending.

4.2.1. Flood Type Awareness and Classification

Based on the results of this study, which classifies flood type, it can be concluded that most emergency responders can be alerted of flood type that is likely to happen in the area. If emergency responders are not local to the area, they may not to be aware of what consequences different flood types can have and what kind of assistance is needed, specific to that flood type. For example, they might only envision damage to infrastructure and overlook building damages. Therefore, improvement measures can be taken to train more qualified staff to reduce the impact flooding may have on homes and other infrastructure.

4.2.2. Preparedness Planning

Depending on the flood type frequently happening to an area, plans can be set to reduce the threat and impact from flooding at the local level. The plan, called the preparedness plan, categorises the roles and responsibilities of each stakeholders i.e., emergency responders, firefighters, police, community and etc. that must take action in case of an emergency situation. By identifying the flood type that is likely to happen, the situation can be monitored and if it reaches an alarming level, a warning for evacuation can be issued to the people local to flood-prone areas. Also, the name of the closest shelter and evacuation route can be provided during an emergency.

4.2.3. Resilience

Bringing human knowledge and AI together is an important way to build resilience. The advantage of this research is to help comprehend, prioritise, and respond to the potential impact of flooding based on flood types and protect the community and environment. Flood type classification based on weather forecast will allow for key decision- makers such as local councils and emergency response agencies to take action to put in place mitigation measures to decrease the potential impact of an oncoming event.

This is achieved through better understanding of flood types and to make a long-term strategic plan to prioritise the need for investment based on flood type risk and consequence to reduce impact on lives, infrastructures, finances, etc. Solutions that are resilient to a variety of flood types can be made, mitigation measures can be implemented, and prioritising locations which are at higher risk can be kept under surveillance leading up to an anticipated flood occurrence.

5. Conclusions

This paper describes a robust evaluation of state-of-the-art ML techniques to classify flood type based on weather forecast, location, days event lasted, begin/end location (name of the place), begin/end latitude and longitude, injuries direct/indirect, death direct/indirect and property and crop damage to classify the flood type. The use of ML on historic data in terms of flood type classification is used for the first time in this study. Extensive historic data has been filtered and used for training and testing purposes. Several models were built and compared using evaluation metrics i.e., RMSE, MAE and confusion matrix. The comparison of the evaluation metrics from the models built suggest that the RF technique outperforms other techniques in terms of RMSE, MAE and confusion matrix (accuracy rate of 80.49%), followed by ANN (accuracy rate of 77.44%). One of the benefits of this work is that the same tools and techniques can be used to classify and estimate many other parameters, which have been used in currently used training set i.e., location, potential financial damage, etc.

This study has focused on flooding as a sub-branch of natural disasters. Nevertheless, there are many other possibilities to apply AI in natural disasters and help build resilience. Data mining can be applied to help insurance companies, estimate level of compensations, estimate damage to crops and buildings, and estimate number of injuries and death in specific areas. By being able to estimate more accurately and learn from past events, many lessons can be learned and applied in building resilience against natural disasters. This will also improve public awareness and preparedness, and save lives, if faced with an adverse natural condition.

A constraint in this study is restrictive access to inclusive data. Most of the big data sets are in the hands of private companies, and there are no principles for data sharing. Accessing and collecting these data is difficult and expensive.

The results from this study show for the first time that ML can be used to analyze datasets from historical disaster events to reveal the most likely events that could occur, should similar events be experienced in the future. From the literature review, to the best of the authors' knowledge there is no equivalent set of data as the NCDC NOAA data from a UK source. It would be advantageous for the UK environmental agency to provide a detailed historic data from past natural disasters similar to the NCDC NOAA. This work has proven that the application of ML concept and if such data is made available from the UK, this ML method can be applied, and more advances can be made within the UK not only for flooding, but any type of natural disaster (based on provided data type) to help preparedness, raise awareness and build resilience in disaster management, especially in areas more prone areas to natural disasters.

The results are highly dependent on data quality and precision. If the data is not reliable or is "bad" data, the ML is trained on wrong information and therefore the results will be completely misleading. Missing information or parameter limitation can also adversely affect the model built.

For further study, building a predictive model for future events will be considered. Furthermore, the use of AI in more natural disaster areas and improving resilience in disaster management, especially in the UK, is strongly suggested.

Author Contributions: Conceptualization, S.S. and R.K.; Data curation, S.S. and D.J.; Formal analysis S.S. and D.J.; Funding acquisition, R.K.; Investigation, S.S.; Methodology, S.S.; Project administration, R.K. and G.F.; Resources, S.S.; Software, S.S.; Supervision, R.K.; Validation, S.S. and D.J.; Visualization, S.S.; Writing–original draft, S.S. and D.J.; Writing–review & editing, S.S.; R.K.; M.R.C.; G.F. and F.M.

Funding: This research was funded by the EPSRC for funding on BRIM (Building Resilience Into Risk Management), Ref: EP/N010329/1

Conflicts of Interest: The authors declare no conflict of interest.

References

1. Arslan, M.; Roxin, A.; Cruz, C.; Ginhac, D. A review on applications of big data for disaster management. In Proceedings of the 2017 13th International Conference on Signal-Image Technology & Internet-Based Systems (SITIS), Jaipur, India, 4–7 December 2017; pp. 370–375. [CrossRef]
2. Sande, C.J.V.D.; Jong, S.M.D.; Roo, A.P.J.D. A segmentation and classification approach of IKONOS-2 imagery for land cover mapping to assist flood risk and flood damage assessment. *Int. J. Appl. Earth Obs. Geoinf.* **2003**, *4*, 217–229. [CrossRef]
3. Wang, Y.; Chen, A.; Fu, G.; Djordjevic, S.; Zhang, C.; Savic, D. An integrated framework for high-resolution urban flood modelling considering multiple information sources and urban features. *Environ. Modell. Softw.* **2018**, *107*, 85–95. [CrossRef]
4. Mathioudakis, M.; Koudas, N. Twitter monitor: Trend detection over the twitter stream. In Proceedings of the 2010 ACM SIGMOD International Conference on Management of Data, ACM, Indianapolis, IN, USA, 6–10 June 2018; pp. 1155–1158.
5. Smith, L.; Liang, Q.; James, P.; Lin, W. Assessing the utility of social media as a data source for flood risk management using a real-time modelling framework. *J. Flood Risk Manag.* **2017**, *10*, 370–380. [CrossRef]
6. Sangeetha, S.; Jayakumar, D. Flash flood forecasting using different artificial intelligence method. *Int. J. Eng. Trends Technol.* **2018**, *3*, 140–144. [CrossRef]
7. Robert, P.; Bella, R.; John, C.; Mark, C. Emergency Situation Awareness: Twitter Case Studies. Available online: https://link.springer.com/chapter/10.1007/978-3-319-11818-5_19 (accessed on 9 May 2019).
8. Lopez-Fuentes, L.; Weijer, J.; González-Hidalgo, M.; Skinnemoen, H.; Bagdanov, D.A. Review on computer vision techniques in emergency situations. *Multimed. Tools Appl.* **2017**, *77*, 17069–17107. [CrossRef]

9. Lai, C.L.; Yang, J.C.; Chen, Y.H. A real time video processing-based surveillance system for early fire and flood detection. In Proceedings of the Instru-mentation and Measurement Technology Conference Proceedings, Warsaw, Poland, 1–3 May 2007; pp. 1–6.
10. Liu, F.; Xu, F.; Yang, S. A flood forecasting model based on deep learning algorithm via integrating stacked autoencoders with BP neural network. In Proceedings of the IEEE International Conference on Multimedia Big Data, Laguna Hills, CA, USA, 19–21 April 2017; pp. 58–61.
11. Martinis, S. Automatic near real-time flood detection in high resolution X-band synthetic aperture radar satellite data using context-based classification on irregular graphs. Ph.D. Thesis, Electronic Theses of LMU Munich, Munich, Germany, 2010.
12. Mason, D.C.; Davenport, I.J.; Neal, J.C.; Schumann, G.J.P.; Bates, P.D. Near real-time flood detection in urban and rural areas using high-resolution synthetic aperture radar images. *Geosci. Remote Sens.* **2012**, *50*, 3041–3052. [CrossRef]
13. Sayers, W.; Savić, D.; Kapelan, A.; Kellagher, R. Artificial intelligence techniques for flood risk management in urban environments. *Proc. Eng.* **2014**, *70*, 1505–1512. [CrossRef]
14. Wu, Z.Y.; Liu, W.; Xu, J.; Feng, S.; Palaiahnakote, T.L. Context-aware attention lstm network for flood prediction. In Proceedings of the 24th International Conference on Pattern Recognition (ICPR), Beijing, China, 20–24 August 2018; pp. 1301–1306. [CrossRef]
15. Witten, I.H.; Eibe, F.; Hall, M.A. *Data Mining: Practical Machine Learning Tools and Techniques*, 3rd ed.; Morgan Kaufmann: San Francisco, CA, USA, 2011; p. 191.
16. Mason, D.C.; Speck, R.; Devereux, B.; Schumann, G.J.P.; Neal, J.C.; Bates, P.D. Flood detection in urban areas using TerraSAR-X. *Geosci. Remote Sens.* **2010**, *48*, 882–894. [CrossRef]
17. Di Martino, G.; Iodice, A.; Riccio, D.; Ruello, G. A novel approach for disaster monitoring: Fractal models and tools. *IEEE Trans. Geosci. Remote Sens.* **2007**, *45*, 1559–1570. [CrossRef]
18. Keiller, N.; Samuel, G.F.; Ícaro, C.D.; de Rafael, O.W.; Javier, A.V.M.; Otávio, A.B.P.; Rodrigo, T.C.; Lin, T.L.; dos Jefersson, A.S.; da Ricardo, S.T. Data-driven flood detection using neural networks. Available online: http://ceur-ws.org/Vol-1984/Mediaeval_2017_paper_39.pdf (accessed on 9 May 2019).
19. Lopez-Fuentes, L.; van de Weijer, J.; Bolanos, M.; Skinnemoen, H. Multi-modal deep learning approach for flood detection. In Proceedings of the MediaEval 2017 Workshop, Dublin, Ireland, 13–15 September 2017.
20. Federal Emergency Management Agency (FEMA), Headquarters: Washington D.C., US. Available online: https://www.fema.gov/data-visualization (accessed on 8 May 2019).
21. National Climatic Data Center (NCDC NOAA), Asheville, North Carolina, US. Available online: https://www.ncdc.noaa.gov/stormevents/ftp.jsp (accessed on 8 May 2019).
22. Hall, M.; Eibe, F.; Geoffrey, H.; Pfahringer, B.; Reutemann, P.; Witten, I.H. The WEKA data mining software: An update. *SIGKDD Explor.* **2009**, *11*, 10–18. [CrossRef]
23. The MathWorks, Inc. *MATLAB and Statistics Toolbox Release 2012b*; The MathWorks, Inc.: Natick, MA, USA, 2012.
24. Demsar, J.; Curk, T.; Erjavec, A.; Gorup, C.; Hocevar, T.; Milutinovic, M.; Mozina, M.; Polajnar, M.; Toplak, M.; Staric, A.; et al. Orange: Data mining toolbox in python. *J. Mach. Learn. Res.* **2014**, *14*, 2349–2353.
25. Sammut, C.; Webb, G.I. *Encyclopedia of Machine Learning and Data Mining*; Springer: Boston, MA, USA, 2011.
26. Harris, D.; Harris, S. *Digital Design and Computer Architecture*, 2nd ed.; Elsevier: San Francisco, CA, USA, 2010; p. 129. ISBN 978-0-12-394424-5.

Communication

Achieving Urban Flood Resilience in an Uncertain Future

Richard Fenner [1,*], Emily O'Donnell [2], Sangaralingam Ahilan [3], David Dawson [4], Leon Kapetas [1], Vladimir Krivtsov [5], Sikhululekile Ncube [5] and Kim Vercruysse [4]

[1] Department of Engineering, University of Cambridge, Cambridge CB2 1PZ, UK; lk411@cam.ac.uk
[2] School of Geography, University of Nottingham, Nottingham NG7 2RD, UK; Emily.O'Donnell@nottingham.ac.uk
[3] Centre for Water Systems, College of Engineering, Mathematics and Physical Sciences, University of Exeter, Exeter EX4 4QF, UK; S.Ahilan@exeter.ac.uk
[4] Faculty of Engineering, University of Leeds, Leeds LS2 9JT, UK; D.A.Dawson@leeds.ac.uk (D.D.); k.vercruysse@leeds.ac.uk (K.V.)
[5] School of Energy, Geoscience, Infrastructure and Society, Heriot-Watt University, Edinburgh EH14 4AS, UK; vladimir.krivtsov@hw.ac.uk (V.K.); s.ncube@hw.ac.uk (S.N.)
* Correspondence: raf37@cam.ac.uk; Tel.: +44-1223-765626

Received: 26 April 2019; Accepted: 22 May 2019; Published: 24 May 2019

Abstract: Preliminary results of the UK Urban Flood Resilience research consortium are presented and discussed, with the work being conducted against a background of future uncertainties with respect to changing climate and increasing urbanization. Adopting a whole systems approach, key themes include developing adaptive approaches for flexible engineering design of coupled grey and blue-green flood management assets; exploiting the resource potential of urban stormwater through rainwater harvesting, urban metabolism modelling and interoperability; and investigating the interactions between planners, developers, engineers and communities at multiple scales in managing flood risk. The work is producing new modelling tools and an extensive evidence base to support the case for multifunctional infrastructure that delivers multiple, environmental, societal and economic benefits, while enhancing urban flood resilience by bringing stormwater management and green infrastructure together.

Keywords: blue-green infrastructure; flood risk management; sustainable; drainage systems; resilience; systems

1. Introduction

Achieving Urban Flood Resilience requires solving a number of interrelated engineering, environmental and socio-political challenges to achieve the transformative change needed in urban stormwater and flood risk management. This will involve dealing with the future uncertainties associated with extreme weather events driven by global warming, and the consequences of increasingly rapid urbanisation. Recognising these constraints, new approaches are urgently needed, which are based on adaptive and flexible designs of a range of traditional grey infrastructure (e.g., underground pipes, detention tanks and lined drainage channels) and innovative blue-green solutions (e.g., swales, rain gardens, wetlands, green roofs and restored urban streams). In moving towards source control techniques which provide infiltration, attenuation and storage that mimic the predevelopment hydrology of an area, the paradigm of urban flood management can be switched from threat to opportunity. This emphasises the resource potential of urban stormwater (for example for water supply and energy generation), as part of a wider "system of systems" of urban infrastructure which can be managed interoperably. However, to deliver these changes planning and adoption barriers must be overcome and solutions must be embraced by the communities they serve.

The Urban Flood Resilience research project was launched in 2016 and comprises a consortium of nine UK Universities supported by government agencies, engineering consultants and planning authorities responsible for managing urban water and flood risk [1] The aim is to provide the methodologies and tools needed to make transformative change possible through adoption of a whole systems approach to urban flood and water management. This is being done through the development of the next generation of hydrosystems models [2] that bridge the interfaces between urban/rural and engineering/natural hydrological systems. Flexible adaptation pathways are being assessed using a multiple benefits approach [3] to determine the most effective mix of blue-green and grey systems for any given location and time. Capturing the resource potential of stormwater through rainwater harvesting [4] and local energy recovery using micro-hydropower [5] is being examined as part of a multifunctional systems approach to urban flood management [6]. The consortium is also considering the importance of the interfaces between planners, developers, engineers and beneficiary communities by investigating citizen's interactions with blue-green infrastructure [7]. The results of the research are being applied and demonstrated in a series of case study locations, including Newcastle and Ebbsfleet, supported by effective Learning Action Alliances which are helping translate these findings into practice [8].

2. Preliminary Results

While the work is ongoing this short communication presents a summary of some of the new research outputs not yet reported elsewhere.

2.1. Contribution of Blue-Green Infrastructure to Improving Natural Capital

The impact of future urban intensification on Natural Capital is being investigated by Heriot Watt, Cambridge and Newcastle Universities, focusing on the London Borough of Sutton. Natural Capital refers to the stock of natural features/assets, e.g., freshwater, land, soil, minerals, air, seas, habitats, biodiversity and processes, which together provide the foundation for the flows of ecosystem services [9]. The work is calculating the extent to which blue-green infrastructure systems such as rain gardens, swales and green roofs can mitigate natural resource depletion associated with new development. This is done using the Natural Capital Planning Tool [10] and a geographic information system (GIS) analysis to assess natural capital indicators, such as flood risk regulation, at different spatiotemporal scales.

Research focuses on the residential area of Carshalton, where recent development has led to the existing drainage network reaching its capacity and extreme storm events have caused local flooding incidents. The local authority is planning to mitigate additional flood risk associated with a plan for developing 3000 new homes with new blue-green infrastructure while also providing a "natural capital uplift". Two development approaches to stormwater management have been tested consisting of a "grey" pipe-based approach and a blue-green approach including green roofs, rainwater harvesting, rain gardens and street swales. The effectiveness of each approach is assessed using the CityCat hydrodynamics model [2].

The Natural Capital Planning Tool calculates the impact score for 10 selected ecosystem services together with an overall aggregated development impact score. The calculated impact scores are based on a range of indicator data such as population density, soil drainage class, size of green space sites and spatial land use information for the pre- and post development state of an area. Such information is automatically translated into impact scores based on an expert informed quantification model embedded in the Natural Capital Planning Tool. The tool also calculates theoretical minimum and maximum possible scores which show the potential of the site to lose or gain natural capital and associated multiple benefits.

Results show an overall negative development impact score of −17.35 resulting from the introduction of housing infrastructure on green field sites in Carshalton, with air quality and local climate regulation particularly affected. Compared to the theoretical minimum possible score of −21.09, this implies that the "grey" pipe-based approach will result in loss of several multiple benefits as this approach replaces most natural capital in the area. With the introduction of different blue-green approaches based on the adoption of sustainable drainage systems (SuDS) the overall development impact score, although still negative, is significantly reduced to −0.12. This is because some blue-green options such as road swales will enhance natural capital in some parts of the study area while other options such as green roofs will negatively impact on predevelopment land uses which have more potential for multiple benefit delivery compared to the green roofs option. A summary of the nature of the ecosystem service impact scores for each individual blue-green/SuDS intervention on the predevelopment land use are shown in Table 1. All blue-green approaches had a positive impact on aesthetic value and local climate regulation, with swales having the most positive impact overall.

Table 1. Natural capital impact scores of blue-green infrastructure options on ecosystem services.

Ecosystem Service	Impact Score		
	Swales	Green Roofs	Rain Gardens
Harvested Products (rainwater)	n/a *	negative	negative
Biodiversity	positive	negative	positive
Aesthetic Values	positive	positive	positive
Recreation	positive	negative	negative
Water Quality Regulation	n/a	negative	n/a
Flood Risk Regulation	n/a	negative	n/a
Air Quality Regulation	positive	negative	negative
Local Climate Regulation	positive	positive	positive
Global Climate Regulation	positive	negative	positive
Soil Contamination	n/a	n/a	n/a

* n/a is not applicable.

For the blue-green/SuDS interventions hydrodynamic modelling showed an increase in water depth in the swales and a decrease in water depths over the northern part of the area. The interventions in the southern area were found to be less useful as this has less connectivity with the downstream parts of the catchment and the outfalls into the Wandle River. Overall the number of floodable properties was reduced by over 65%. The results highlight the trade-offs and synergies in multiple benefits associated with different combinations of blue-green options. It was found that grey development options increased the flood risk downstream compared to a three times reduction using blue green options. Overall such Natural Capital assessments can help practitioners, local authorities, planning agencies and developers understand the interdependencies between the natural and urban environments, highlight the different environmental and social benefits that strategies may deliver and provide insight into the relative performance of a range of flood risk management measures.

2.2. Adaptation Pathways for Drainage Infrastructure Planning

Again using Carshalton to demonstrate new assessment procedures, Cambridge University has developed a roadmap for urban drainage adaptation over the next 40 years, shown conceptually in relation to changing climate inputs Figure 1.

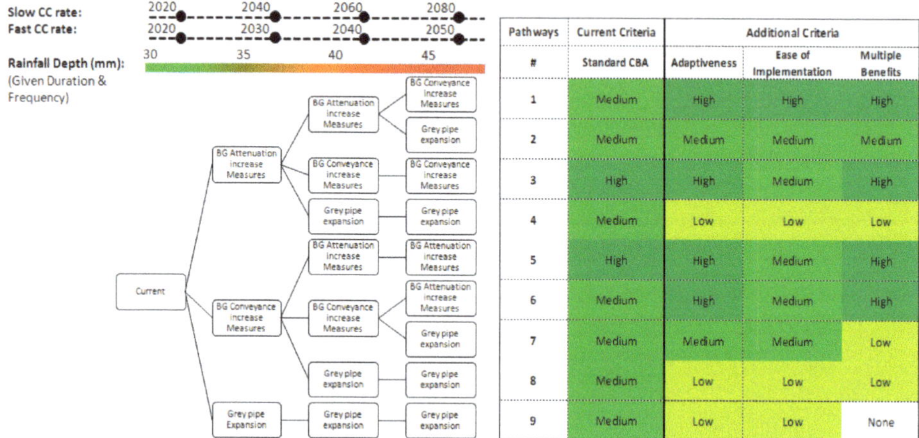

Pathways	Current Criteria	Additional Criteria		
#	Standard CBA	Adaptiveness	Ease of Implementation	Multiple Benefits
1	Medium	High	High	High
2	Medium	Medium	Medium	Medium
3	High	High	Medium	High
4	Medium	Low	Low	Low
5	High	High	Medium	High
6	Medium	High	Medium	High
7	Medium	Medium	Medium	Low
8	Medium	Low	Low	Low
9	Medium	Low	Low	None

Figure 1. Assessment of adaptation pathways to meet uncertainties in future climate.

In comparing each of these possible future pathways additional criteria are used to complement conventional cost–benefit analysis, such as indicators of adaptiveness, flexibility, ease of implementation and a monetised and spatial evaluation of the wider multiple benefits that can be delivered by SuDS and blue-green options. The procedures are using the SWMM dynamic rainfall-runoff-routing simulation model (as developed by US Environment Protection Agency (EPA)) to identify when service thresholds are exceeded, triggering the need for further interventions. The integration of a variety of appraisal techniques offers a new perspective to help inform the timing and placement of blue-green interventions, while acknowledging future uncertainties. As an envelope of possible climatic and urbanisation rates are considered, a viable planning horizon becomes evident. The approach provides a pragmatic response to changing drivers of urban flooding allowing real options evaluation techniques to help determine the scale of interventions that are required and when they should be implemented.

2.3. Interoperability

Leeds University have introduced the concept "interoperability" to guide transition from local multifunctionality to city-scale multisystem flood management, through actively managing connections between infrastructure systems to convey, divert and store flood water. They define interoperability as "The ability of any water management system to redirect water and make use of other system(s) to maintain or enhance its performance function during exceedance events" [6].

The work is focusing on achieving a better understanding of how flood prone areas are linked to flood source areas within urban catchments and how opportunities for capturing or transferring stormwater along these pathways can be identified. A source to impact analysis is using a transferable modelling approach based on the CityCAT hydrodynamic model to systematically identify locations contributing most to flood hazard within a catchment. This is reported using a spatial analysis framework by synthesising and combing spatial data on (i) flood hazard, (ii) intervention efficiency and (iii) opportunities for interoperability at the catchment scale. Figure 2 illustrates the application of these ideas in the case study city of Newcastle.

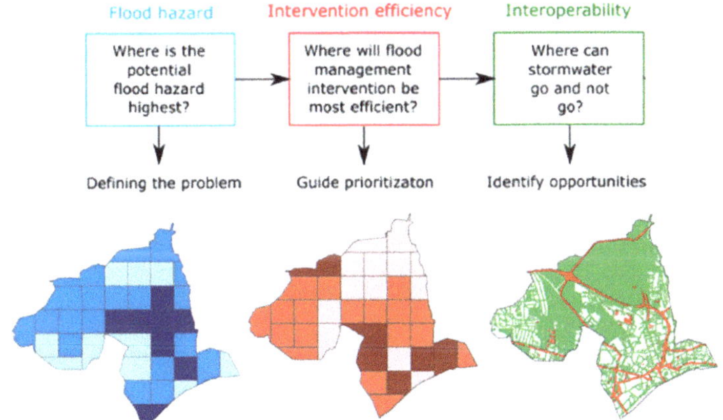

Figure 2. Geographic information system (GIS) analysis of flood impacts across Newcastle.

The first image (flood hazard) depicts the potential flood hazard from a 1 in 50 year event mostly in the middle to lower part of the city centre. The second image (intervention efficiency) shows flood source areas, which are located mostly in the upper and lower part of the catchment. The third image (interoperability) illustrates the infrastructure systems that can potentially act as a water management asset (green) and systems where additional flood water should be avoided.

This approach can help identify different types of flood source areas and, when combined with information on infrastructure systems, can guide the selection of appropriate flood management solutions from a catchment perspective. Linking flood hazard to flood source areas provides insights into the hydrological processes and spatial interactions within the urban catchment, and can help prioritise locations for flood management intervention.

2.4. Uncovering Implicit Perceptions of Blue-Green Interventions for Flood Resilience

The Universities of Nottingham and the West of England are working to better understand the attitudes and preferences towards blue-green and grey infrastructure in public open space, moving beyond stated preference studies where attributes are openly articulated (e.g., through questionnaires) to developing new Implicit Association Tests (IATs) that measure hidden perceptions and subconscious attitudes.

Explicit and implicit preferences for SuDS in public greenspace were investigated using, respectively, a Likert scale test and an IAT, based on the method presented by Greenwald, McGhee, and Schwartz [11] and using the FreeIAT software (Adam W.Meade, North Carolina State University, Raleigh, CA, USA) [12]. In the IAT, participants responded to photographs (target concepts) illustrating public greenspace with and without SuDS and to positive and negative words representing evaluative attributes. Implicit preferences were calculated based on reaction times. Key findings from a trial with 44 participants in Bristol revealed significant differences between implicit and explicit preferences for public greenspace with and without SuDS. Overall, respondents tended to explicitly prefer greenspace without SuDS (70%, compared with 7% who explicitly preferred greenspace with SuDS) (Figure 3). In contrast, more respondents implicitly preferred public greenspace with SuDS (41%, compared with 30% who implicitly preferred greenspace without SuDS). This suggests that the respondents' subconscious attitudes are more favourable towards SuDS, rather than just greenspace. The combined explicit–implicit tests provide insight into perceptions of attractiveness, safety and tidiness associated with SuDS and public greenspace which may help blue-green infrastructure be designed in ways that may improve public acceptability of features.

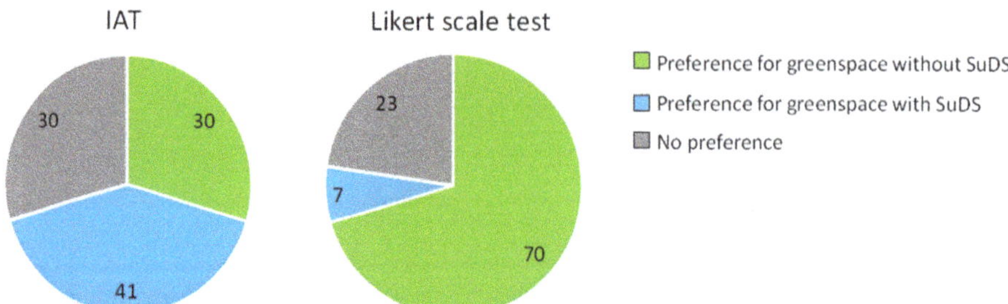

Figure 3. Attitudes towards sustainable drainage systems (SuDS) in public greenspace evaluated via an Implicit Association Tests (IAT) and Likert scale test (explicit measure). Sample population consists of 44 respondents in Bristol, UK, who completed the tests in May–July 2018.

In related work, the Heriot-Watt University team are investigating the potential for SuDS retrofitting at the Houston Industrial Estate, Scotland, assessing the public awareness of SuDS technology, relevant regulations and barriers to retrofit using a questionnaire survey. Most companies were unaware of the Scottish Environmental Protection Agency's General Binding Rules, which provide statutory controls over certain low risk activities that may affect the water environment in Scotland (e.g., diffuse pollution), although most companies claimed familiarity with some SuDS techniques such as permeable paving and gravel filter drains. Many of the potential plot scale techniques, such as detention basins and larger scale rain gardens, were unfamiliar to most companies and there was a lot of confusion in the understanding of SuDS features. A site survey confirmed the presence or absence of these features on individual plots and compared them to the claims of the site owners. Approximately 40 SuDS features, including detention basins, swales, filter strips and gravel filter drains (and more), were claimed by premises on the industrial estate and yet were not identified as present when the research team conducted their survey. Conversely permeable block paving was ubiquitous across the development but not always recognised as present by the respondents. This work suggests that there is a clear need for sustained engagement and education to raise awareness of SuDS technology and overcome barriers to their adoption in areas where they need to be retrofitted.

2.5. Urban Metabolism Modelling in Ebbsfleet Garden City

Urban metabolism refers to the combination of the technical and socioeconomical processes that occur in cities and that result in growth, production of energy and elimination of waste. The metabolism-based modelling approach led by Exeter University overcomes issues commonly encountered by independent modelling and management of urban water systems (water supply, wastewater and surface water collection) by providing an integrated approach that considers the interconnections and interdependencies of all urban water subsystems. This approach increases resilience to extreme events, from floods to droughts, by enabling the integrated modelling of future water management intervention strategies, such as water harvesting and grey water recycling, and their impact on the performance of downstream infrastructure such as stormwater systems.

The study is using WaterMet^2, a mass-balance conceptual urban metabolism modelling tool [13] to evaluate the sustainability performance of the urban water systems in the Ebbsfleet Garden City, over a predefined long-term planning horizon and for a range of future possible intervention strategies. The integrated evaluation of the water and wastewater management master plans of the local water companies (Thames and Southern Water) through WaterMet^2 simulations will aid long-term decision making by illustrating the impact of different sustainability strategies, including options for wastewater treatment, on the urban water system.

The ultimate aim is to couple this work with a semi-quantitative system dynamics model (currently being coproduced by the Ebbsfleet Water Forum) to further assess sustainable water management options for the Ebbsfleet Garden City. The Ebbsfleet Water Forum, established by the Urban Flood Resilience consortium in 2017 and based on a 'Learning and Action Alliance' framework [8], is coordinated by Open University team members and the Ebbsfleet Development Corporation. The vision of the Forum is to incorporate blue-green infrastructure into development design from the outset, and champion urban flood resilience to encourage the realistic delivery of sustainable urban water management. The system dynamics model investigates sustainable water management options for the Garden City, including the reduction of residential potable water use, increased blue-green space and SuDS, and rainwater harvesting. An early causal loop diagram is presented in Figure 4. The system dynamics model is currently being codeveloped iteratively by the project team and Ebbsfleet stakeholders, and will be run under a range of future climate and socioeconomic conditions, and including a range of policy incentives, to enhance the capacity of local stakeholders to influence policy in a more sustainable direction.

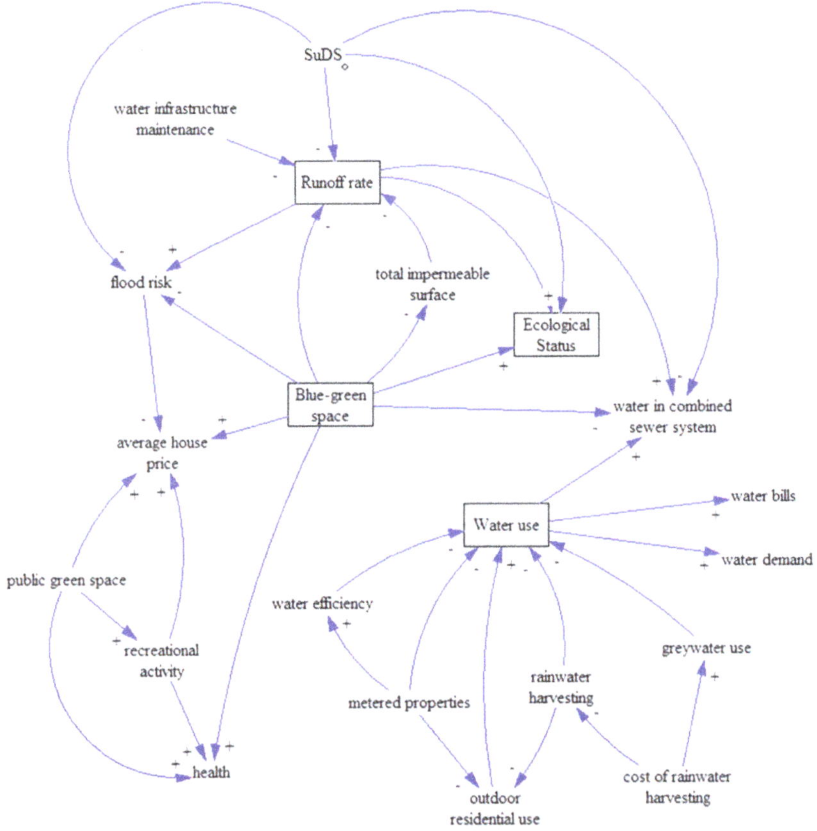

Figure 4. Early version of the causal loop diagram created by the project team and members of the Ebbsfleet Water Forum illustrating variables and linkages in response to a discussion of sustainable water management options for the Ebbsfleet Garden City.

2.6. Further Research

Further research of the consortium not reported here includes work by Newcastle University on developing a new comprehensive model of urban hydrosystems, a study of suspended particulate matter and water quality by Heriot-Watt University illustrating ecosystem functioning of SuDS ponds, challenges associated with implementing SuDS through the strengthened English planning system led by the Open University, effective approaches to blue-green infrastructure community engagement by the University of the West of England and Exeter University's development of Rainwater Management Systems that concurrently reduce stormwater discharges and potable water consumption.

3. Conclusions

Achieving urban flood resilience is a multifaceted problem which requires integrated solutions across a range of disciplines: From advances in modelling the hydrodynamic performance of combined grey and blue-green interventions, and flexible engineering design, to social insights into public perceptions and community acceptability. The work reported in this paper has shown that adopting blue-green approaches which rely on storage and infiltration through vegetated surfaces can contribute to wider Natural Capital in areas subject to rapid urbanisation. Adaptive design pathways that help identify the right balance between grey pipe-based approaches and sustainable drainage systems, and how these can be managed interoperably across urban catchments, are key outputs to date, in addition to research into SuDS perceptions and sustainable water management options through system dynamics modelling, that generate insight into how the public and businesses understand the value of blue-green systems.

Author Contributions: Conceptualization, Writing—Review & Editing: R.F. and E.O.; Section 2.1: S.N., Table 1, Figure 1 and Section 2.2: L.K. and R.F., Section 2.3 and Figure 2: K.V. and D.D., Section 2.4: E.O. and V.K., Section 2.5: S.A.

Funding: This research was performed as part of an interdisciplinary project undertaken by the Urban Flood Resilience Research consortium (www.urbanfloodresilience.ac.uk). This work was supported by the Engineering and Physical Sciences Research Council (grant numbers EP/P004180/1, EP/P003982/1, EP/P004210/1, EP/P004237/1, EP/P004261/1, EP/P004296/1, EP/P004318/1, EP/P004334/1 and EP/P004431/1).

Acknowledgments: The authors acknowledge the contributions of these peoples (Colin Thorne, Scott Arthur, Stephen Birkinshaw, David Butler, Brian D'Arcy, Glyn Everett, Vassilis Glenis, Chris Kilsby, Jessica Lamond, Greg O'Donnell, Karen Potter, Tudor Vilcan, Nigel Wright) in developing the above work.

Conflicts of Interest: The authors declare no conflict of interest. The funders had no role in the design of the study; in the collection, analyses, or interpretation of data; in the writing of the manuscript, and in the decision to publish the result

References

1. Urban Flood Resilience Project (2019), Key Project Outputs Report. Available online: http://www.urbanfloodresilience.ac.uk/publications/publications.aspx (accessed on 26 April 2019).
2. Glenis, V.; Kutija, V.; Kilsby, C. A fully hydrodynamic urban flood modelling system representing buildings, green space and interventions. *Environ. Modell. Softw.* **2018**, *109*, 272–292. [CrossRef]
3. Fenner, R. Spatial evaluation of multiple benefits to encourage multi-functional design in blue green cities. *Water* **2017**, *9*, 953. [CrossRef]
4. Ahilan, S.; Melville-Shreeve, P.; Kapelan, Z.; Butler, D. The influence of household rainwater harvesting system design on water supply and stormwater management efficiency. In *New Trends in Urban Drainage Modelling*; Mannina, G., Ed.; Springer: Cham, Swizerland, 2018; pp. 369–374. ISBN 33199986762018.
5. Costa, J.; Fenner, R.; Kapetas, L. A Screening Tool to Assess the Potential for Energy Recovery from the Discharge of Stormwater Runoff. In Proceedings of the Institution of Civil Engineers-Engineering Sustainability, London, UK, 20 December 2018; ICE Publishing: London, UK, 2018. [CrossRef]
6. Vercruysse, K.; Dawson, D.; Wright, N. Interoperability: A conceptual framework to bridge the gap between multi-functional and multi-system urban flood management. *J. Flood Risk Manag.* **2019**. [CrossRef]
7. Everett, G.; Lamond, J. Considering the value of community engagement for (co-) producing Blue-Green infrastructure. In *Urban Water Systems & Floods II. WIT Transactions on the Built Environment*; WIT Press: Southampton, UK, 2018; Volume 184, pp. 1–14.
8. O'Donnell, E.; Lamond, J.; Thorne, C. Learning and Action Alliance framework to facilitate stakeholder collaboration and social learning in urban flood risk management. *Environ. Sci. Policy* **2018**, *80*, 1–8. [CrossRef]

9. Guerry, A.D.; Polasky, S.; Lubchenco, J.; Chaplin-Kramer, R.; Daily, G.C.; Griffin, R.; Ruckelshaus, M.; Bateman, I.J.; Duraiappah, A.; Elmqvist, T.; et al. Natural capital and ecosystem services informing decisions: From promise to practice. *Proc. Natl. Acad. Sci. USA* **2015**, *112*, 7348–7355. [CrossRef] [PubMed]

10. Holzinger, O.; Sadler, J.; Scott, A. Natural Capital Planning Tool. 2018. Available online: http://ncptool.com/ressources/ (accessed on 12 April 2019).

11. Greenwald, A.G.; McGhee, D.E.; Schwartz, J.L. Measuring individual differences in implicit cognition: The implicit association test. *J. Person. Soci. Psychol.* **1998**, *74*, 1464. [CrossRef]

12. Meade, A.W. FreeIAT: An open-source program to administer the implicit association test. *Appl. Psychol. Meas.* **2019**, *33*, 643. [CrossRef]

13. Behzadian, K.; Kapelan, Z. Modelling metabolism based performance of an urban water system using Water MET 2. *Resour. Conserv. Recycl.* **2015**, *99*, 84–99. [CrossRef]

Article

A Conceptual Time-Varying Flood Resilience Index for Urban Areas: Munich City

Kai-Feng Chen * and Jorge Leandro

Chair of Hydrology and River Basin Management, Department of Civil, Geo and Environmental Engineering, Technical University of Munich, Arcisstrasse 21, 80333 Munich, Germany; jorge.leandro@tum.de
* Correspondence: kaifeng.chen@tum.de; Tel.: +886 927-211-971

Received: 9 March 2019; Accepted: 16 April 2019; Published: 19 April 2019

Abstract: In response to the increased frequency and severity of urban flooding events, flood management strategies are moving away from flood proofing towards flood resilience. The term 'flood resilience' has been applied with different definitions. In this paper, it is referred to as the capacity to withstand adverse effects following flooding events and the ability to quickly recover to the original system performance before the event. This paper introduces a novel time-varying Flood Resilience Index (FRI) to quantify the resilience level of households. The introduced FRI includes: (a) Physical indicators from inundation modelling for considering the adverse effects during flooding events, and (b) social and economic indicators for estimating the recovery capacity of the district in returning to the original performance level. The district of Maxvorstadt in Munich city is used for demonstrating the FRI. The time-varying FRI provides a novel insight into indicator-based quantification methods of flood resilience for households in urban areas. It enables a timeline visualization of how a system responds during and after a flooding event.

Keywords: flood resilience index; flood resilience analysis; urban floods; flood risk assessment; flood inundation modelling

---

1. Introduction

According to worldwide evidence of the last decades, the frequency and severity of extreme flooding events in urban areas are increasing [1–3]. The characteristics of an urban environment, such as the high portion of impervious area and increased population density, raise the vulnerability to flooding [4,5]. Traditional engineering measures face great challenges in providing sufficient flood protection when facing a more severe and frequent flooding condition [6,7]. In response, current flood protection strategies move away from measures to increase flood proofing towards flood resilience [8]. Various approaches to improve resilience to urban flooding have been proposed recently across different continents, such as the Best Management Practices (BMPs), Low Impact Development (LID), or Sustainable Urban Drainage Systems (SUDS). Furthermore, policies for improving public awareness of flood risk, advocating flood insurance, automated warning systems, etc. have also been advocated [9]. These approaches aim to mitigate the flooding impacts in cities by maintaining a high level of system performance during flooding events and facilitating the recovery stage of the system after flooding, i.e., its resilience. This study aims to develop a novel methodology to assess the flood resilience level of households within an urban area with time during and after a flooding event by incorporating physical, social, and economic factors.

There are numerous studies evaluating the benefits of flood resilience-enhancing strategies. However, many of them focus on flood impact reduction instead of resilience. Indeed, various flood impact assessment techniques have been formulated according to a wide diversity of research purposes, availability of data, and accessibility of resources [10]. On the contrary, the assessment of flood resilience faces many challenges, including its definition, dimensions used (e.g., social, economic, or physical aspects), and methods of quantification [11,12]. Nevertheless, there is a growing number of research projects and studies aiming at quantifying flood resilience using integrated [13] or multi-criteria [14] approaches, assessing climate variability [15] or the impact of infrastructure [16,17] while considering socioeconomic aspects [18]. Governance strategies for improving flood resilience have also been studied [19].

The study of resilience was originated in the field of ecology [20], where Holling defined it as the measure of the ability of an ecosystem to absorb changes and persist [21]. Since then, variations of the resilience concept started to emerge in different research fields. In the context of flood risk and flood management, various definitions have been introduced recently [22–27]. According to the literature, the definitions of flood resilience differ from each other. However, they generally comprise two major elements: 1. The coping capacity in the face of flooding, and 2. the recovery capacity after flooding. In this paper, these two major elements are adopted. Flood resilience is thus defined as "the capacity to withstand adverse effects following flooding events and the ability to quickly recover to the original system performance before the event".

Resilience assessment can be used to evaluate flood risk management strategies at a city scale [28–30]. However, there still exists no consensus on how to measure flood resilience [31]. One commonly applied approach to quantify resilience is to utilize indicators that measure the characteristics of a system facing urban flooding. De Bruijn defined a set of indicators for flood resilience quantification, which covers three aspects: The amplitude of reaction, the graduality of the increase of the reaction with increasingly severe flood waves, and the recovery rate [32]. These three aspects describe the state of system performance when facing flooding events. In addition, the value of indicators reflects the physical, social, and economic factors regarding flood risk management. Batica and Gourbesville developed an urban flood vulnerability and resilience assessment tool with indicators providing a comprehensive overview of vulnerability and resilience of a city and its community [33]. An index is proposed to describe resilience level by assigning grades (0 to 5) to different indicators according to the availability levels to various urban services when facing a 100-year flooding event. Mugume et al. quantified the resilience of urban drainage systems in the UK by applying the utility performance function combined with the depth–damage data for residential properties that relates the overall performance of a drainage system to flood depths [25]. Analogously, Lee and Kim proposed a resilience index for urban drainage systems in Korea based on flooding damage that resulted from damage functions calculated by multi-dimensional flood damage analysis [34]. However, both studies on urban drainage systems lack socioeconomic aspects when estimating resilience. Bertilsson and Wiklund developed a spatialized index to measure and visualize flood resilience changes in an urban area of Rio [35], incorporating five dimensions: Flood level, exposed population, susceptibility, material recovery, and flood duration.

Despite the already existing studies on flood resilience quantification, there is a lack of methods for assessing how a system's resilience level is affected during and after flooding. As discussed, most existing studies are not time-dependent. Therefore, the aim of this study is to propose a time-dependent method for quantifying flood resilience of households in urban areas.

Section 2 introduces the study area of Maxvorstadt in Munich city. In Section 3, the structure of the Flood Resilience Index (FRI) and the computation of each parameter are explained in detail. In Sections 4 and 5, the inundation and FRI modelling results for Maxvorstadt as well as the sensitivity analysis of the applied reference parameters are provided and discussed. Finally, in Section 6, the conclusion highlights the main advantage and limitation of the proposed FRI method and consideration for future work.

2. Study Area and Data

The study area of Maxvorstadt is one of the 25 boroughs within Munich city, located at the city center. The borough contains an area of 429.79 ha, and is composed of 69% of buildings, 7% of recreation area, and 24% of road surface [36]. The geographical range of the study site is trimmed alongside the roads at the boundary of the administrative area of Maxvorstadt to exclude buildings crossing over multiple boroughs. Maxvorstadt consists of nine districts, which are Königsplatz, Augustenstraße, St. Benno, Marsfeld, Josephsplatz, Am alten nördlichen Friedhof, Universität, Schönfeldvorstadt, and Maßmannbergl (see Figure 1a). Table 1 shows the number of buildings and area within each district. Figure 1b illustrates the population and age distribution of each district. The population of Maxvorstadt lies mainly between 20 to 30 years old [36]. Like other urban areas, the majority of the surface area within Maxvorstadt is sealed. However, there are parks, cemeteries, and lawns composing 7% of the total area as green spaces. Furthermore, some buildings are constructed with green roofs or roof-top gardens, making up more green surface areas. Figure 1c shows the land use map of the study site.

The average yearly precipitation from 1981 to 2010 of Munich City was 944 mm [37]. Regarding rainfall events, the German Meteorological Office (Deutscher Wetterdienst, DWD) provides a dataset storing grids of return periods of heavy rainfall over Germany (Koordinierte Starkniederschlagsregionalisierung und -auswertung des DWD, KOSTRA-DWD). The dataset contains statistical rainfall intensity values as a function of the duration and return period. It is often applied to assess damages caused by severe design rainfalls with regard to their return period [38]. This paper applies the latest version of the dataset, the KOSTRA-DWD-2010R, which encompasses the time period from 1951 to 2010 and focuses on the 15 min duration rainfall events for various return periods (see Figure 1d).

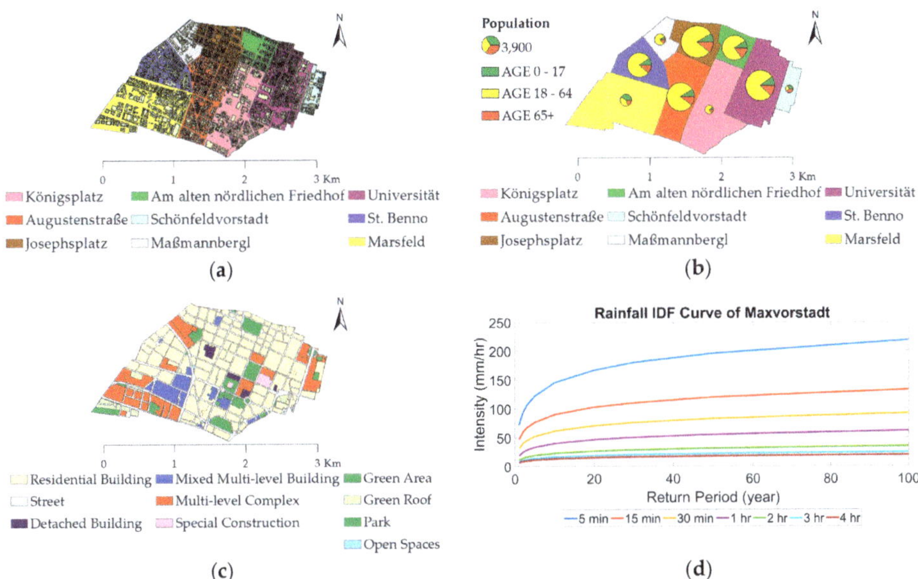

Figure 1. (**a**) Location of the nine districts and buildings within Maxvorstadt; and (**b**) demographic structure with age distribution of Maxvorstadt. The size of the age pie chart is proportional to the population amount. (**c**) Land use map of Maxvorstadt; and (**d**) Rainfall Intensity Duration Frequency curve of Maxvorstadt. Data retrieved from KOSTRA-DWD-2010R database (Grid no. 92049) containing information from 1951 to 2010. This paper applies the 15 min duration rainfall events for various return periods.

Table 1. The amount of buildings and area for each district within Maxvorstadt.

District Name	N [2]	% of Total N	Area [ha]	% of Total Area	D [3]
Königsplatz	417	7.48%	62.43	16.83%	6.68
Augustenstraße	1196	21.46%	51.88	13.98%	23.05
St. Benno	620	11.13%	32.52	8.76%	19.07
Marsfeld	630	11.30%	75.96	20.47%	8.29
Josephsplatz	787	14.12%	31.30	8.44%	25.14
Am a- n- Friedhof [1]	437	7.84%	21.35	5.75%	20.47
Universität	1195	21.44%	64.84	17.48%	18.43
Schönfeldvorstadt	168	3.01%	13.91	3.75%	12.08
Maßmannbergl	123	2.21%	16.83	4.54%	7.31
Maxvorstadt	5573	100%	371.02	100%	15.02

[1] Am alten nördlichen Friedhof. [2] Number of buildings. [3] Building density [building/ha].

3. Methods

3.1. Time-Varying Flood Resilience Index: FRI

3.1.1. Structure of the Flood Resilience Index

A time-varying Flood Resilience Index (FRI) is developed to quantify the resilience level of households in Maxvorstadt, ranging from 0 to 1 as the minimum and maximum value, respectively. The FRI quantifies the capacity to withstand the adverse effects during flooding and the ability to quickly recover from them at each timestep. Indicators reflecting physical, social, and economic dimensions are considered for computing the FRI.

The evaluation of the FRI is split into two phases: The event phase and the recovery phase, depending on the indoor water depth (see Figure 2). In the event phase, physical indicators from flood modelling, i.e., water depth, accumulated water depth, flooding duration, and water accumulation rate are incorporated to assess the flooding impacts. It is assumed that after each flooding event, when the indoor water depth recedes to zero, the recovery phase is initiated. Aside from the physical indicators, social indicators (i.e., percentage of households with children and percentage of elderly population) and economic indicators (i.e., household income) are considered to evaluate the recovery capacity, which facilitates the system to bounce back to the original performance level before flooding (FRI = 1). A description of each indicator and its computation will be introduced in the following section.

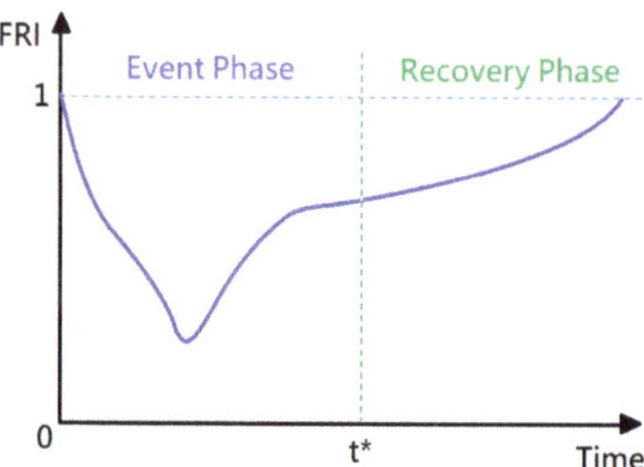

Figure 2. Illustration of the Flood Resilience Index (FRI) structure. t* stands for the timestep when the indoor water depth returns to zero, which separates the event and recovery phases.

3.1.2. Event Phase Indicators

When the indoor water depth is larger than zero, it is considered as an event phase. In this case, four physical indicators are considered for calculating the FRI: Water depth ($I_h(t)$), accumulated water depth ($I_{AWD}(t)$), flooding duration ($I_D(t)$), and water accumulation rate ($I_{WAR}(t)$).

The water depth indicator indicates the severity of flooding at each timestep. The higher the water depth is, the more a household, human, and items are affected, and thus the less resilient the system becomes. A value is assigned to a reference parameter, which indicates the maximum water depth that the building can withstand (h_{ref} [m]). The resilience level decreases as the indoor water depth rises, and once the indoor water depth exceeds the reference parameter, the water depth indicator becomes zero. Equation (1) computes the water depth indicator ($I_h(t)$), where variable $h_{in}(t)$ [m] is the indoor water depth at time t. The reference parameter of water depth h_{ref} [m] is assigned a value of 0.5 m.

$$I_h(t) = \begin{cases} 1 - \frac{h_{in}(t)}{h_{ref}}, & if \ h_{ref} \geq h_{in}(t) \\ 0, & otherwise \end{cases} \tag{1}$$

The water depth indicator shows the severity of the flooding at a certain timestep. However, it is also important to investigate the full scope of the impact that the flooding event has caused. Hence, the accumulated water depth indicator is developed. A reference parameter is inserted, stating the maximum accumulated water depth that the building can withstand (AWD_{ref} [m]). Equation (2) calculates the accumulated water depth indicator ($I_{AWD}(t)$) at every time step (every 10 s), where t_s [s] is the starting time of the flooding event. AWD_{ref} [m] is assigned a value of 3 m.

$$I_{AWD}(t) = \begin{cases} 1 - \frac{\sum_{t_s}^{t} h_{in}(t)}{AWD_{ref}}, & if \ AWD_{ref} \geq \sum_{t_s}^{t} h_{in}(t) \\ 0, & otherwise \end{cases} \tag{2}$$

The duration of the flooding event plays an important role in evaluating the FRI. The longer the flood lasts, the higher damage it will cause. Young points out several impacts that the long-lasting floods could bring to human health, including toxic chemical exposure, growing mold causing respiratory problems, and mosquitos carrying a variety of diseases [39]. In addition, financial damage, social losses, and impacted level of well-being, such as breakdowns of factories and transportations, can have a large impact on the society leading to a lower resilience level [40–42]. These adverse effects become more significant when the flood duration increases. The flooding duration indicator ($I_D(t)$) is calculated by Equation (3), where $D(t)$ [min] stands for the flooding duration until time t. The reference parameter D_{ref} [min] presents the maximum flooding duration that a household can withstand, which is assigned a value of 800 min.

$$I_D(t) = \begin{cases} 1 - \frac{D(t)}{D_{ref}}, & if \ D_{ref} \geq D(t) \\ 0, & otherwise \end{cases} \tag{3}$$

The rising rate of the floodwater is one of the most influential factors which determines the damage magnitude caused by flooding events. For instance, the evacuation procedure should be executed within a limited time span. If the rising rate of the floodwater is high, the evacuation might be incomplete or executed with a reduced efficiency. Facilities with higher vulnerability to fast-rising water, such as schools and nursing homes, will then have a much lower resilience level. In this paper, a water accumulation rate indicator is considered at the rising stage of a flood. Equation (4) computes the water accumulation rate indicator ($I_{WAR}(t)$), where $r_{rise}(t)$ [cm/min] stands for the water rising rate during time t and $t-1$. The reference parameter WAR_{ref} [cm/min] represents the highest water rising rate that can be tolerated, which is assigned a value of 5 cm/min.

$$I_{WAR}(t) = \begin{cases} 1 - \frac{r_{rise}(t)}{WAR_{ref}}, & if \ WAR_{ref} \geq r_{rise}(t) \\ 0, & otherwise \end{cases} \tag{4}$$

3.1.3. Recovery Phase Indicators

When the indoor water depth recedes to zero, the recovery phase is initiated. In this case, not only the physical indicators, but the social and economic ones are applied for calculating a recovery factor in order to enhance the FRI after flooding. The four physical indicators include flood severity (I_{fs}), total flooding depth (I_{TFD}), total flooding time (I_{TFT}), and maximum water accumulation rate (I_{WRAmax}). The two social indicators are households with children (I_C) and elderly population (I_E). The economic indicator is household income (I_I). Seven indicators in total comprise the recovery factor, which is a product of seven exponential terms.

The concepts of the physical indicators in the recovery phase are similar to those in the event phase. However, there is a slight difference at the evaluation time frame. Instead of taking values for the numerators at current time steps, the maximum or the accumulated values during the previous event phase are considered. For flood severity (I_{fs}) and maximum water accumulation rate indicators (I_{WARmax}), the maximum value of the indoor water depth and the water accumulation rate within the previous event phase are considered, respectively. As for total flooding depth (I_{TFD}) and total flooding time indicators (I_{TFT}), a cumulative value of the indoor water depth and total flooding duration within the previous event phase are considered. Equations (5)–(8) show the calculation of the four physical indicators, respectively. Variables t_s and t_e represent the starting and ending timesteps of flooding in the previous event phase.

$$
I_{fs} = \begin{cases} e^{\left(1-\frac{\max\limits_{t \in [t_s,t_e]} h_{in}(t)}{h_{ref}}\right)}, & if\ h_{ref} \geq \max\limits_{t \in [t_s,t_e]} h_{in}(t) \\ 1, & otherwise \end{cases}
\tag{5}
$$

$$
I_{TFD} = \begin{cases} e^{\left(1-\frac{\sum_{t_s}^{t_e} h_{in}(t)}{AWD_{ref}}\right)}, & if\ AWD_{ref} \geq \sum_{t_s}^{t_e} h_{in}(t) \\ 1, & otherwise \end{cases}
\tag{6}
$$

$$
I_{TFT} = \begin{cases} e^{\left(1-\frac{D(t_e)}{D_{ref}}\right)}, & if\ D_{ref} \geq D(t_e) \\ 1, & otherwise \end{cases}
\tag{7}
$$

$$
I_{WARmax} = \begin{cases} e^{\left(1-\frac{\max\limits_{t \in [t_s,t_e]} r_{rise}(t)}{WAR_{ref}}\right)}, & if\ WAR_{ref} \geq \max\limits_{t \in [t_s,t_e]} r_{rise}(t) \\ 1, & otherwise \end{cases}
\tag{8}
$$

Social and economic indicators are assigned to evaluate the recovery capacity from flooding for each household according to different districts within Maxvorstadt. The demographic and social–economic characteristics, such as race, gender, age, and income are principal drivers of a population's ability to recover from damaging flooding events [43–45]. The more children and elderly people within a district, the higher vulnerability to flooding and lower recovery strength the community has. Equations (9) and (10) show the calculation for the indicators of households with children (I_C) and elderly population (I_E), respectively. Furthermore, household income straightforwardly reflects the recovery strength from a flooding event. The more a household earns, the easier and faster it can recover from flooding by repairing or replacing the damaged goods. Equation (11) computes the income indicator (I_I).

$$
I_C = \begin{cases} e^{\left(1-\frac{C}{C_{ref}}\right)}, & if\ C_{ref} \geq C \\ 1, & otherwise \end{cases}
\tag{9}
$$

$$
I_E = \begin{cases} e^{\left(1-\frac{E}{E_{ref}}\right)}, & if\ E_{ref} \geq E \\ 1, & otherwise \end{cases}
\tag{10}
$$

$$
I_I = \begin{cases} e^{\frac{I}{I_{ref}}}, & if\ I_{ref} \geq I \\ e^{1}, & otherwise \end{cases}
\tag{11}
$$

C [%] and E [%] stand for the percentage of households with children and elderly population, respectively, in the district that the household is seated in. Reference parameters, C_{ref} [%] and E_{ref} [%], are assigned values 20% and 12%, respectively, which provide the thresholds that the recovery capacity decreases as C and E increase. I [€] represents the annual household income, and reference parameter I_{ref} [€], assigned 80,000€, represents the threshold that the recovery capacity increases as I increases.

3.1.4. Time Series of the Flood Resilience Index

Once the indicators for evaluating FRI in the event phase and the recovery factor in the recovery phase are calculated, the time series of FRI can be computed. Like the calculation of the indicators, the computation of the FRI time series should be divided into event and recovery phases.

In the event phase, the calculated indicators of water depth ($I_h(t)$), accumulated water depth ($I_{AWD}(t)$), flooding duration ($I_D(t)$), and water accumulation rate ($I_{WAR}(t)$) are applied to evaluate the FRI at time t following Equation (12). WF stands for the weighting factor for each indicator, which determines the relative level of significance among the indicators. WF_h, WF_{AWD}, WF_D, and WF_{WAR} are assigned values of 3, 1, 3, and 2, respectively. t_s and t_e represent the starting and ending timesteps of flooding in the event phase.

$$
\begin{aligned}
FRI(t) &= \\
&\left[\frac{WF_h \cdot I_h(tt) + WF_{AWD} \cdot I_{AWD}(t) + WF_D \cdot I_D(t) + WF_{WAR} \cdot I_{WAR}(t)}{(WF_h + WF_{AWD} + WF_D + WF_{WAR})} \right] \\
&if\ t \in [t_s, t_e]
\end{aligned}
\tag{12}
$$

In the recovery phase, a recovery factor (RF) is calculated based on the physical characteristics of flooding in the previous event phase, and the social and economic indicator values of a household and its corresponding district. Equation (13) computes the recovery factor, applying the indicators of flood severity (I_{fs}), total flooding depth (I_{TFD}), total flooding time (I_{TFT}), maximum water accumulation rate (I_{WARmax}), households with children (I_C), elderly population (I_E), and household income (I_I). Weighting factors WF_{fs}, WF_{TFD}, WF_{TFT}, WF_{WARmax}, WF_C, WF_E and WF_I are assigned values of 3, 1, 2, 1, 1, 2, and 3, respectively. A value of 0.001 is assigned as a scaling constant. At last, the FRI at time t is computed as the product of the recovery factor and the FRI at the previous timestep $t - 1$ (see Equation (14)). Note that the recovery phase will last until the FRI value reaches 1.

$$
x = \{fs,\ TFD,\ TFT,\ WARmax,\ C,\ E,\ I\}
$$
$$
RF = \left[\prod (I_x)^{WF_x} \right]^{\frac{0.001}{\sum WF_x}}
\tag{13}
$$

$$
FRI(t) = FRI(t - 1) \times RF,\ if\ t \notin [t_s, t_e]
\tag{14}
$$

3.2. Indoor Water Depth Modelling

The indoor water depth modelling can be conducted based on a one-way coupling computation given the inundation modelling result from the 2D surface runoff model Parallel Diffusive Wave (P-DWave). It is assumed that floodwater flows into the buildings through doors with known location and it follows the fluid mechanics of discharge over a rectangular weir. Furthermore, the width of the door for every building is assumed to be 75 cm.

Figure 3 diagrammatizes the flow dynamic of water coming into the building. The flow should be analyzed by the upper and lower portion. At the upper portion of the flow, the income discharge is calculated by Equation (15), which describes a free discharge under a head of water equal to $(h_{out} - h_{in})$. At the lower portion of the flow, the income discharge is calculated by Equation (16), which describes a submerged discharge under a head of water equal to h_{in}. Q_u [m³/s] and Q_l [m³/s] represent the upper and lower portions of the discharge, respectively. C_d stands for the discharge coefficient, which in this case is assigned a value of 1. L (m) is the width of the door, assumed to be 0.75 m. Variable h_{out} (m) and h_{in} (m) represent the outdoor and indoor water level, respectively.

$$Q_u = \frac{2}{3} C_d \times L \times \sqrt{2g} \times (h_{out} - h_{in})^{\frac{3}{2}} \tag{15}$$

$$Q_l = C_d \times L \times h_{in} \times \sqrt{2g \times (h_{out} - h_{in})} \tag{16}$$

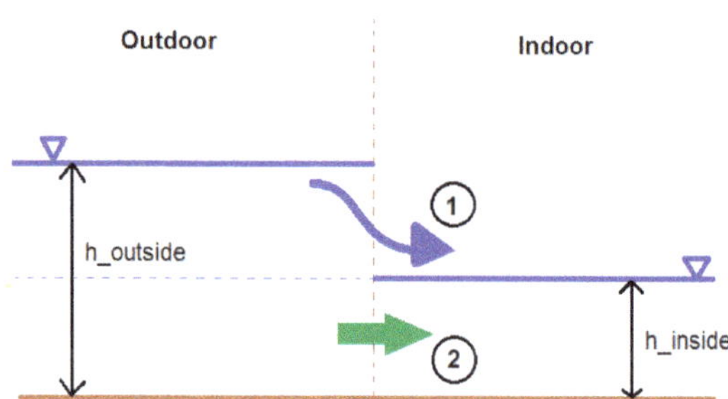

Figure 3. One-way coupling computation of the flow dynamic of water coming into the building. The upper portion of the flow, in which the water level outside the building is higher than the water depth inside the building, is marked as 1, and the lower portion, in which the water level outside the building is equal to the water depth inside the building, is marked as 2. h_outside and h_inside stand for the water depth outside and inside the building, respectively. When the indoor water depth is higher than that of the outdoor surroundings, the computation remains the same, whereas the discharge becomes negative as the flow direction faces the opposite direction.

The total discharge, Q_t [m³/s], is calculated by summing up Q_u and Q_l. When the outdoor water level recedes, $h_{outside}$ becomes lower than h_{in}, hence, Q_u and Q_l become negative. From that point, the water is flowing outwards, shown as a negative value of Q_t. After $Q_t(t)$ is calculated at time t, the water volume entering or exiting the building at time t, $V_{flux}(t)$ [m³], can be calculated by Equation (17). Then, the water volume inside the house at time t, $V(t)$ [m³], can be computed by Equation (18). At last, the indoor water depth at time t, $h_{in}(t)$ [m], is computed by Equation (19). Δt [s] represents the computation time interval, which in this case is 10 s. *Area* [m²] stands for the building area.

$$V_{flux}(t) = Q_t(t) \cdot \Delta t \tag{17}$$

$$V(t) = V(t-1) + V_{flux}(t) \tag{18}$$

$$h_{in}(t) = \frac{V(t)}{Area} \tag{19}$$

3.3. Parallel Diffusive Wave Model: P-DWave

The Parallel Diffusive Wave Model, P-DWave, is the surface runoff model applied for flood inundation modelling in this study. It is a first-order finite volume explicit discretization scheme that takes the conservative form of the 2D Shallow Water Equations into account and neglects the inertial terms (see Equations (20) and (21)). h is the water depth [m] and t is the time [s]. Velocity is defined by $u^2 = u_x^2 + u_y^2$. $u = \begin{bmatrix} u_x & u_y \end{bmatrix}^T$ stands for the depth-averaged flow velocity vector [-], in which u_x is the flow velocity in the x direction [m/s] and u_y is the flow velocity in the y direction [m/s]. R represents the source/sink term (e.g., rainfall, inflow, surcharge, drainage [m/s]). z is the bed elevation [m]. The bed friction is approximated by Manning's formula (Equation (22)), in which $S_f = \begin{bmatrix} S_{fx} & S_{fy} \end{bmatrix}^T$ stands for the bed friction vector [-]. S_{fx} is the bed friction slope in the x direction [-] and S_{fy} is the bed friction slope in the y direction [-]. n is the Manning's roughness coefficient [s/m1/3]. The modulus of the depth-averaged flow velocity vector is given by Equation (23), where $S_{wx} = d(h+z)/dx$ is the water level gradient in the x direction [-] and $S_{wy} = d(h+z)/dy$ is the water level gradient in the y direction [-]. Further details of the P-DWave model can be found elsewhere [46]. Note that neither the sewer system nor the infiltration processes are considered in this study. As such, not all water could be drained from the surface and the event phase could not be considered complete. Therefore, in order to enable the start of the recovery phase, we assume that the outdoor water depths after the end of the 60-min simulation time return automatically to zero. The one-way coupling indoor water depth simulation for each building is then conducted following this assumption.

$$\frac{dh}{dt} + \nabla(uh) = R \tag{20}$$

$$g\nabla(h+z) = gS_f \tag{21}$$

$$\begin{bmatrix} S_{fx} \\ S_{fy} \end{bmatrix} = \begin{bmatrix} \frac{n^2|u|u_x}{h^{4/3}} \\ \frac{n^2|u|u_y}{h^{4/3}} \end{bmatrix} \tag{22}$$

$$|u| = \frac{h^{2/3}\left(S_{wx}{}^2 + S_{wy}{}^2\right)^{1/4}}{n} \tag{23}$$

4. Results

4.1. Flood Inundation Modelling

The maximum inundation modelling of the 15-min rainfall applying the data from the KOSTRA-DWD-2010R database is conducted for various return periods, providing the information of locations that are more likely to encounter severe flooding conditions in Maxvorstadt. (see Figure 4). Table 2 shows the average inundation depth on streets versus different return periods. By applying the one-way coupling indoor water depth computation, the maximum indoor water depth for each building can be seen in Figure 5, also showing the information of the percentage of buildings facing different levels of indoor flooding.

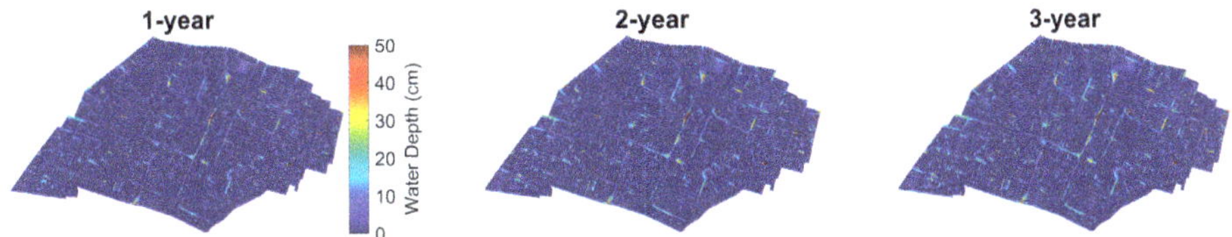

Figure 4. *Cont.*

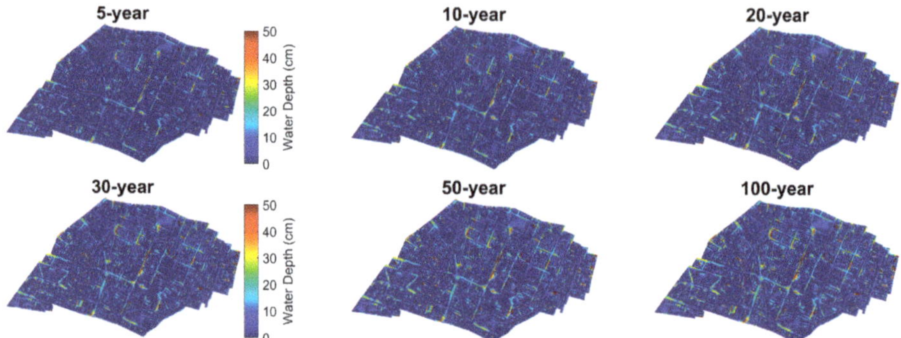

Figure 4. Maximum inundation map in Maxvorstadt with various return periods, applying rainfall data from the KOSTRA-DWD-2010R database. Luisen Street, located at the center of Maxvorstadt, faces the most extreme flooding condition with the range crossing over two blocks and maximum water depth over 50 cm in the case of a 100-year flooding event.

Table 2. Average inundation depth on streets versus return periods in Maxvorstadt.

Return Period (year)	1	2	3	5	10	20	30	50	100
Average Water Depth on Streets (cm)	4.11	5.19	5.84	6.60	7.67	8.71	9.33	10.08	11.11

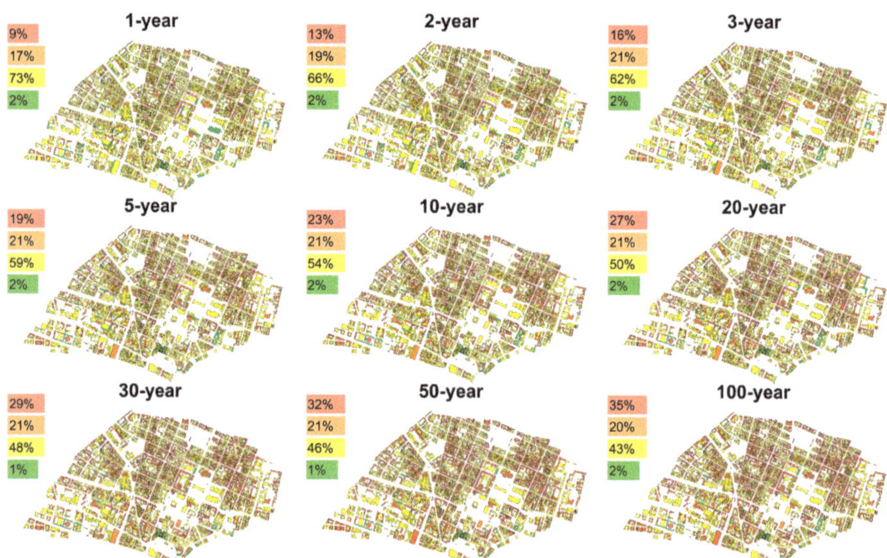

Figure 5. Maximum indoor water depth modelling given flooding events with various return periods by applying the one-way coupling computation. The indoor water depth in buildings with color green: 0 cm, yellow: 0–5 cm, orange: 5–10 cm, and red: above 10 cm. Percentage stands for the amount of building in each level of indoor flooding.

4.2. Parameter Sensitivity Analysis

A sensitivity analysis for the reference parameters is conducted for the building which encounters the most severe indoor inundation caused by a 100-year flooding in district Königsplatz. The alteration of each indicator by changing its corresponding reference parameter is examined by comparing the case of adding and subtracting 50% from the original values of the reference parameter. The differences can be detected by comparing the case applying the original set of reference parameters (shown in blue dashed lines) and the upper edges of the areas. The results of the sensitivity analysis in the event (see Figure 6) and recovery phase (see Figure 7) are shown below.

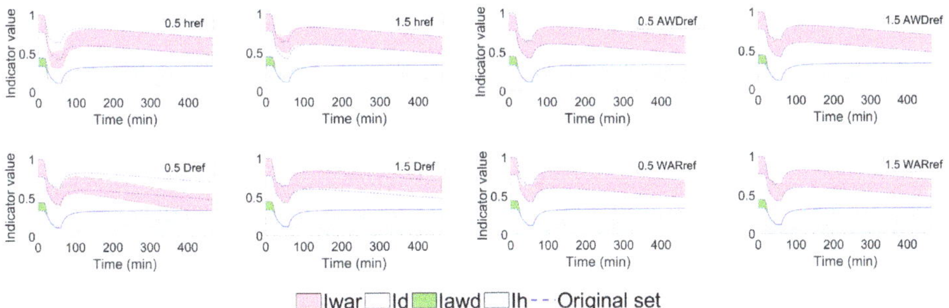

Figure 6. Sensitivity analysis of the four reference parameters for each physical indicator in the event phase by comparing the case of adding and subtracting 50% from the original reference parameters. Blue dashed line stands for the case when applying the original set of reference parameters. Annotation in each graph illustrates which reference parameter is examined by either adding or subtracting 50% from its original value.

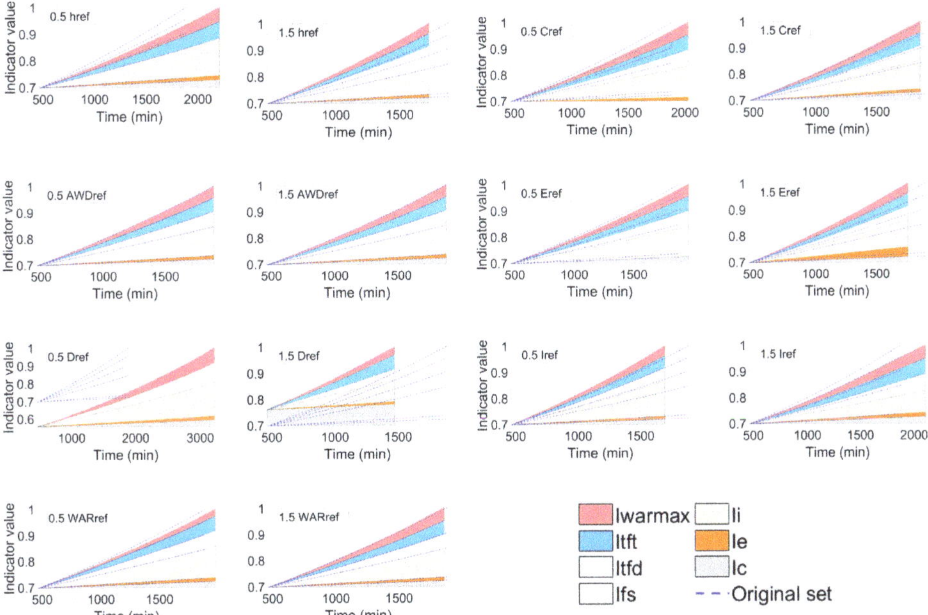

Figure 7. Sensitivity analysis of the seven reference parameters for each physical, social, and economic indicator in the recovery phase by comparing the case of adding and subtracting 50% from the original reference parameters. Blue dashed line stands for the case when applying the original set of reference parameters. Annotation in each graph illustrates which reference parameter is examined by either adding or subtracting 50% from its original value.

4.3. Flood Resilience Index

Figure 8 shows the mean FRI curves as an aggregated result for every household in Maxvorstadt according to different reference parameters (either by adding or subtracting 50% from their original values) considered in the sensitivity analysis. In addition, the standard deviation curves are provided to illustrate the level of dispersion of the FRI at each timestep.

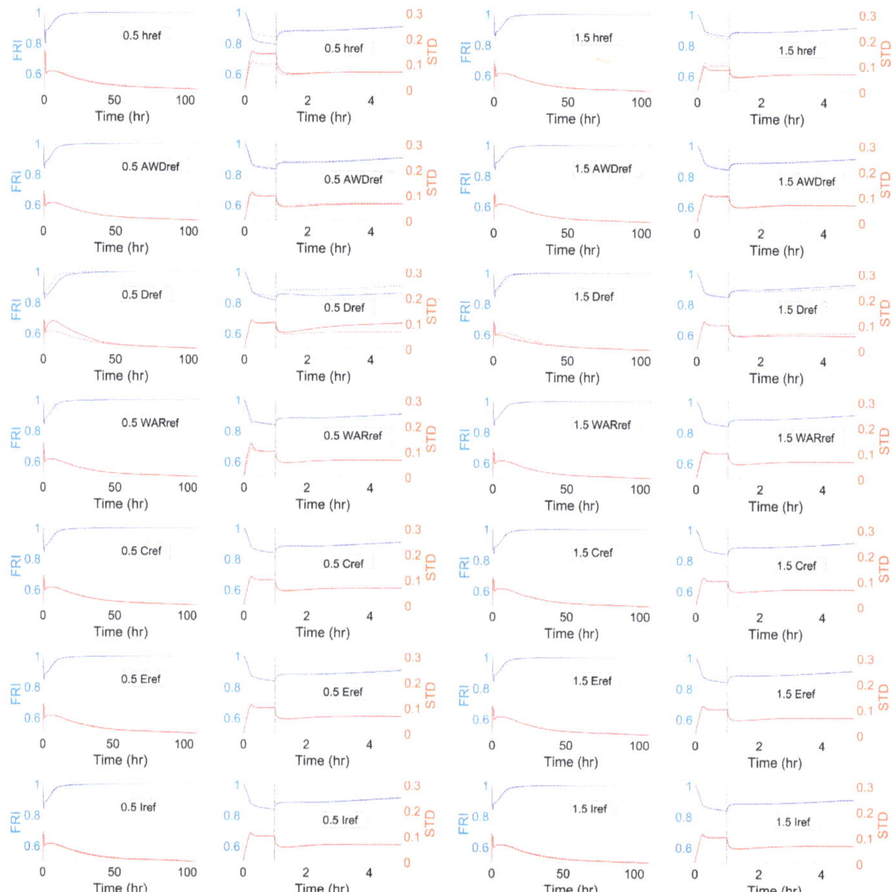

Figure 8. Mean FRI curves as aggregated results for every household in Maxvorstadt (blue lines) and standard deviation curves (red lines) showing the level of dispersion of the FRI at each timestep in face of a 100-year flooding. The simulation considers different multiplication factors of the reference parameters applied in the sensitivity analysis. Solid lines represent the considered case (either with a 50% increment or decrement of the original reference parameter) and the dashed lines correspond to the case applying the original set of reference parameters. Black dashed lines at $t = 1$ h in the zoom-in graphs (right-hand side) illustrate the end of the outdoor inundation modelling, for which the outdoor water depth is set to zero to enable the start of the recovery phase.

5. Discussion

5.1. Flood Inundation Modelling

According to the results of the maximum inundation modelling (see Figure 4), it can be shown that the scenarios with higher return period as well as rainfall intensity return larger inundation areas and higher water depths. The average water depth on streets also shows an increasing trend as the return period increases (see Table 2), which corresponds to the shape of the considered 15 min rainfall intensity–frequency curve in Figure 1d. In the maximum inundation maps, there are several spots showing a small area of higher water depths. The reason is that these are areas enclosed by buildings or tunnels, which have a lower elevation than the surroundings. The surface runoff modelling applying the P-DWave model only includes the overland surface routing of rainwater. Hence, in this case, it is not possible to model the underground drainage of water accumulated at the low-laying areas. According to the maximum indoor water depth results (see Figure 5), the percentage of buildings which face a more severe indoor flooding rises as the return period increases.

5.2. Parameter Sensitivity Analysis

According to the results of parameter sensitivity analysis, different reference parameters have an effect on the corresponding FRI indicator and total FRI. In the event phase (see Figure 6), the higher the four physical reference parameters are, the higher the indicators and FRI values will be. However, the sensitiveness of changing different reference parameters differs according to the assigned weighting factors and the original values of the reference parameters. Not surprisingly, reference parameters of water depth and flooding duration, which are assigned with the greatest weighting factors, have the highest impact. The water depth reference parameter controls the water depth indicator when the building is facing an indoor flooding. As shown in Figure 6, the sensitiveness of altering such a parameter is obvious within the timeframe from 0 to 100 min, in which the building is encountering the peak water depth. By increasing this reference parameter, the water depth indicator shows a 0.1 increment at the lowest point of the indicator curve. In contrast, by decreasing this reference parameter, the water depth indicator decreases and drops to zero during the peak water depth. The sensitiveness of altering the reference parameter of flooding duration is evident at the tail end of the event phase. Like the water depth indicator, the flooding duration indicator rises when the reference parameter increases and falls down to zero at the timestep at 400 min when the reference parameter decreases. Note that changing the reference parameter of flooding duration will lead to different endpoints of the FRI in the event phase, which makes different starting points for the recovery phase. The alteration of the water accumulation rate indicator only appears at the front end of the event phase, which corresponds to the rising limb of the indoor hydrograph. The altering of the water accumulation indicator is more evident when the reference parameter decreases. However, when the indoor water is receding, the water accumulation rate becomes ineffective, and thus changing this reference parameter does not affect the water accumulation rate indicator. Finally, the reference parameter of accumulated water depth in this study is set as an extreme case (3 m) to show that by either increasing or decreasing it by 50%, the indicator will fall down to zero. The difference between these two cases is only at which timestep the indicator drops to zero. According to Figure 6, the difference between increasing and decreasing the reference parameter of accumulated water depth is not significant in this case study.

The seven indicators included in the recovery phase constitute the recovery factor which is responsible for the system to return to the original state (FRI = 1). Similar to the indicators in the event phase, the effect of reference parameters in the recovery phase depends on the assigned weighting factors and the original values of the reference parameters (see Figure 7). All indicators, with exception for the total flooding depth and income indicators, increase as their reference parameters increase. Below is a short description of the impact of each reference parameter/indicator on the recovery phase:

1. Flood severity indicator—when its reference parameter decreases, the maximum water depth in the event phase becomes greater than its reference parameter, and thus the indicator is no longer

contributive to the recovery factor and does not appear in the sensitivity analysis graph in the recovery phase. In this case, the system requires a longer recovery time (approximately 300 min longer) with the smaller recovery factor. On the contrary, when the reference parameter of flood severity increases, the indicator contribution to the recovery factor increases and the recovery of system is faster (200 min).

2. Total flooding depth indicator—there is no difference between increasing and decreasing the reference parameter of total flooding depth as the indicator does not appear in both graphs. The reason is that the altered reference parameter remains in any case below the total water depth during the event phase, and thus the indicator has no contribution to the recovery factor.

3. Total flooding time indicator—changes in the reference parameter of total flooding time show different starting points for the FRI curve at the beginning of the recovery phase. This is due to different endpoints in the event phase, as mentioned in the previous section. In this case, there will be a higher starting point in the recovery phase according to an increased reference parameter of total flooding time, and vice versa. When the reference parameter decreases, the total flooding time in the event phase exceeds the threshold, and thus the indicator is not contributive to the recovery factor and does not appear in the sensitivity analysis graph for the recovery phase. In contrast, when the reference parameter increases, not only the starting point of the recovery phase raises, but the system is able to bounce back to the original state of performance approximately 500 min faster.

4. Maximum water accumulation rate indicator and households with children indicator—due to the low weighting factors assigned to both indicators, the differences between increasing or decreasing their reference parameters is relatively small. However, it can be seen that the degree of changing is larger when their reference parameters are decreased.

5. Elderly population indicator—this indicator has a slightly higher impact than the two previous indicators. This is due to a larger weighting factor assigned for it. In addition, the effect of the indicator is larger when its reference parameter is increased.

6. Income indicator—decreasing/increasing its reference parameter increases/decreases the indicator impact on the recovery factor. The reference parameter of income could be considered as the threshold that defines whether the income amount reaches the maximum recovery strength, at which the income indicator equals to e^1. As a result, if the reference parameter decreases, the threshold decreases, and the household will either reach the maximum recovery strength or have a larger income indicator.

The sensitivity analysis of the reference parameters provides detailed information of the FRI composition, which is a powerful tool for decision makers to decide which aspect requires instant improvement through visual comparison among indicators. The results also allow a better understanding on how external influencing factors affect the FRI. For instance, when a house is equipped with water-proof furniture, the reference parameter of water depth should raise, standing for a higher resilience to flooding depth (i.e., higher FRI).

5.3. Flood Resilience Index

According to the results of the FRI simulation (see Figure 8), the mean FRI curve drops together with an increment of the standard deviation curve at the beginning. Within the one-hour simulation duration, buildings within Maxvorstadt experience indoor flooding, and thus the mean FRI decreases in the event phase. The indoor flooding hydrograph from every building differs quite significantly. Hence, the standard deviation curve climbs, illustrating a higher level of dispersion of the FRI values. At the timestep of 1 h, the outdoor water depth is set to recede to zero (see assumptions in Section 3.3.) and the indoor water depth of each building faces a significant drop according to the one-way coupling computation, leading to a sudden increase of the mean FRI curve and a sharp drop of the standard deviation curve. Note that the recovery phase does not start at the timestep of 1 h. It starts at the timestep when the indoor water depth recedes to zero, so it differs for each building (see Figure 2). After the timestep at 1 h, the mean FRI curve climbs following the recession of indoor water depth, and gradually returns to one during the recovery phase. The standard deviation curve then returns back to zero along with the increment of the FRI values, reaching one for every building. The impact of each reference parameter on the FRI is now shortly summarized:

- Increasing/decreasing the reference parameter of water depth (event phase) and flood severity (recovery phase) increases/decreases the mean FRI curve along with a decreasing/increasing standard deviation curve.
- The altered reference parameter of accumulated water depth (event phase) and total flooding depth (recovery phase) slightly changes the mean FRI and standard deviation curve but only at the beginning of the simulation, at which the accumulated water depth does not exceed the reference parameter in the event phase.
- The altered reference parameter of flooding duration (event phase) and total flooding time (recovery phase) makes the most significant changes on the mean FRI and standard deviation curve among all considered reference parameters.
- The changes on the mean FRI and standard deviation curve due to the altered reference parameter of water accumulation rate (event phase) and maximum water accumulation rate (recovery phase) appear within the timestep at 1 h, which lies at the rising limb of all indoor hydrographs for every building. Aside from this, the effect of changing such a reference parameter is not significant.
- The altered reference parameters for the social and economic indicators can only affect the recovery phase, in which these indicators are taken into account. They are highly sensitive to the assigned weighting factors and the original set of reference parameters.
- Regarding the social and economic indicators, the altered reference parameter of income has the greatest impact on the mean FRI and standard deviation curve, while that of households with children has the least.

The summary table (see Table 3) provides information of the FRI duration (time lasting from the system being hit by flooding to a full recovery, including the event and recovery phase) and the minimum value of the mean FRI considering the effect of the altered reference parameters of each indicator. The alteration of the reference parameter of income has the highest impact considering the FRI duration, which is a 7.2 h difference comparing increasing and decreasing the reference parameter by 50%. The minimum value of the mean FRI is caused by the physical impacts from flooding and thus lies within the event phase, hence the altered reference parameters of the households with children, elderly population, and income indicators, which are only considered in the recovery phase and cannot have an effect on it. Among the four physical indicators, the alteration of the reference parameter of water depth has the largest effectiveness on the minimum mean FRI.

The aggregated FRI results provide the information of the flood resilience level within the study area (regarding it as a whole system). Based on this information, it is possible to verify whether: (a) The severity of the flooding impact hits the system and induces a significant drop of the FRI value, or (b) the system is undergoing a slow or a fast recovering process. Furthermore, the dispersiveness of the FRI curves also provides the information of how homogenously the urban components react to a certain event. If the standard deviation value is high, it means that the urban components react differently and some districts will need more assistance during low FRI periods than their neighboring zones, which have a higher FRI value.

Table 3. Summary table for the FRI simulation considering different multiplication factors of the reference parameters.

Ref. [1]	Original Value	Multiplication Factor	FRI Duration [2] (h)	Lowest Mean FRI
href	0.5 m	0.5	110.4	0.79
		1.5	106.2	0.85
AWDref	3 m	0.5	107.2	0.83
		1.5	107.2	0.84
Dref	800 min	0.5	107.2	0.81
		1.5	107.2	0.84
WARref	5 cm/min	0.5	107.4	0.83
		1.5	107.1	0.83
Cref	20 %	0.5	107.7	0.83
		1.5	105.9	0.83
Eref	12 %	0.5	109.8	0.83
		1.5	106.3	0.83
Iref	80,000 €	0.5	103.8	0.83
		1.5	110.6	0.83
Original	-	1	107.2	0.83

[1] Reference parameter. [2] Total duration including the event and recovery phase.

6. Conclusions

In this paper, we developed an indicator-based flood resilience quantification method by introducing the time-varying Flood Resilience Index (FRI). The FRI is able to quantify the flood resilience level for households within an urban area, divided into event and recovery phases. Therefore, the new FRI embodies the definition of flood resilience as the capacity to withstand adverse effects following flooding events and the ability to quickly recover to the original system performance before the event. During the flooding event, the FRI is estimated based on physical indicators, namely the water depth, accumulated water depth, flooding duration, and water accumulation rate. During the recovery phase, the FRI is estimated based on social indicators, i.e., the percentage of households with children and that of elderly population, as well as an economic indicator, i.e., annual household income.

The sensitivity analysis of the parameters (and indicators) provided a useful tool to understand better how external influencing factors affect the FRI. The aggregated FRI results allow the identification of fragilities in the urban household as part of a system. It is easy to identify which households have a slow-recovering process or which are being hit severely by the event. Furthermore, the dispersiveness of the FRI curves also provides the information of how homogenously the urban components of the system react to a certain event.

The novel time-varying FRI therefore provides a novel insight into the indicator-based quantification method of flood resilience level for households in an urban area. The time-dependent characteristic of the proposed method contributes to advancing the research field by enabling a quantifiable characterization and visualization of how a system responds during and after a flooding event. Therefore, the introduced FRI could become a valuable tool for urban planning and public communication, and promote a better flood risk management plan. Future work will see the inclusion of the sewer network and possible extension of the urban area of Maxvorstadt, which is considered at the moment isolated from other boroughs in Munich City.

Author Contributions: Conceptualization, K.-F.C. and J.L.; Data curation, K.-F.C.; Formal analysis, K.-F.C.; Investigation, K.-F.C.; Methodology, K.-F.C. and J.L.; Project administration, K.-F.C.; Resources, J.L.; Software, K.-F.C.; Supervision, J.L.; Visualization, K.-F.C.; Writing—original draft, K.-F.C.; Writing—review & editing, J.L.

Funding: This research received no external funding.

Acknowledgments: The authors are grateful to Professor Stephan Pauleit from the Centre for Urban Ecology and Climate Adaptation, TUM for providing GIS data of Maxvorstadt applied in this study.

Conflicts of Interest: The authors declare no conflict of interest.

References

1. Barroca, B.; Bernardara, P.; Mouchel, J.M.; Hubert, G. Indicators for identification of urban flooding vulnerability. *Nat. Hazards Earth Syst. Sci.* **2006**, *6*, 553–561. [CrossRef]
2. Campbell, J.; Douglas, I.; Mclean, L.; Mcdonnell, Y.; Alam, K.; Maghenda, M. Unjust waters: Climate change, flooding and the urban poor in Africa. *Environ. Urban.* **2008**, *20*, 187–205.
3. Few, R. Flooding, vulnerability and coping strategies: Local responses to a global threat. *Prog. Dev. Stud.* **2003**, *3*, 43–58. [CrossRef]
4. Adelekan, I.O. Vulnerability of poor urban coastal communities to flooding in Lagos, Nigeria. *Environ. Urban.* **2010**, *22*, 433–450. [CrossRef]
5. Huong, H.; Pathirana, A. Urbanization and climate change impacts on future urban flooding in Can Tho city, Vietnam. *Hydrol. Earth Syst. Sci.* **2013**, *17*, 379–394. [CrossRef]
6. Hallegatte, S. Strategies to adapt to an uncertain climate change. *Glob. Environ. Chang.* **2009**, *19*, 240–247. [CrossRef]
7. Jones, H.P.; Hole, D.G.; Zavaleta, E.S. Harnessing nature to help people adapt to climate change. *Nat. Clim. Chang.* **2012**, *2*, 504–509. [CrossRef]
8. Leandro, J. Title of Special Issue: Towards more Flood Resilient Cities. *Urban Water J.* **2015**, *12*, 1–2. [CrossRef]
9. Robadue, D., Jr. Understanding resistance to resilience in coastal hazards and climate adaptation: Three approaches to visualizing structural and process obstacles, opportunities and adaptation responses. In Proceedings of the 52nd Hawaii International Conference on System Sciences, Maui, HI, USA, 8–11 January 2019.
10. Messner, F.; Penning-Rowsell, E.; Green, C.; Meyer, V.; Tunstall, S.; van der Veen, A. *Evaluating Flood Damages: Guidance and Recommendations on Principles and Methods*; FLOODsite: Wallingford, UK, 2007.
11. Schelfaut, K.; Pannemans, B.; van der Craats, I.; Krywkow, J.; Mysiak, J.; Cools, J. Bringing flood resilience into practice: The FREEMAN project. *Environ. Sci. Policy* **2011**, *14*, 825–833. [CrossRef]
12. Jones, L. Resilience isn't the same for all: Comparing subjective and objective approaches to resilience measurement. *Wiley Interdiscip. Rev. Clim. Chang.* **2019**, *10*, e552. [CrossRef]
13. Allen, T.R.; Crawford, T.; Montz, B.; Whitehead, J.; Lovelace, S.; Hanks, A.D.; Christensen, A.R.; Kearney, G.D. Linking Water Infrastructure, Public Health, and Sea Level Rise: Integrated Assessment of Flood Resilience in Coastal Cities. *Public Work. Manag. Policy* **2019**, *24*, 110–139. [CrossRef]
14. Moghadas, M.; Asadzadeh, A.; Vafeidis, A.; Fekete, A.; Kötter, T. A multi-criteria approach for assessing urban flood resilience in Tehran, Iran. *Int. J. Disaster Risk Reduct.* **2019**, *35*, 101069. [CrossRef]
15. Zhang, Y.; Li, W.; Sun, G.; King, J.S. Coastal wetland resilience to climate variability: A hydrologic perspective. *J. Hydrol.* **2019**, *568*, 275–284. [CrossRef]
16. Karamouz, M.; Taheri, M.; Khalili, P.; Chen, X. Building Infrastructure Resilience in Coastal Flood Risk Management. *J. Water Resour. Plan. Manag.* **2019**, *145*, 4019004. [CrossRef]
17. Murdock, H.; de Bruijn, K.; Gersonius, B. Assessment of Critical Infrastructure Resilience to Flooding Using a Response Curve Approach. *Sustainability* **2018**, *10*, 3470. [CrossRef]
18. Walsh, B.; Hallegatte, S. *Measuring Natural Risks in the Philippines: Socioeconomic Resilience and Wellbeing Losses*; World Bank Group: Washington, DC, USA, 2019.
19. Driessen, P.; Hegger, D.; Kundzewicz, Z.; van Rijswick, H.; Crabbé, A.; Larrue, C.; Matczak, P.; Pettersson, M.; Priest, S.; Suykens, C.; et al. Governance Strategies for Improving Flood Resilience in the Face of Climate Change. *Water* **2018**, *10*, 1595. [CrossRef]
20. Hammond, M.J.; Chen, A.S.; Djordjević, S.; Butler, D.; Mark, O. Urban flood impact assessment: A state-of-the-art review. *Urban Water J.* **2015**, *12*, 14–29. [CrossRef]
21. Holling, C.S. Resilience and Stability of Ecological Systems. *Annu. Rev. Ecol. Syst.* **1973**, *4*, 1–23. [CrossRef]
22. de Bruijn, K.M. Resilience and flood risk management. *Water Policy* **2004**, *6*, 53–66. [CrossRef]
23. Tourbier, J. A Methodology to Define Flood Resilience. In Proceedings of the EGU General Assembly, Vienna, Austria, 22–27 April 2012.

24. United Nations Office for Disaster Risk Reduction. Terminology on Disaster Risk Reduction. Available online: https://www.unisdr.org/we/inform/terminology#letter-r (accessed on 15 December 2018).

25. Mugume, S.; Gomez, D.; Butler, D. Quantifying the resilience of urban drainage systems using a hydraulic performance assessment approach. In Proceedings of the 13th International Conference on Urban Drainage, Sarawak, Malaysia, 7–12 September 2014.

26. Perfrement, T.; Lloyd, T. Identifying and Visaulising Resilience to Flooding via a Composite Flooding Disaster Resilience Index. In Proceedings of the 56th Floodplain Management Australia National Conference, Nowra, Australia, 17–20 May 2016.

27. Keating, A.; Campbell, K.; Szoenyi, M.; McQuistan, C.; Nash, D.; Burer, M. Development and testing of a community flood resilience measurement tool. *Nat. Hazards Earth Syst. Sci.* **2017**, *17*, 77–101. [CrossRef]

28. Bizzotto, M. Resilient Cities Report 2018. In Proceedings of the 9th Global Forum on Urban Resilience and Adaptation, Bonn, Germany, 26–28 April 2018.

29. The Rockefeller Foundation and ARUP. *City Resilience Index*; ARUP: London, UK, 2014.

30. Fischer, K.; Hiermaier, S.; Riedel, W.; Häring, I. Morphology Dependent Assessment of Resilience for Urban Areas. *Sustainability* **2018**, *10*, 1800. [CrossRef]

31. Winderl, T. *Disaster Resilience Measurements: Stocktaking of Ongoing Efforts in Developing Systems for Measuring Resilience*; United Nations Development Programme: New York, NY, USA, 2014.

32. de Bruijn, K.M. Resilience indicators for flood risk management systems of lowland rivers. *Int. J. River Basin Manag.* **2004**, *2*, 199–210. [CrossRef]

33. Gourbesville, P.; Batica, J. Flood Resilience Index—Methodology and Application. In Proceedings of the 11th International Conference on Hydroinformatics, New York, NY, USA, 17–21 August 2014.

34. Lee, E.H.; Kim, J.H. Development of Resilience Index Based on Flooding Damage in Urban Areas. *Water* **2017**, *9*, 428. [CrossRef]

35. Bertilsson, L.; Wiklund, K.; de Moura Tebaldi, I.; Rezende, O.M.; Veról, A.P.; Miguez, M.G. Urban flood resilience—A multi-criteria index to integrate flood resilience into urban planning. *J. Hydrol.* **2018**, in press. [CrossRef]

36. Landeshauptstadt München. *Statistisches Taschenbuch 2018—München und seine Stadtbezirke*; Landeshauptstadt München: Munich, Germany, 2018.

37. Landeshauptstadt Munchen. *Munich Facts and Figures 2018*; Landeshauptstadt Munchen: Munich, Germany, 2018.

38. Junghänel, T.; Ertel, H.; Deutschländer, T. *Bericht zur Revision der koordinierten Starkregenregionalisierung und -auswertung des Deutschen Wetterdienstes in der Version 2010*; Deutscher Wetterdienst: Offenbach am Main, Germany, 2017.

39. Young, G. From Chemical Exposure and Mold to Mosquitos, Problems Don't Stop When the Rain Does. Available online: https://today.ttu.edu/posts/2015/06/flooding-long-term-impact-health-environment (accessed on 20 November 2018).

40. Unterberger, C. How Flood Damages to Public Infrastructure Affect Municipal Budget Indicators. *Econ. Disasters Clim. Chang.* **2018**, *2*, 5–20. [CrossRef]

41. Ten Brinke, W.; Knoop, J.; Muilwijk, H.; Ligtvoet, W. Social disruption by flooding, a European perspective. *Int. J. Disaster Risk Reduct.* **2017**, *21*, 312–322. [CrossRef]

42. Milojevic, A.; Armstrong, B.; Wilkinson, P. Mental health impacts of flooding: A controlled interrupted time series analysis of prescribing data in England. *J. Epidemiol. Community Heal.* **2017**, *71*, 970–973. [CrossRef]

43. Rufat, S.; Tate, E.; Burton, C.G.; Maroof, A.S. Social vulnerability to floods: Review of case studies and implications for measurement. *Int. J. Disaster Risk Reduct.* **2015**, *14*, 470–486. [CrossRef]

44. Teo, M.; Goonetilleke, A.; Ziyath, A.M. An integrated framework for assessing community resilience in disaster management. In Proceedings of the 9th Annual International Conference of the International Institute for Infrastructure Renewal and Reconstruction, Risk-informed Disaster Management: Planning for Response, Recovery and Resilience, Brisbane, Australia, 7–10 July 2013.

45. Saja, A.; Goonetilleke, A.; Teo, M.; Ziyath, A. A critical review of social resilience assessment frameworks in disaster management. *Int. J. Disaster Risk Reduct.* **2019**, *35*, 101096. [CrossRef]

46. Leandro, J.; Chen, A.; Schumann, A. A 2D parallel diffusive wave model for floodplain inundation with variable time step (P-DWave). *J. Hydrol.* **2014**, *517*, 250–259. [CrossRef]

Article

A Model-Based Engineering Methodology and Architecture for Resilience in Systems-of-Systems: A Case of Water Supply Resilience to Flooding

Demetrios Joannou [1,*], Roy Kalawsky [1], Sara Saravi [1], Mónica Rivas Casado [2], Guangtao Fu [3] and Fanlin Meng [3]

[1] Advanced VR Research Centre, Loughborough University, Loughborough, Leicestershire LE11 3TU, UK; r.s.kalawsky@lboro.ac.uk (R.K.); s.saravi@lboro.ac.uk (S.S.)
[2] School of Water, Energy and Environment, Cranfield University, Cranfield, Bedfordshire MK43 0AL, UK; m.rivas-casado@cranfield.ac.uk
[3] Centre for Water Systems, College of Engineering, Mathematics and Physical Sciences, University of Exeter, Exeter, Devon EX4 4QF, UK; G.Fu@exeter.ac.uk (G.F.); M.Fanlin@exeter.ac.uk (F.M.)
* Correspondence: d.joannou@lboro.ac.uk; Tel.: +44-(0)-1509-635674

Received: 5 February 2019; Accepted: 3 March 2019; Published: 8 March 2019

Abstract: There is a clear and evident requirement for a conscious effort to be made towards a resilient water system-of-systems (SoS) within the UK, in terms of both supply and flooding. The impact of flooding goes beyond the immediately obvious socio-aspects of disruption, cascading and affecting a wide range of connected systems. The issues caused by flooding need to be treated in a fashion which adopts an SoS approach to evaluate the risks associated with interconnected systems and to assess resilience against flooding from various perspectives. Changes in climate result in deviations in frequency and intensity of precipitation; variations in annual patterns make planning and management for resilience more challenging. This article presents a verified model-based system engineering methodology for decision-makers in the water sector to holistically, and systematically implement resilience within the water context, specifically focusing on effects of flooding on water supply. A novel resilience viewpoint has been created which is solely focused on the resilience aspects of architecture that is presented within this paper. Systems architecture modelling forms the basis of the methodology and includes an innovative resilience viewpoint to help evaluate current SoS resilience, and to design for future resilient states. Architecting for resilience, and subsequently simulating designs, is seen as the solution to successfully ensuring system performance does not suffer, and systems continue to function at the desired levels of operability. The case study presented within this paper demonstrates the application of the SoS resilience methodology on water supply networks in times of flooding, highlighting how such a methodology can be used for approaching resilience in the water sector from an SoS perspective. The methodology highlights where resilience improvements are necessary and also provides a process where architecture solutions can be proposed and tested.

Keywords: architecture modelling flood resilience; resilience engineering; system-of-systems water systems

---

1. Introduction

Water is essential to people, businesses and the environment. The sustainability of water resources is fundamental to life as we know it and access to clean water is essential to society. A generalised depiction (rich picture [1]) of the water "system-of-systems" (SoS) can be seen in Figure 1, which illustrates the complexity and high-level connectivity between stakeholders and constituent systems. An SoS is a large set of interconnected systems which collaborate to bring about behaviours and capabilities greater than the sum of the individual parts.

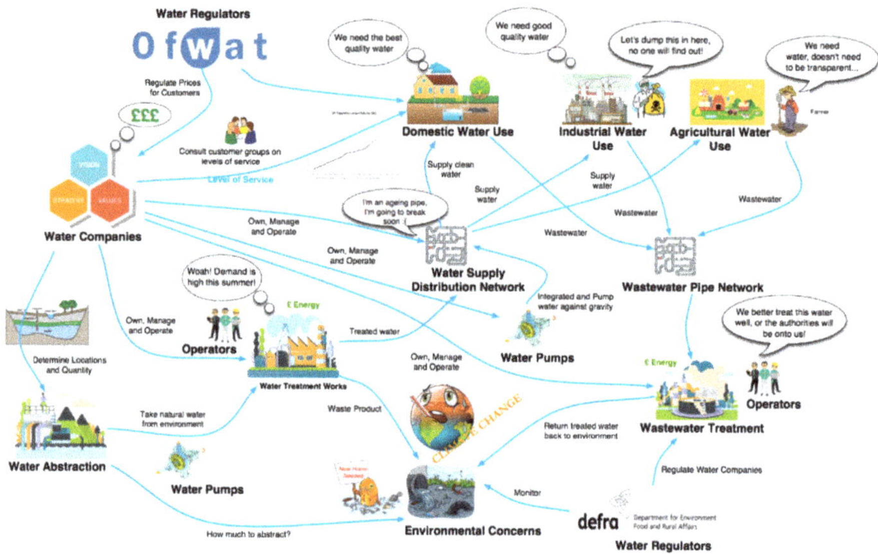

Figure 1. Rich picture of UK water sector.

Climate projections (UKP09) [2] suggest that the change in climate [3] and the rise in demand for water will increase the strain on UK water resources. General trend projections indicate warmer and drier summers that will result in less water being available for all uses in the coming decades. Additionally, what is expected is that winters will generally be wetter with heavier downpours of rain which will increase the risk of flooding [4]. To this end, the water needs of the UK population, industry and associated dependencies is expected to increase due to forecast population growth [5].

The risks posed by flooding on supply water to demand points ought to be considered just as much as the immediately obvious and devastating impact flooding has on the lives of people, buildings and the environment. This would imply that water is successfully supplied to all demand regions (domestic, industrial and agricultural) whatever the circumstances are of the environment and that the water SoS holistically demonstrates levels of resilience in different operational scenarios. This entails both the aforementioned scenario of less water availability and increased temperatures, and also in crisis scenarios such as natural and man-made disasters, where water may become unavailable or available only at a depleted level, for a certain period of time. Therefore, when attempting to "engineer resilience" [6,7] into such systems, there needs to be an understanding of levels of performance, acceptable levels of performance, and also recovery strategies to mitigate against loss of service.

The objective of this paper is to demonstrate how the resilience methodology created can explore where resilience improvements are necessary and then increase the resilience of water supply systems during times of flooding, by adopting an SoS-based approach and applied model-based processes. A key contribution is the inclusion of a resilience viewpoint within an existing architecture framework to allow engineers to model certain aspects of SoS resilience using different views and model types.

1.1. Resilience in the Water Sector

'Resilience' is a term that has been used for a long time and in many different domains [6,8,9]. The popularity of the concept is currently on the rise [10] and has attracted great interest from engineers [11] and safety analysts [12] alike. Many variations of the definition exist but the core connotation remains constant with the Latin origin of the word "resilire", to spring back or rebound [13] i.e., to recover. Traditionally, resilience has been closely linked with safety and risk management strategy planning, taking a hindsight and "lessons-learnt" approach from past incidents to improve existing and future systems. For knowledge purposes, there are many advantages in studying past failures and disasters to gain an insight into why things go wrong, how response systems performed, taking the positives and negatives from a given instance to inform better decision-making in the future. That said, a more effective approach is a proactive one, to test possible future scenarios and see how the system would respond, in order to develop systems with the capability to deal with change prior to an incident and mitigating the severity of the potential impact.

For the purpose of this research, the water system and the occurrence of a flooding event was considered as a system-of-systems (SoS) [14] as a means to appreciate the complexity of water supply systems and associate systems in times of flooding. This permitted the development of model-based methodology and to propose methods to design future phases of water systems that ensure resilience strategies are included early on in their lifecycle phases. In this specific case, the methodology supports increasing the overall resilience of the systems delivering water to consumers in times of flooding.

Literature shows resilience frameworks being developed from an "adaptation" perspective [14] which is interesting from the standpoint of existing infrastructure having to cope with new threats. Nelson et al. [14] provide a definition for systems which are resilient as being able to "undergo change and still retain the same function and structure while maintaining the options to develop." This suggests that systems are able to support a degraded level of performance and provide the core functions required to meet the system's goals. The notion of "surprises" is introduced by Hollangel [15] in his definition of resilience which suggests the idea of unanticipated events, making the challenge for engineering for certain types of events more difficult because they are less predictable. Certain definitions highlight the importance of time and costs whilst a system is in the recovery stages of an incident [16] and the significance of reducing these as optimally as possible. With all definitions, the implications of resilience are all context dependent, therefore defining what is meant by resilient in any given scenario is a critical part of engineering for resilience. This is equally true in our scenario of continuing the supply of water during extreme weather events like flooding. Legacy systems (pumps, for example) may not be able to cope with the changes in flood frequency and flood severity, and evolutions to existing systems should to be made, say by replacing aging systems, ensuring sustainability in the UK's water supply SoS.

The concept of resilience in this instance would attempt to mitigate against this risk of failure and to ensure the availability of water, even in times of severe conditions, i.e., infrastructure resilience. The resilience curve [17,18] is a common representation of the responsiveness of a system to undergo a disturbance and to recover from it in terms of performance; it has four main stages; reliability, unreliability, recovery, and, recovered steady state [8], as seen in Figure 2a.

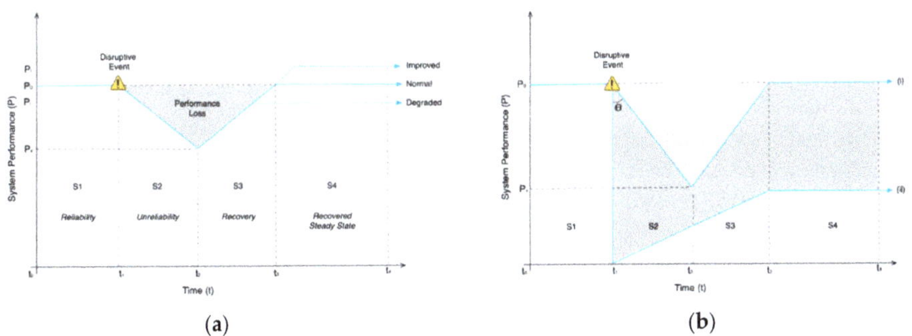

Figure 2. (**a**) Generic resilience curve; (**b**) different impacts on resilience.

The objective when considering the resilience of the water system, and the constituent systems which enable the supply of water is that when the occurrence of a flood for example occurs, then the resilience curve is more like profile (i) in Figure 2b, than profile (ii). Profile (ii) shows complete failure of the system and recovers to a state that is substantially below its original level. Where ideally, there would be no loss in performance, however profile (i) shows an "acceptable loss" in performance, and a full recovery to its normal operational state over a period of time. Of course, this is context dependent, but describing the system resilience in this way enables engineers and architects to apply the resilience methodology and test future designs against variables which are of importance.

Having a succinct methodology which enables engineers, architects and decision makers in the water domain to assess the resilience of current systems, and to then propose and test future architectural designs would be invaluable. The methodology which has been created within this article demonstrates a workflow which defines the scope of interest, and then models the elements within a reference architecture using systems modelling languages and architecture frameworks. Additionally, these static architectures can be examined by migrating them into simulation environments to assess their performance. Favourably, alternative designs and architectures can be modelled, simulated and tested to see which provides the most resilient solutions in terms of water availability and continual supply.

The value of adding the resilience viewpoint within the methodology is to specifically address aspects of resilience like criticality and vulnerability between certain nodes of the network and to identify key risks in certain scenarios. This allows the architect to think about strategies which enable the operational systems to mitigate against certain incidents and remain reliable and resilient at all times. Assigning metrics to individual constituents within a system is seen to be a step in the right direction for developing a set of metrics to measure the resilience in the water sector. This is seen to be another feature of the resilience viewpoint, an ongoing development process and research initiative.

1.2. Systems-of-Systems (SoS) in the Context of Water Supply/Flooding

Several definitions of what constitutes an SoS have been provided by Jamshidi 2005, 2008 [19], though the most favourable definition from the book is "systems of systems are large-scale integrated systems that are heterogeneous and independently operable on their own, but are networked together for a common goal," [20]. Maier [21,22] claimed that the term 'system-of-systems' did not have a distinct and recognised definition, although he did acknowledge that the SoS idea is widely accepted and generally recognised. He cited a number of examples such as integrated air defence networks, the internet, intelligent transport systems, and enterprise information networks, which are an emergent class of systems comprising large-scale systems in their own right. This has led to the categorisation of an SoS into five important characteristics:

1. Operational independence of the elements: The constituent systems (CSs) can operate independently.
2. Managerial independence of the elements: The constituent systems are acquired separately by different managerial entities.
3. Evolutionary development: An SoS evolves over time, developing its capabilities as the constituent systems are changed, added or removed.
4. Emergent behaviour: The SoS itself offers additional services above and beyond the capabilities of the constituent systems. However, it can also exhibit unexpected and potentially damaging behaviours.
5. Geographic distribution: The geographic extent of the constituent systems is large.

Accordingly, a definition of resilience has been created by the authors in the context of SoS, such as that of the water supply system, which is of topic in this case; "The dynamic ability of the SoS to re-adjust and recover when faced with change and disruption, at both the SoS and constituent system level. To continue to provide operational capacity at a certain level of (degraded) functionality and performance."

There is a clear and evident requirement for a conscious effort to be made towards a resilient water SoS within the UK, in terms of both supply and in terms of flooding. This effort is as stated, an 'SoS-wide' effort which must be supported by a range of stakeholders, including; water companies, emergency response agencies, wastewater companies, regulators, and consumers of water, to name a few. Natural disasters (including flooding) increase the probability of the water infrastructure being damaged which also could jeopardize the supply of water to customers of all types (domestic, industry, agriculture).

2. Materials and Methods

The SoS Resilience Framework/Methodology

The SoS resilience methodology is an encompassing set of methods and processes which helps decision makers, engineers and other stakeholders "design" resilience into SoS. The framework has been developed with the goal of evaluating resilience in the 'as is' state of the system and then implementing changes to target resilience in certain areas of the system which could be improved. The water scenario is a very useful case to test the methodology as it is representative of these types of mega-systems. Modelling and simulation (M&S) methods form the basis of the methodology, however it is a tool independent framework and can be replicated in a different set of modelling languages and tools, if required.

The methodology is systematic and takes a holistic approach to engineering future resilient systems. Stakeholder interaction is a fundamental characteristic of the methodology and subject matter experts (SMEs) from the water sector have been consulted throughout the application of the methodology at all phases. An overview of the developed framework is shown in Figure 3.

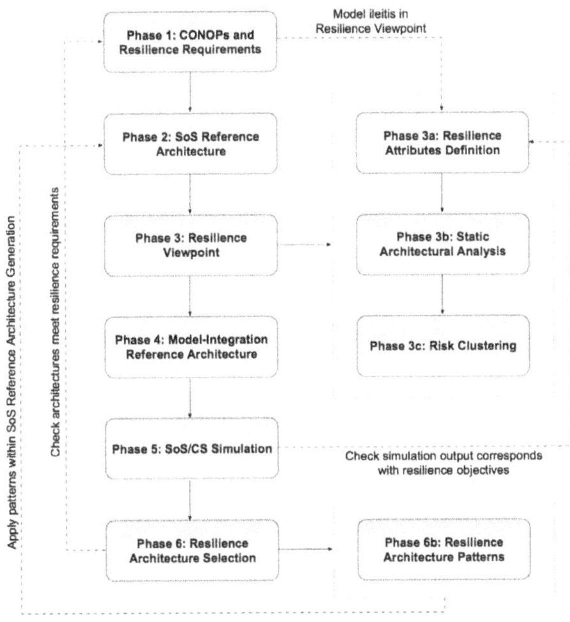

Figure 3. System-of-systems (SoS) resilience methodology.

Phase 1. CONOPs and Resilience Requirements

Phase 1 commences with deriving a set of requirements specific to the area of concern and understanding the concept of operations (CONOPs) of the operational systems involved. The tools applied here are rich picture diagrams—a soft systems method [1]—and causal loop models [23,24] to understand the dynamic interactions between system elements. Subsequently a reference architecture [1,4,25] is created using systems modelling languages such as SysML and the Department of Defense Architecture Framework (DoDAF) [26] which is common language used to create systems architecture descriptions of systems and enterprises.

Phase 2. SoS Reference Architecture

The reference architecture forms the backbone to the methodology as it defines the interactions between constituent systems (CS) and sub-elements within the water SoS. A reference architecture, as the name suggests is a source of reference when regarding systems from different architectural perspectives. A set of views that form the reference architecture are shown in Figure 4. The architecture is a resource which is typically shared amongst project stakeholders, and engineering teams in the development phases of software, a product, a service, a system or an SoS. A common set of information is one of the key benefits of model-based systems engineering (MBSE) [27–29] and the purpose of a reference architecture, which is an MBSE practice.

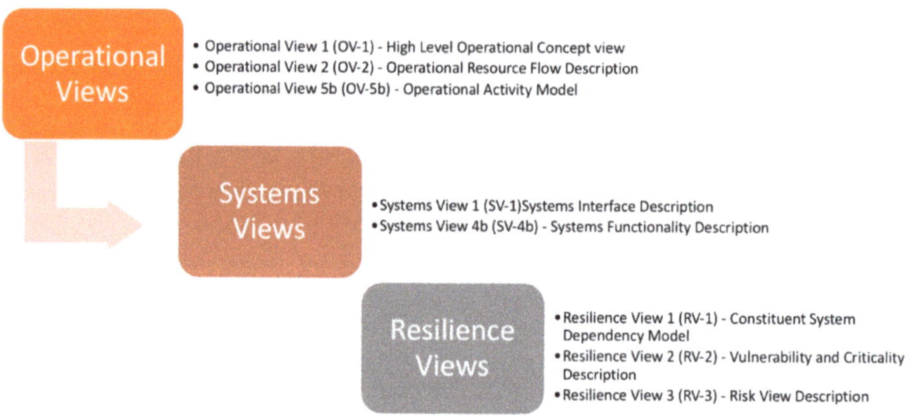

Figure 4. Architecture modelling views process.

Phase 3. Resilience Viewpoint

A novel viewpoint within the framework has been created solely focused on the resilience aspects of the architecture, and proposes three model types (Figure 4) to address; static structural analysis of the nodes within a network of systems and to explore the cascading effects of a failure or disruption; risks and those who are responsible for mitigating against risks, and; resilience attributes which are being designed into the architecture, for example, non-functional features referred to as "ilities" [30,31] such as flexibility, robustness, security, and availability. These non-functional properties are assessed through observation from the subsequent simulation models created in the next phases of the methodology.

Phase 4. Model-Integration Reference Architecture

The model-integration architecture provides the "big picture" view of the model architecture and puts it in context with the rest of the stakeholder's model portfolio, showing how the project's models and model platforms fit together. The importance of a modelling and the simulation integration framework cannot be overlooked, as it provides details at both high and low levels of granularity, conditional of the MBSE approach employed. This architecture specifies the tools used to develop models, data types, and the exchange of models and data between platforms and engineering teams. This process is highly recommended for when translating architecture to an executable model as it defines exactly what data is required for setting the parameters in the model. The integration framework is independent from the simulation tool; therefore, the engineer can select tools which are apt at a later date from when the architecture was created. The integration framework is beneficial in communicating requirements between stakeholders, and the "chief architect" and has the responsibility of gathering the essential information from individual model owners to generate a complete picture of the integration framework.

Phase 5. SoS/CS Simulation

Determining the resilience of an architecture is not a trivial task. A predominant reason for this is the nature of the architecture. Architectural designs are essentially static representations of the SoS whereas the real SoS is a dynamic entity whose behaviour is created from the interactions of the constituent elements. For successful analysis of SoS resilience to be conducted the SoS framework must consist of dynamic simulations to explore emergent behaviours which may arise from alternative architectures. The SoS reference architecture strongly supports the depiction of the SoS and its constituent systems, but it is desirable to represent the models that can be simulated in suitable simulation environments. In this case, the platform Simulink (MATLAB_R2018b, MathWorks, Cambridge, UK) was used—in conjunction with IBM Rhapsody which was used to create the architecture in phases 2, 3 and 4—to run simulations of supply and demand scenarios within a particular region of the UK.

Phase 6. Resilience Architecture Selection

This process is illustrated by Figure 5, where IBM Rhapsody architectures are manually translated into Simulink for simulation, and subsequent analysis in an additional visual analytics tool, in this case Tableau software (Tableau Desktop 10.5, Seattle, WA, USA). Simulink was selected as it permits the use of sliders and other parameter controls as the simulation runs in real-time. This enables decision makers to explore different parameters and immediately visualize the outputs.

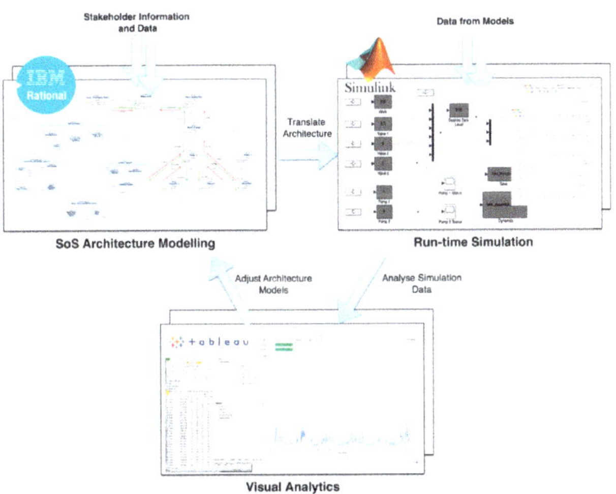

Figure 5. Modelling and simulation workflow.

Although an overall SoS resilience metric is desirable, it gives little information of the resilience at the local, CS level (or lower) which could be problematic when resilience is required in a specific area of the SoS. Uncertainty and unknown parameters are likely to make the task of achieving overall SoS resilience extremely difficult for the architect, hence why targeting resilience at the CS level is probably more logical. Metrics associated to executable architectures provide insight to the performance of technical and social systems which assist in the design space exploration process for more resilience solutions.

3. Case Study Results

3.1. Case Study Overview

This case study considers a scenario where flooding has resulted in pump failure (directly from flood water or from power failure), where the pump is a constituent system that forms part of the SoS. If we consider this case and assume that the flood has created a redundancy problem as the pump supplies multiple demand points within a network, it becomes evident that this challenge is greater than a redundancy problem and we must consider alternative resilience strategies to overcome the difficulties. Let us consider demand centres which are supplied by a water source of some kind (e.g., primary service reservoir) and utilise a pump to feed water up against gravity, to a set of demand centres via a number of service reservoirs. Of the four demand centres as seen in Figure 6, it is assumed (for demonstration purposes) that two are for domestic use, one for industrial and one for agriculture.

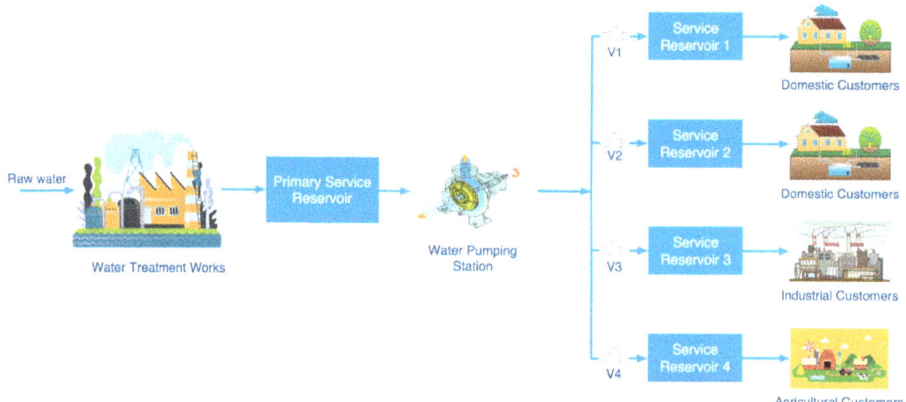

Figure 6. Case study scenario.

As previously mentioned, the resilience is the capacity of the water supply system to meet demands during emergency situations like floods, and in this case during the failure of a pump in the network. As it can be seen from Figure 6, this would cease supply to all four service reservoirs, enabling supply to the demand points until those service reservoirs completely depleted. Multiple strategies can be implemented, and numerous architectural variants can be suggested to increase the resilience here, however before getting to these, the upcoming results section shows the application of certain phases of the SoS resilience methodology.

3.2. Case Study Results

Following the phases outlined previously in Figure 3, a great understanding of the water SoS is achieved through applying CONOPs methods such as rich picture diagrams (Figures 1 and 6) and causal loop models (Figure 7). The advantages of causal loop diagrams are that it helps analyse complex systems and helps identify key dynamic variables for later simulations. Cause and effect diagrams are a crucial aspect of designing resilience systems, particularly for understanding the dynamics between key variables. The loops which are created show an overall effect on the system and this is one of the elements of a causal loop diagram as illustrated in Figure 7, where the overall effect is a decrease in water available due to damaged infrastructure in times of a flood event. Although this seems a basic construct, reducing to this level of complexity is beneficial in exploring potential solutions, and determining the types of data sets required for simulating later in the process. Thus, from the causal loop it can be determined that the solution that needs exploring is the flexibility of the current systems in place to search for strategies and also architectural solutions which regulate the availability of water and hence, the supply of water to those that require it.

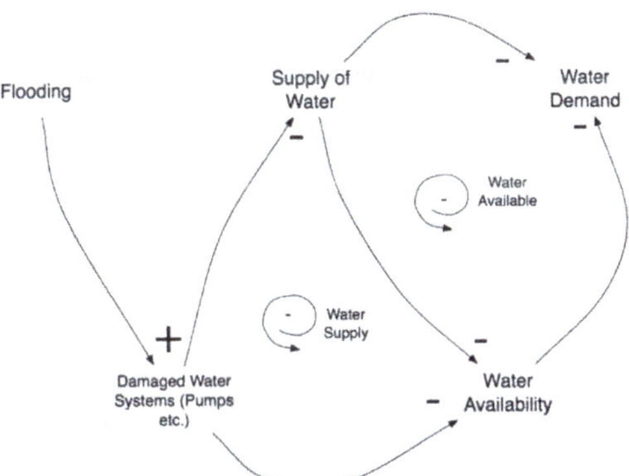

Figure 7. Causal loop of water problem.

The rich picture and causal loop models helped define the requirements for this case study, and eliciting requirements for resilience in this case was done in collaboration with a range of stakeholders via small workshops, including; water companies, water consultancies and academics interested in water applications.

A subset of architecture models has been provided here to illustrate the type of modelling, but the reality is there would be numerous instances of some models from different stakeholders' standpoints to describe the CSs and to elicit further resilience requirements. The two models here show a general Operational Viewpoint (OV-5b) of the water processing procedure and a Systems Viewpoint (SV-1) of the water companies' systems arrangement and connectivity (Figures 8 and 9, respectively). Figure 8 shows a high-level process model of taking raw water and passing it through a series of stages such as initial water storage, water screening, filtering etc. prior to being supplied for water customers and consumption. Figure 9 shows four high level components of a water system; a water company, distribution network, customers and consumers and waste water distribution network. Again, these are then populated with further subsystems which must be present to make the systems functional. Multiple instances of each model type can be created to model specific system structures, but these are for demonstration purposes.

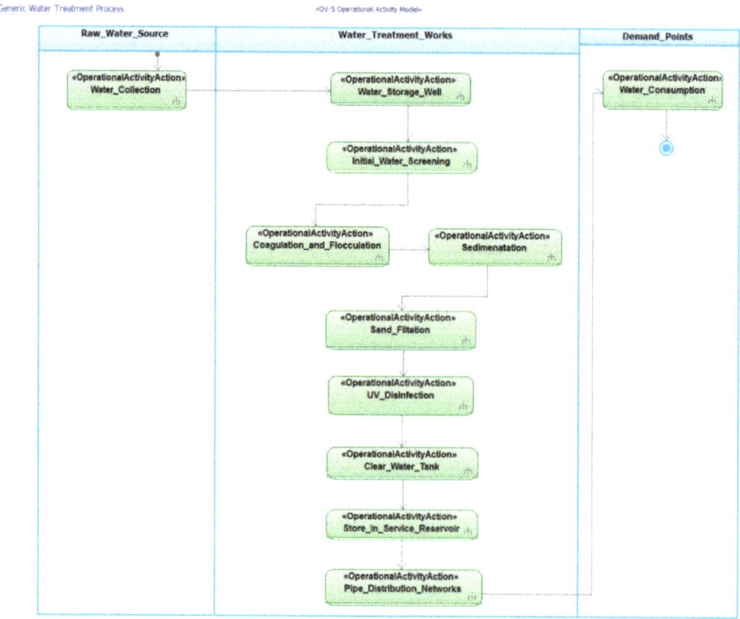

Figure 8. OV-5b—Operational view of high-level water treatment process.

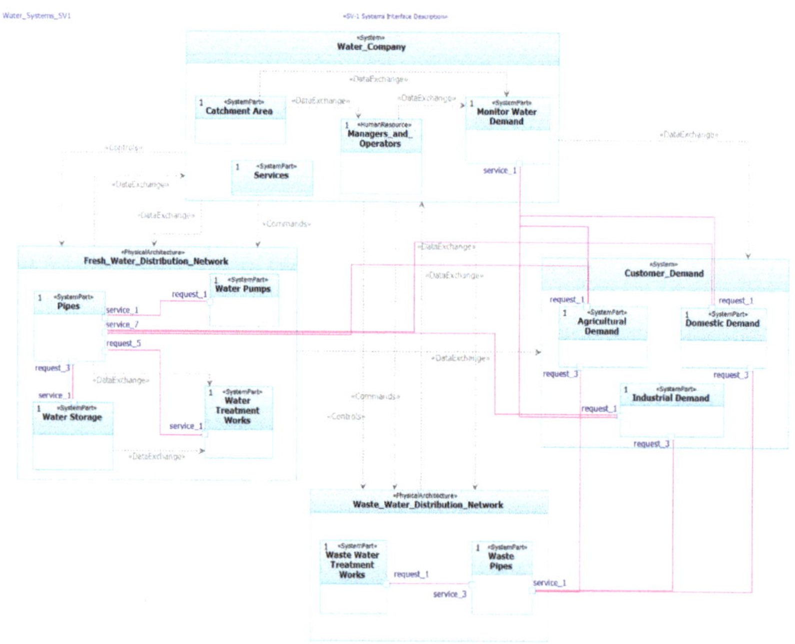

Figure 9. SV-1 Water Company systems.

Evolutionary development of the SoS architecture should be addressed by iteratively updating the specific constituent systems with the overall SoS in mind. The reference architecture is only as good as the quality of the information it holds and therefore, it is crucial that quality information is captured in its views/models. These static architectural representations go a long way in the design phases, allowing effective communication with stakeholders who have an influence on shaping the CS and the requirements definition process for future phases of the SoS evolution.

Subsequently, modelling the system using the resilience viewpoint (more specifically, Resilience View 2—RV-2) allows the engineer to highlight which nodes are critically dependent on each other and also to show which nodes are vulnerable if failure was to occur at some point in the network. For instance, modelling the case study example, shown in Figure 6, the viewpoint shows the four demand points being critically dependent on the functioning of the pump, and vice versa, the demand points being vulnerable on the pump failing. Similarly, the water storage component, or the primary service reservoir becomes vulnerable if the pump is no longer working, because that clean water has a set period which it can be stored for before becoming unsafe to distribute. Figure 10 shows the capability of RV-2 to model further details of each SoS component, for instance, the main supply pipeline has certain attributes which can be assigned to that element. These may include important parameters such as flow rates, flow direction, capacity, and others which are important in later simulation phases.

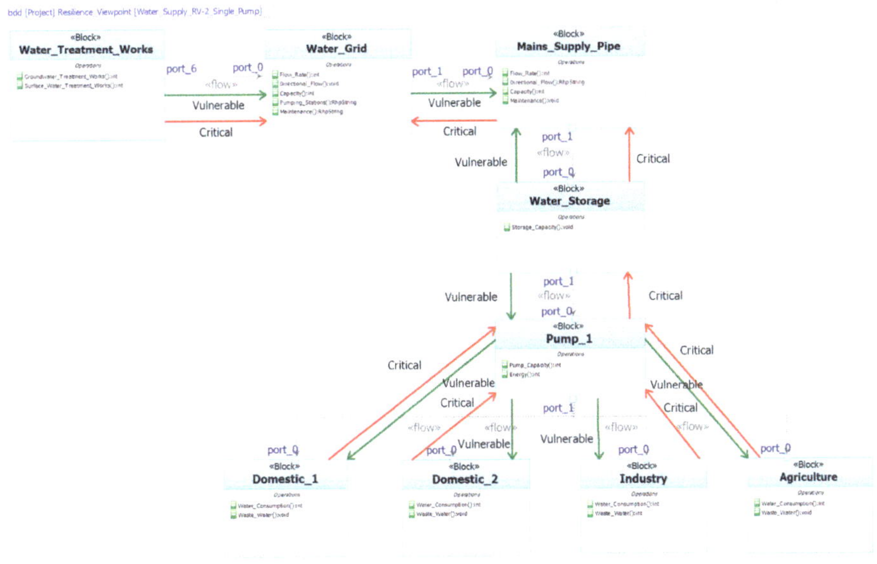

Figure 10. Resilience View 2 (RV-2) pump failure.

How would this look on the resilience curve? Referring back to Figure 2, the outage of a pump in a single pump supply system could result in curve 2b (ii), the performance dropping off to zero until that pump was restored. Alternatively, an architectural solution to the problem could be to add a secondary pump (Figure 11), that could be turned on immediately after failure, and performance would again increase back to a lower level of performance, or back to ideal performance P0. On the other hand, a decrease in performance could be avoided altogether if the failure of Pump 1 was anticipated, and Pump 2 switched on prior to failure.

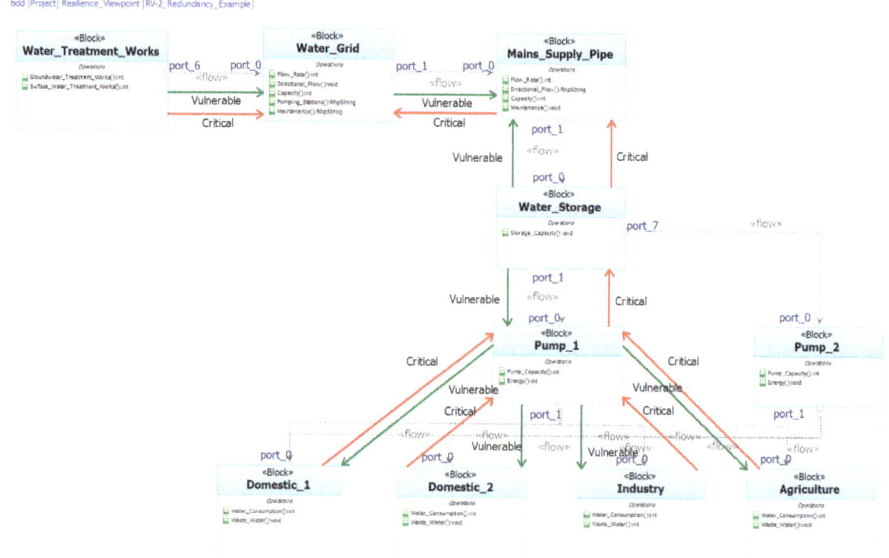

Figure 11. RV-2—Two pump architecture solution.

At first glance, this may seem like an obvious problem with some obvious solutions to increase the resilience of supplying water to the consumers. Resilience here is considered as a redundancy issue. However, resilience can be a set of operational strategies implemented during a time of disruption to solve the supply problem. For example, prioritizing which demand points are more critical, decision-makers can set constraints on where to supply water and for which certain periods of time. For instance, it may be acceptable to supply the two domestic areas from say 6.00 am to 10.00 am and then from 3.00 pm to 10.00 pm in attempt to cut back on water consumption and resultant water waste. Another solution would be to decrease the pressure of which water is supplied to these points which would decrease the overall volume of water available to be used. Furthermore, depending on what is prioritised, the system may gracefully degrade its capacity to supply the two domestic points and the industrial point, reducing its supply to agriculture completely for a given period of time. This is one resilience strategy, although there will be many like this which should be explored architecturally and simulated to evaluate the feasibility of each one. These are just a few of many strategies which could be implemented to resolve the problem and, in the short-term at least, solve some issues of increasing resilience against flood scenarios.

From a risk perspective, the resilience viewpoint offers a model (Resilience View 3, RV-3) to capture risks which can be communicated with all stakeholder groups. Additionally, the risk view as illustrated in Figure 12, permits the architect or engineer to allocate the risks to constituent systems or stakeholders within the water SoS. This allocation of risks establishes responsibility of each identified risk and provides accountability to whom or to what is responsible for mitigating against that risk. The advantage from an architectural perspective is that risks can be avoided if mitigation strategies are embedded within the design of constituent systems. An example of an RV-3—Risk View Description—is shown in Figure 12. The example shows the risk of failure (or partial failure) of the water pump from the above example. Although this a simple example for demonstration purposes, the number of risks in a scenario like this are likely to be in the tens or hundreds, and this is a valued way to manage these risks as the complexity can be captured in multiple instances. This view (RV-3) within the resilience view package is a novel inclusion to the architecture framework which historically does not consider risks (in this way) as part of the architecture modelling phases of systems lifecycle modelling or development.

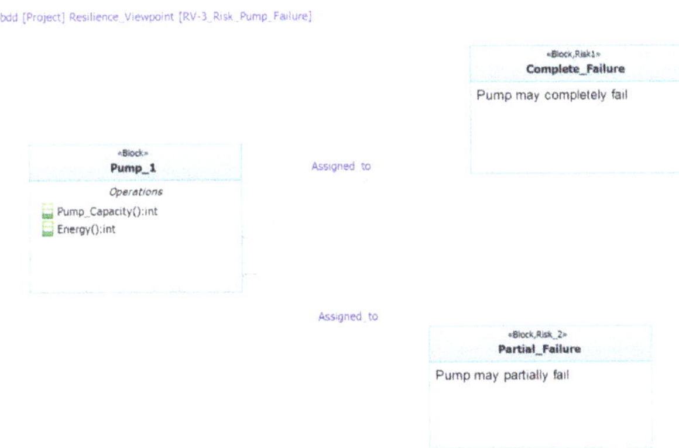

Figure 12. Resilience View 3 (RV-3)—Risk view description.

The final phase to be applied is simulation of architectures. Due to the nature of an SoS such as the water SoS, it is extremely important to simulate as much of the SoS and with as many of the CSs included as possible. The purpose of simulation in our case is to determine whether the overall resilience goals of the SoS can be satisfied by evaluating the interactions between the constituent systems and to understand where the current system falls short in terms of resilience. Since we cannot always rely on the CSs being modelled with the same tools, special techniques must be used to transform the architecture into a set of system models that can be executed within the same co-simulation [32] environment. Whilst it seems convenient for all CSs to be modelled using the same tool this is not always possible, nor desirable, due to commercial restrictions i.e., different water companies use different modelling tools and languages to design future capabilities. Also, it is highly probable that certain existing CSs have already been created and tested using other tools and provision should be made to use these wherever possible.

System dynamics (SD) [33] is an approach applied to understand the behaviour of complex, dynamic, nonlinear systems. Defining variables is a key step in creating an SD model, as these are assigned to specific model elements to simulate how changes in the system occur over time, and thus provide understanding of the basic structure of a system and the rationale for its behaviours. A positive of employing such a method in the SoS context comes from the need to understand the behaviour of the whole through understanding the behaviour of the interconnected parts. Numerical values can be assigned to the model and the dynamics can be simulated through execution of the variables. This study used the software tool, Simulink to create the model and simulation for our case study that stemmed from the specifications defined within IBM Rhapsody.

The Simulink model as seen in Figure 13 shows the element that reflects the problem scenario outlined in Figure 6, which looks at the use of pumps and valves to carry water from a primary reservoir, against gravity, to four service reservoirs which supply different types of customers (domestic, industry and agricultural). The model enables the user to set desired tank levels and the prescribed levels and to then manipulate the pumps and valves to direct water to specific service reservoirs depending on the resilience strategy in place. This specific strategy focussed on keeping the supply high to the two domestic customer groups i.e., service reservoirs 1 and 2, and to drop the amount of water being supplied to agriculture for the short term in order to maintain water to the domestic and industry customers. Preliminary results for this simulation run can be seen in Figure 14 where the domestic customers and their respective service reservoirs remain constantly at a high water capacity and the agricultural service reservoir (service reservoir 4) remains low. Service reservoir 3, the industry service reservoir fluctuates depending on the demand.

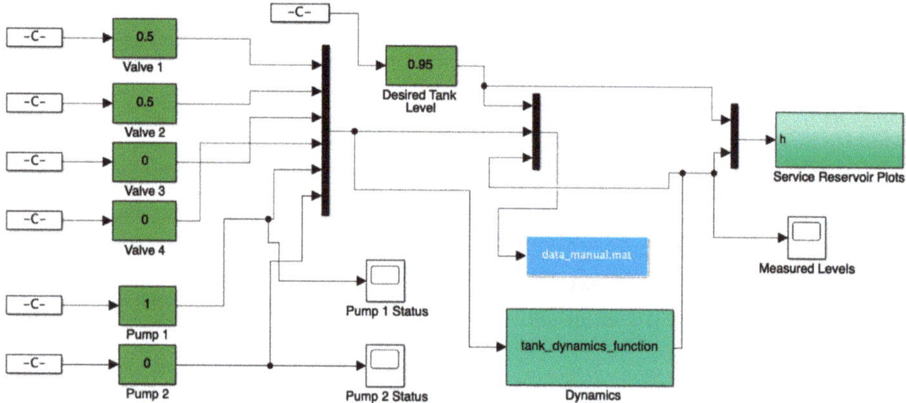

Figure 13. Simulink model.

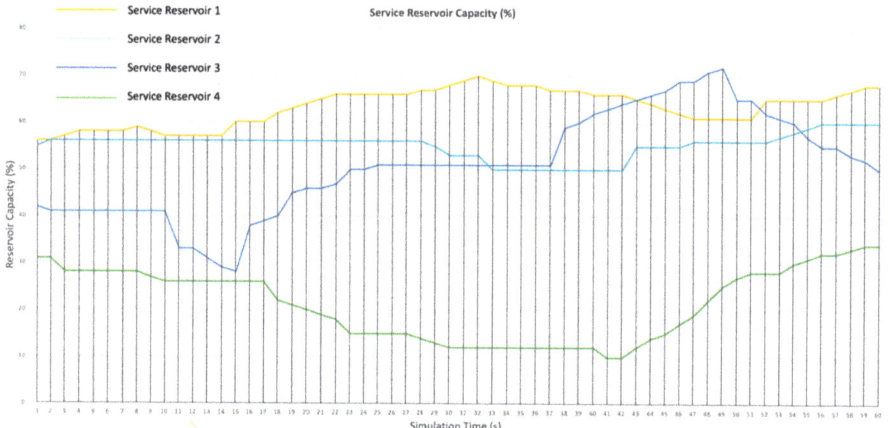

Figure 14. Service reservoir capacity from simulation run.

In order to test the model further, real data from the demand of customer groups and the volume and capacity of real reservoirs from specific locations will enable more accurate simulations to be run that are reflective of incidents in times of flooding. A great advantage of running a model like this is Simulink, in that it enables the engineer or decision maker to manipulate the parameters and variables using sliders and other input types to assess the response of the system in real time. Thus, allowing trade offs to be made in real time with respect to the events of a flood and the resilience strategies which are being tested and implemented.

4. Discussion

The methodology presented provides a means to explore resilience in SoS and to engineer resilience into systems such as the water SoS at multiple points within its lifecycle. Engineering resilience is a systematic endeavour and the model-based engineering methodology shown enables resilience to be evaluated in the 'as is' state and to explore future states to proactively mitigate against certain risks within particular contexts. Overall, the methodology will stipulate many outputs which are regarded as tools for engineering resilience in SoS. Such artefacts include; information about the water SoS which is stored within a reference architecture; architectural designs which can be stored as architecture patterns [34] for later re-use in design activities; simulation data which reflects the performance of architectures and helps inform decisions about architecture design; and finally; a detailed and shared understanding of key resilience challenges and issues, within the water sector, across a broad range of stakeholders.

In order to avoid commercially sensitive data, generic data was presented in this paper that is based on real water data to test the methodology. A significant result is the creation of a resilience viewpoint within an architecture framework that solely focusses on resilience. The anticipated future results would be to simulate some of the water architectures suggested to see how they perform using a resilience curve and to assess the responsiveness of certain designs. This would be done using the workflow as suggested in Figure 5, and data would be analysed to infer resilience measures for future SoS evolutions.

The intrinsic characteristic of any SoS is its state of constant flux and evolution. This means the reference architecture and subsequent simulations must be updated on a regular basis, especially when simulation results depart from the observed operation of the SoS. Attempting to perform this evaluation on the actual SoS is fraught with danger, because surprise emergent behaviours may be highly detrimental or unsafe. The reference architecture (blueprint) becomes an essential asset when considering the evolution of an SoS or where the SoS begins to exhibit unpredicted behaviour. However, we must always guard against the differences between the real world and the simulated SoS. Nevertheless, having a reference benchmark is extremely important. Similarly, maintaining a library of patterns for the constituent systems is very helpful not only in representing complex characteristics of the constituent systems but also in their reuse.

It was clear from the engagement with water companies and consulting companies that asset resilience is of high priority to water companies, in ensuring an aging infrastructure can meet the demands of present and future trends. To solve the challenges of modernizing infrastructure and maintaining waste water networks, there is an evident need for an interdisciplinary approach involving experts from water supply companies, systems engineers, climate experts, and key customers (domestic, industry and agriculture). By treating the water sector as an SoS, it should be possible to begin to address the more challenging aspects of ensuring resilience in water supply for the future; for example, the application of a reference architecture for mapping the current resources, surveying the current condition, analysing the problem areas, and understanding the nexus within all the stakeholders like geographical and economic implications. Strengthening the links between several elements from academic research and industry could make an important contribution to solving these challenges via the application of an SoS reference architecture.

Future work would include generating architectures that can be assessed quantifiably allowing resilience metrics to be developed for different types of SoS while measuring the performance of architectures can iteratively be done in modelling and simulation environments bespoke to these kinds of investigations. To also link the resilience curve with specific data from water supply failures would provide a way forward for quantifying resilience using some important metrics.

Where a pre-existing model of a constituent system exists, it is not always possible to guarantee that it will be compatible with the simulation environment. Therefore, care must be taken to understand the limitations of the model and its context of use. Such requirements place specific demands on the co-simulation environment, which must be a component-based simulator where system models are incorporated as the composition of a set of hierarchical modules. Consequently, the reference architecture must be structured to support a composition of constituent system models. Whilst this task is straightforward at the reference architectural modelling level, this requirement restricts the choice of simulation environment and depends on what simulations are needed. However, underlying this is a complex process of transforming system architectures into executable models [32], and something which should be looked into in future work.

5. Conclusions

This paper presents a methodology to explore resilience in the water sector from multiple perspectives, by adopting an SoS approach to problem solving. Water supply to flooding has been considered from an SoS standpoint, and the creation of a methodology allows resilience to be explored in many water scenarios to assess resilience and propose future designs. Resilience is seen as a new paradigm to risk management. A proactive approach is needed to explore future scenarios which may cause problems for systems with high dependencies and raise concerns with regards to system performance. Resilience engineering is seen as an important topic in the domain of systems engineering and certainly in sectors such as water, where the importance of successful operations is critical to human life and the environment.

Model-based systems engineering, and a structured framework or methodology allows decision makers and engineers to; (i) evaluate the resilience of current systems through static and dynamic methods, and; (ii) to discover future system states and resilience strategies through architecture exploration and implement them through a tested methodology. The water supply and flooding example shown within this paper illustrates the practicality of applying a set of MBSE methods to increase the resilience of systems which deliver water to consumers and demands points within a supply network. Water systems may suffer impairments and disruptions during times of severe weather, e.g., floods and droughts, or even failure due to legacy systems aging and failing naturally. The paper has shown how water may still be made available during a flood scenario, through architectural alternatives to deliver water prior and during a disruption. However, there are multiple variations of strategies and architectural alternatives which could provide a solution to this specific issue. Furthermore, water availability data of a specific region would help validate this process via informed simulation models, developed in conjunction with subject matter experts from a water company, however the methodology is a strong step in the right direction to explore resilience in SoS and the water sector during times of flooding.

Author Contributions: Conceptualization, D.J., R.K.; Data curation, D.J. and S.S.; Formal analysis, D.J. and S.S.; Funding acquisition, D.J. and R.K.; Investigation, D.J.; Methodology, D.J.; Project administration, D.J., R.K., M.R.C. and G.F.; Resources, D.J.; Software, D.J. and S.S.; Supervision, R.K.; Visualization, D.J.; Writing—original draft, D.J. and S.S.; Writing—review & editing, D.J., R.K., G.F., M.R.C. and F.M.

Funding: The authors would like to thank the EPSRC for the funding on BRIM (EP/N010329/1).

Conflicts of Interest: The authors declare no conflict of interest.

References

1. Monk, A.; Howard, S. Methods & tools: The rich picture: A tool for reasoning about work context. *Interactions* **1998**, *5*, 21–30.
2. Jenkins, G.J.; Murphy, J.M.; Sexton, D.M.H.; Lowe, J.A.; Jones, P.; Kilsby, C.G. *UK Climate Projections: Briefing Report*; Met Office Hadley Centre: Exeter, UK, June 2009; ISBN 9781906360023.
3. *HM Government UK Climate Change Risk Assessment*; HM Government: London, UK, 2017; p. 24.
4. Alessandra Scotto di Santolo UK Weather Warning: Expert Warns UK Heatwave is First of Many as Summers Will Get HOTTER. *EXPRESS*. 2018. Available online: https://www.express.co.uk/news/weather/997034/UK-weather-warning-climate-change-summer-heatwave-latest (accessed on 15 August 2018).
5. Office for National Statistics. *Overview of the UK Population*; Overv. UK Popul.; Office for National Statistics: Newport, UK, July 2017; pp. 1–17.
6. Hosseini, S.; Barker, K.; Ramirez-Marquez, J.E. A review of definitions and measures of system resilience. *Reliab. Eng. Syst. Saf.* **2016**, *145*, 47–61. [CrossRef]
7. Tran, H.T.; Balchanos, M.; Domerçant, J.C.; Mavris, D.N. A framework for the quantitative assessment of performance-based system resilience. *Reliab. Eng. Syst. Saf.* **2017**, *158*, 73–84. [CrossRef]
8. Bhamra, R.; Dani, S.; Burnard, K. Resilience: The concept, a literature review and future directions. *Int. J. Prod. Res.* **2011**, *49*, 5375–5393. [CrossRef]

9. Woods, D. Four Concepts for resilience and the Implications for the Future of Resilience Engineering. *Reliab. Eng. Syst. Saf.* **2015**, *141*, 5–9. [CrossRef]

10. Righi, A.W.; Saurin, T.A.; Wachs, P. A systematic literature review of resilience engineering: Research areas and a research agenda proposal. *Reliab. Eng. Syst. Saf.* **2015**, *141*, 142–152. [CrossRef]

11. Adjetey-Bahun, K.; Birregah, B.; Châtelet, E.; Planchet, J.L. A model to quantify the resilience of mass railway transportation systems. *Reliab. Eng. Syst. Saf.* **2016**, *153*, 1–14. [CrossRef]

12. Bergström, J.; Van Winsen, R.; Henriqson, E. On the rationale of resilience in the domain of safety: A literature review. *Reliab. Eng. Syst. Saf.* **2015**, *141*, 131–141. [CrossRef]

13. Patriarca, R.; Falegnami, A.; Costantino, F.; Bilotta, F. Resilience engineering for socio-technical risk analysis: Application in neuro-surgery. *Reliab. Eng. Syst. Saf.* **2018**, *180*, 321–335. [CrossRef]

14. Nelson, D.R.; Adger, W.N.; Brown, K. Adaptation to Environmental Change: Contributions of a Resilience Framework. *Annu. Rev. Environ. Resour.* **2007**, *32*, 395–419. [CrossRef]

15. Hollnagel, E.; Paries, J.; Woods, D.D.; Wreathall, J. *Resilience Engineering in Practice: A Guidebook*; Ashgate Publishing Limited: Farnham, UK, 2011; ISBN 978-1-4094-1035-5.

16. Haimes, Y.Y. On the definition of resilience in systems. *Risk Anal.* **2009**, *29*, 498–501. [CrossRef] [PubMed]

17. Nan, C.; Sansavini, G. A quantitative method for assessing resilience of interdependent infrastructures. *Reliab. Eng. Syst. Saf.* **2017**, *157*, 35–53. [CrossRef]

18. Gama Dessavre, D.; Ramirez-Marquez, J.E.; Barker, K. Multidimensional approach to complex system resilience analysis. *Reliab. Eng. Syst. Saf.* **2016**, *149*, 34–43. [CrossRef]

19. Jamshidi, M. *Systems of Systems Engineering: Principles and Applications*, 1st ed.; Jamshidi, M., Ed.; CRC Press: Boca Raton, FL, USA, 2008; ISBN 978-1-4200-6589-3.

20. Jamshidi, M. *System Of Systems Engineering: Innovations for the 21st Century*; Jamshidi, M., Ed.; WILEY: Hoboken, NJ, USA, 2009; ISBN 978-0-470-19590.

21. Maier, J.; Eckert, C.; Clarkson, P. Model Granularity and Related Concepts. In Proceedings of the International Design Conference—Design 2016, Cavtat, Croatia, 16–19 May 2016; pp. 1327–1336.

22. Maier, M. *Art of Systems Architecting*, 3rd ed.; CRC Press: Boca Raton, FL, USA, 2000; ISBN 9780849304408.

23. Inam, A.; Adamowski, J.; Halbe, J.; Prasher, S. Using causal loop diagrams for the initialization of stakeholder engagement in soil salinity management in agricultural watersheds in developing countries: A case study in the Rechna Doab watershed, Pakistan. *J. Environ. Manag.* **2015**, *152*, 251–267. [CrossRef] [PubMed]

24. Binder, T.; Belyazid, S.; Haraldsson, H.V.; Svensson, M.G.; Kennedy, M.; Winch, G.W.; Langer, R.S.; Rowe, J.I.; Yanni, J.M. Developing System Dynamics Models from Causal Loop Diagrams. In Proceedings of the 22nd International Conference on Neural Information Processing Systems, Vancouver, BC, Canada, 7–10 December 2004.

25. US Dept. of Defence/Office of the DoD CIO. *Reference Architecture Description*; US Department of Defense: Washington, DC, USA, 2010.

26. Hause, M. The unified profile for DoDAF/MODAF (UPDM) enabling systems of systems on many levels. In Proceedings of the 2010 IEEE International Systems Conference, San Diego, CA, USA, 5–8 April 2010; pp. 426–431.

27. Ibarra-zannatha, J.M.; Limón, R.C.; Hernandez, W.E.C.; Electronico, I.; Estefan, J.A.; Obaidat, M.S.; Boudriga, N.A.; van der Aalst, W.M.P.; Voorhoeve, M.; Page, E.H.; et al. Survey of Model-Based Systems Engineering (MBSE) Methodologies 2. Differentiating Methodologies from Processes, Methods, and Lifecycle Models. *Environment* **1994**, *32*, 397–438.

28. Piaszczyk, C. Model Based Systems Engineering with Department of Defense Architectural Framework. *Syst. Eng.* **2011**, *14*, 305–326. [CrossRef]

29. Kalawsky, R.S.; O'Brien, J.; Chong, S.; Wong, C.; Jia, H.; Pan, H.; Moore, P.R. Bridging the gaps in a model-based system engineering workflow by encompassing hardware-in-the-loop simulation. *IEEE Syst. J.* **2013**, *7*, 593–605. [CrossRef]

30. Ricci, N.; Fitzgerald, M.E.; Ross, A.M.; Rhodes, D.H. Architecting systems of systems with ilities: An overview of the SAI method. *Procedia Comput. Sci.* **2014**, *28*, 322–331. [CrossRef]

31. Ross, A.M.; Rhodes, D.H. Towards a prescriptive semantic basis for change-type ilities. *Procedia Comput. Sci.* **2015**, *44*, 443–453. [CrossRef]

32. Blochwitz, T.; Otter, M.; Arnold, M.; Bausch, C.; Clauß, C.; Elmqvist, H.; Junghanns, A.; Mauss, J.; Monteiro, M.; Neidhold, T.; et al. The Functional Mockup Interface for Tool independent Exchange of Simulation Models. In Proceedings of the 8th International Modelica Conference, Dresden, Germany, 20–22 March 2011; pp. 173–184.

33. Salzano, E.; Di Nardo, M.; Gallo, M.; Oropallo, E.; Santillo, L.C. The application of System Dynamics to industrial plants in the perspective of Process Resilience Engineering. *Chem. Eng. Trans.* **2014**, *36*, 457–462.
34. Kalawsky, R.S.R.S.; Joannou, D.; Tian, Y.; Fayoumi, A. Using Architecture Patterns to Architect and Analyze Systems of Systems. *Procedia Comput. Sci.* **2013**, *16*, 283–292. [CrossRef]

 *water*

Article

Automated Floodway Determination Using Particle Swarm Optimization

Huidae Cho [1,*,†], **Tien M. Yee** [2,†] **and Joonghyeok Heo** [3]

[1]   Institute for Environmental and Spatial Analysis, University of North Georgia, Oakwood, GA 30566, USA
[2]   Department of Civil and Construction Engineering, Kennesaw State University, Marietta, GA 30060, USA; tyee@kennesaw.edu
[3]   Physical Sciences Department, University of Texas Permian Basin, Odessa, TX 79762, USA; heo_j@utpb.edu
*   Correspondence: hcho@isnew.info
†   These two authors contributed equally to this work.

Received: 4 September 2018; Accepted: 6 October 2018; Published: 10 October 2018

Abstract: The floodway plays an important role in flood modeling. In the United States, the Federal Emergency Management Agency requires the floodway to be determined using an approved computer program for developed communities. It is a local government's interest to minimize the floodway area because encroachment areas may be permitted for human activities. However, manual determination of the floodway can be time-consuming and subjective depending on the modeler's knowledge and judgments, and may not necessarily produce a small floodway especially when there are many cross sections because of their correlation. Very little work has been done in terms of floodway optimization. In this study, we propose an optimization method for minimizing the floodway area using the Isolated-Speciation-based Particle Swarm Optimization algorithm and the Hydrologic Engineering Center's River Analysis System (HEC-RAS). This method optimizes the floodway by defining an objective function that considers the floodway area and hydraulic requirements, and automating operations of HEC-RAS. We used a floodway model provided by HEC-RAS and compared the proposed, manual, and default HEC-RAS methods. The proposed method consistently improved the objective function value by 1–40%. We believe that this method can provide an automated tool for optimizing the floodway model using HEC-RAS.

Keywords: floodway; optimization; particle swarm optimization; HEC-RAS; flood mitigation; hydraulic modeling

1. Introduction

The floodplain and floodway are an essential part of hydrologic and hydraulic studies of riverine flooding. The floodplain shows any area that will be covered by water when a flood event occurs. In the United States, the Federal Emergency Management Agency (FEMA) requires the 100-year and 500-year floodplains as part of the National Flood Insurance Program (NFIP) [1]. For developed communities, FEMA also requires a floodway to be determined within the 100-year floodplain using a computer program approved by them [2]. FEMA approved the Hydrologic Engineering Center's River Analysis System (HEC-RAS) [3] because it is widely accepted and used for floodplain mapping and flood risk modeling around the world [4–8]. HEC-RAS was developed by the US Corp of Engineers [3] and is available to the public at no cost. However, since this computer program is not open source, its source code is not available to the research community and implementing improvements within HEC-RAS is very difficult, if at all possible. To address this challenge, HEC-RAS provides the Application Programming Interface (API) called the HECRASController that allows the user to control its user interface non-interactively [9].

The floodway is defined by vertical encroachments on both sides of the main channel area within the 100-year floodplain as shown in Figure 1 and conveys flood water without raising the water surface elevation by a regulated threshold, typically set to be 0.305 m (1 ft) in the United States by FEMA unless a lesser rise criterion is imposed by the state [10]. Figure 1 shows a schematic of the 100-year flood elevation and the floodway elevation before and after encroachment, respectively. Ideally, development within the floodplain should be avoided. However, in dense urban areas, development that encroaches into the floodplain is unavoidable because an increase in human activities often pushes development closer to rivers and streams, and even into floodplains. At the same time, for safety reasons, many regulations do exist to prohibit excessive encroachments into natural streams because such encroachments can cause rise in the 100-year flood elevation and result in severe flooding upstream. The floodway then becomes a boundary to prevent encroachments from excessively impeding the conveyance of flow and hence excessive rise in the flood elevation. Since encroachment areas on both sides are often used for human activities such as business, leisure, parks, etc., it is one of the major interests for landowners and local governments to maximize these areas. To satisfy their interests in a safe manner, the width of the floodway can be reduced during flood modeling as long as the hydraulic requirements are met to minimize flood hazards. By optimizing the floodway, local governments can achieve more sustainable land use planning, better risk and safety assessment, and will be able to mitigate legal issues due to subjective floodway interpretations. However, floodway optimization is a difficult task because the floodway boundary can be established in many different ways [11].

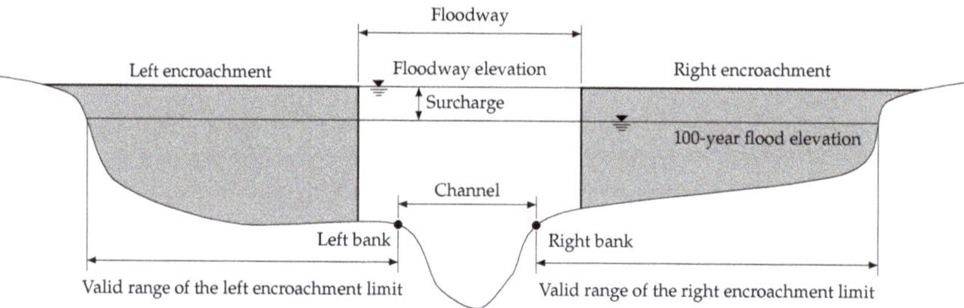

Figure 1. Schematic of the 100-year flood elevation and floodway.

Protocols in creating the floodway boundary are not well defined [12,13] and are primarily left to the practitioner [13]. One major problem without well-defined methods of floodway determination is that floodway boundaries produced by different modelers may vary significantly. Depending on the modeler's motivation, experience level, and understanding of the subject matter, different modelers may produce different floodway boundaries even if given the identical hydraulic model. This subjectivity may not necessarily result in optimal floodway boundaries and hence highlights the need and advantage of a computer-aided optimization method.

Selvanathan and Dymond [14] developed an ArcGIS [15] tool that can run HEC-RAS, post-process the results, and visualize and smooth the floodway from within an ArcGIS environment. Their tool also allows the modeler to adjust encroachments visually inside ArcGIS and has a limited optimization routine that tries to satisfy the surcharge requirement. However, the focus of their research was mainly on automation of iterative HEC-RAS runs and post-processing of its outputs. Franz and Melching [11] introduced the Full Equations Utilities (FEQUTL) model that uses an iterative trial and error procedure to determine the left and right encroachment limits. Their model was not developed to perform reach-wide optimization of the floodway, but it does provide the reader with a glimpse of the technique used for floodway determination. Majority of focus in floodway modeling revolves around the criteria and methods used in floodway modeling [12,13,16,17]. Most discussions are primarily focused on modeling techniques, recommendations to modeling standards and procedures, and evaluating the practicality and feasibility of applying one uniform standard for all floodways. Thomas and Golaszewski [17] suggested an improved iterative procedure that involves a consideration of non-steady section-averaged velocity, variation of velocity-depth product, topographic and geomorphic features, control of hydraulic structures, flow conveyance and results of hydraulic models. They also acknowledged that an experienced practitioner is required to delineate and assess floodways using the improved iterative procedure.

Lots of effort in optimization of floodplains and floodways have been devoted towards the system operations of flow control structures within the floodway [18–20] and floodplain management related issues such as flood risk assessment and cost benefit analysis of different floods and structures [21–31]. Szemis et al. [18] introduced an optimization framework for scheduling environmental flow management alternatives using ant colony optimization [32]. Bogardi and Balogh [19] developed a model that calculates the probability of levee failures and optimizes floodway operations. Luke et al. [20] studied the impact on the floodway and levee damages in the New Madrid floodway in Missouri of the detonation control during May of 2011 and concluded that passive control would have greatly reduced the costs of repairing those hydraulic structures. Lund [21] used linear programming to develop an approach that minimizes the expected flood damages and costs. Shafiei et al. [22] examined different genetic algorithms (GA) [33] for optimizing the levee encroachment design and concluded that GAs are acceptable tools for solving the levee design problems while non-GA-based optimization techniques may not be able to find the global optimum of such problems because of the non-linear nature of the objective function surface. Mori and Perrings [23] developed a model for finding optimal floodplain development decisions. Yazdi and Neyshabouri [24] used the non-dominated sorting GA to find the optimal Pareto solutions of two objective functions minimizing flood mitigation costs and potential damages to the floodplain. Lu et al. [25] proposed an inexact sequential response planning approach for optimal management of floodplains. Porse [26] used linear programming to evaluate decisions for urban floodplain development and assess potential flood damages. Woodward et al. [27,28] developed a decision support system that generates effective mitigation measures and optimizes their performance using a multi-objective optimization algorithm. Lopez-Llompart and Kondolf [29] and Kondolf and Lopez-Llompart [30] studied how floodways in the Mississippi River had been affected by land use conflicts and management. Czigáni et al. [31] used the MIKE 21 model [34] for multi-purpose floodway zoning and floodway delineation along the lower Hungarian Drava section. However, very little has been done to produce an optimized floodway boundary for the entire stream [35] and an extensive literature review revealed Froehlich's [36] and Yu's [35] works.

Froehlich [36] suggested that delineation of the floodway boundary be done in a fair way. He pointed out legal issues including violation of the constitution with floodplain regulations and related those issues towards a need for a just and fair way to delineate a floodway. He further hinted that a reach-wide optimal solution would be an impartial way to define regulatory floodway boundaries. Froehlich's proposed approach uses dynamic programming to delineate the floodway boundary of the steady-state energy balance equation using the standard step method. His research is significant in the way that he realized the importance of reach-wide optimization of the floodway and attempted to use dynamic programming as an optimization tool. However, since the work was done using hydraulic code that is not approved by FEMA, his floodway optimization technique cannot be used to generate floodplains and floodways for FEMA.

Yu [35] used a GA to calculate the floodway encroachment limits within HEC-RAS. For the objective function, he used the sum of the absolute difference of simulated and desired water surface elevations within the floodway for all cross sections. He attempted to find floodway encroachment limits that keep the surcharge for all cross sections within an acceptable range, but he did not consider minimizing the floodway area nor maintaining a subcritical flow state reach-wide. His approach was not able to produce better results than the default methods in HEC-RAS, but the idea of using a heuristic algorithm to determine the floodway aligns well with this study.

In one-dimensional riverine modeling, cross sections are extracted from terrain data along the river where there are hydraulically important features. For each cross section, the left and right encroachment limits are assigned, which define the floodway boundaries. In flood modeling, the flood elevations are interrelated between cross sections. When modeling streams that consist of only a few cross sections, it may be feasible to manually find optimal encroachment limits based on engineering judgments, but as the number of cross sections increases, seeking feasible floodway boundaries becomes more complex and time-consuming. This difficulty usually discourages the modeler from repeatedly running the model using different combinations of encroachment limits to determine the best feasible reach-wide solution that would meet the surcharge and hydraulic requirements. Even if the modeler is willing to put forth the trial and error effort, the floodway determined in this manner may not be optimized. Further, there is no one structured and uniform procedure that exists in defining the floodway area [12,13,16], let alone finding the best feasible one. The default methods built in HEC-RAS are generally used as a starting point [13] and should not be considered a method for reach-wide optimization. Optimization of the floodway in a reach-wide manner involves generating the left and right encroachment limits in each cross section that will result in a regulatory rise in the water surface elevation along the entire stream. At the same time, these encroachment limits should satisfy certain requirements set by the modeler. As we mentioned above, our interest is to minimize the footprint area of the floodway to maximize the land use of both encroachments by local governments while ensuring the hydraulic safety from potential flood hazards.

In this study, we developed software using the HECRASController to communicate with HEC-RAS and implemented an optimization method for reach-wide floodway determination using a heuristic algorithm called Isolated-Speciation-based Particle Swarm Optimization (ISPSO) [37] and the HEC-RAS hydraulic model. In Section 2, we defined the floodway optimization problem and formulated engineering judgments as an objective function to remove any subjectivity. We also briefly introduced ISPSO, and combined the objective function and ISPSO to elaborate on the development of the proposed method called the Automated Floodway Optimizer for HEC-RAS (AFORAS). In Section 3, we conducted a case study using a universally available floodway model from HEC-RAS and discussed its results. Finally, in Section 4, we highlighted the main advantages and limitations of AFORAS and concluded this study with a consideration of future work.

2. Materials and Methods

2.1. Study Area

A floodway model called FLODENCR included in the HEC-RAS 4.1.0 installation was used for this study. This model was chosen at random to demonstrate the effectiveness of the optimization method. In addition, since input files for the case study are freely available from the HEC-RAS installation, researchers may duplicate the simulation if desired. Although this case study may not generalize all other cases, it serves as a simple but yet applicable case scenario where the floodway can be obtained using the optimization method and manually for cross-validating if the optimization method is indeed providing results that mimic the behavior of a manually obtained floodway. This model represents a 1.59 km of Beaver Creek near Kentwood, Louisiana, and consists of total 12 cross sections with a bridge structure as shown in Figure 2. Figure 3 shows the 9-pier bridge structure on Highway 1049 located at 8.69 km. The upstream area of the bridge is mostly covered by grass while its downstream area is dominated by forests. Two cross sections just upstream and downstream of the bridge define the geometries of the upstream and downstream faces of the structure, respectively. The opening under the low chord of the bridge is 60.05 m wide and its deck is 12.19 m wide. On both sides of the opening, ineffective areas are defined to model the embankment area that effectively blocks the water.

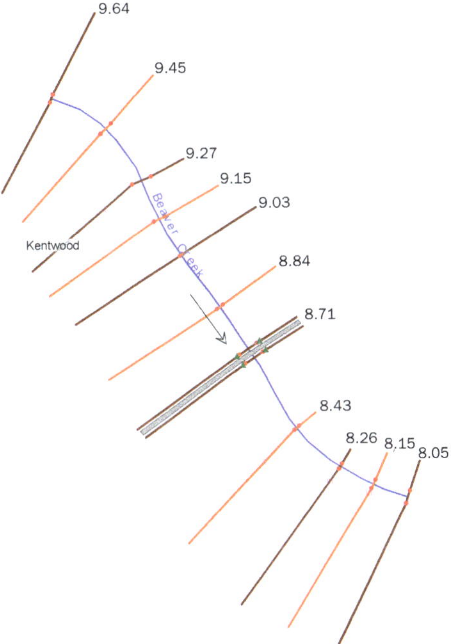

Figure 2. River and cross section geometries for Beaver Creek near Kentwood, Louisiana, in the FLODENCR model. The arrow indicates the direction of flow. A bridge structure is located at 8.69 km.

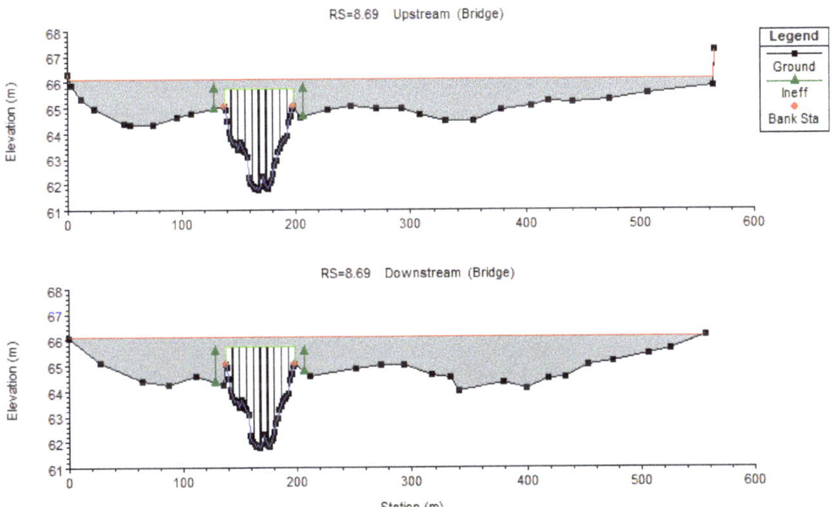

Figure 3. Bridge structure on Highway 1049 located at 8.69 km.

2.2. Mathematical Representation of the Problem

In order to optimize the floodway in HEC-RAS using an optimization algorithm, an objective function needs to be defined that evaluates the fitness of the model parameters objectively because there cannot be the modeler's intervention during an optimization run. The objective function should reflect the modeler's knowledge about the model and their engineering judgments about the model outputs.

Figure 4 depicts a floodway in a river with cross sections, channel banks, and 100-year floodplain extents. The hatched polygon representing the floodway is created by connecting the left and right encroachment limits in each cross section. The area of the hatched polygon indicates the floodway footprint area denoted by A_{fw} in this section. In a similar way, the maximum and minimum floodway areas ($A_{fw,max}$ and $A_{fw,min}$, respectively) can be calculated by connecting the 100-year floodplain extents (diamonds) and channel banks (dots), respectively. Our interest is to minimize the floodway area A_{fw} while satisfying hydraulic requirements.

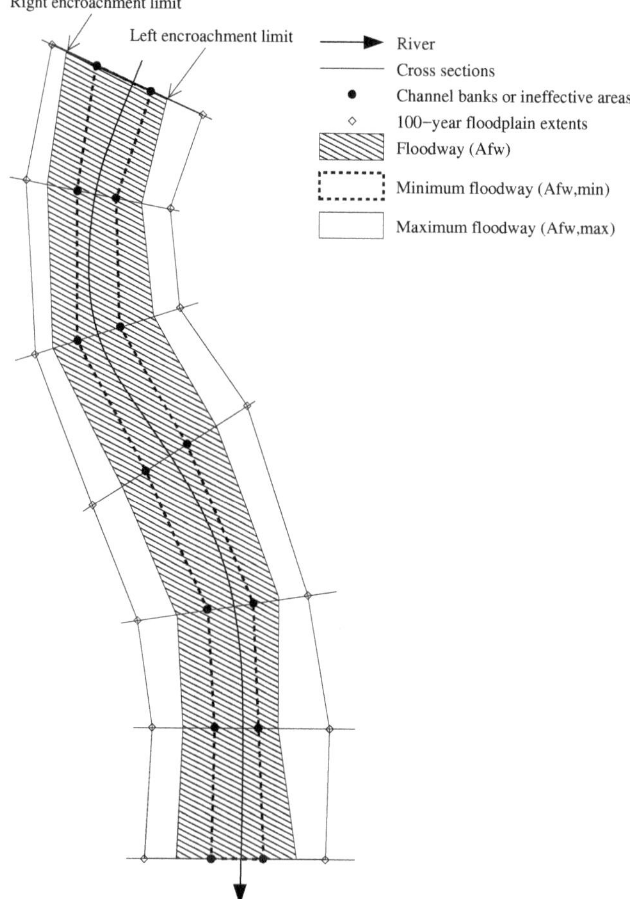

Figure 4. Plan view of a floodway along with a river, cross sections, channel banks, and 100-year floodplain extents.

In floodway determination, the objective function should consider three criteria: (1) Floodway surface area, (2) surcharge, and (3) flow state indicating whether or not the flow is subcritical. The floodway surface area should be minimized while the surcharge is kept within acceptable limits and the subcritical flow state is maintained. There are many different hydraulic parameters in HEC-RAS. However, both models for the 100-year floodplain and floodway are required to share the same hydraulic parameters including the channel bottom geometry, Manning's roughness coefficient, structural parameters, etc. because hydraulic modeling for the floodway should simulate the same hydraulic conditions as for the 100-year floodplain. Otherwise, the floodway model would not simulate the 100-year flood condition anymore. The only exception is the left and right encroachment limits which defines the floodway area. In other words, the optimization algorithm only adjusts the shape of the hatched polygon in Figure 4 and evaluates the model outputs to calculate the objective function. The model outputs from HEC-RAS provide the surcharge and cross-sectional Froude number [38], which can be used to evaluate violations of criteria (2) and (3) stated above. Ideally, the surcharge should be between 0 and an allowable limit mandated by the FEMA or the state government while the Froude number should be less than 1 for flow to be subcritical.

Most of the streams in the United States flow at a subcritical state except in the mountainous area, which is most likely undeveloped and underpopulated. Since FEMA only requires the floodway analysis for developed communities, the objective function in this study is formulated to handle subcritical flow conditions. However, if the flow state changes to supercritical, the modeler would have to choose the flow condition to model within HEC-RAS. In cases where the flow turns from subcritical to supercritical, HEC-RAS will default to the critical depth [2] to proceed with its calculation even though the water surface elevation may be erroneous. There are options within HEC-RAS for simulation of mixed-type or supercritical flows, but either the trial-and-error method or a priori knowledge of the flow state is required to be able to select these options. While the optimization algorithm itself does not discriminate between the different flow states, encroaching a stream that is flowing at a critical or supercritical flow state will, most often than not, result in a decrease in the water surface elevation and an excessive increase in the flow velocity. Since a negative surcharge is not allowed by FEMA [39], encroachment analysis is not necessary for the supercritical flow state in most of the case.

There are also two constraints: (1) The floodway should not be narrower than the area defined by the left and right channel banks because it should not affect the effective flow conveyance of the channel; and (2) the floodway limits cannot fall outside the 100-year floodplain by definition. The first constraint reserves the channel area between the left and right channel banks to carry the flood water by not impeding this area by floodway encroachment. The second constraint does not allow the floodway encroachment limits on a non-inundated dry area, which effectively leads to no encroachment at all in the river. In Figure 4, the left encroachment limit is constrained between the left channel bank (dots) and the left extent of the 100-year floodplain (diamonds). The same constraints apply to the right encroachment limit. These valid ranges of the left and right encroachment limits are shown in the cross-sectional view in Figure 1. From these two constraints, the minimum and maximum floodway bounds can be defined so that the objective function can compare different floodways by evaluating how close solutions are to the minimum possible floodway area. Given that the above two hydraulic conditions including the surcharge and Froude number are satisfactory, the floodway with the minimum surface area is deemed the optimal floodway.

By formulating the above three criteria and two constraints, the following objective function can be defined:

$$f(\mathbf{M}(\mathbf{g}(\mathbf{x}))) = \begin{cases} \frac{A_{\text{fw}} - A_{\text{fw,min}}}{A_{\text{fw,max}} - A_{\text{fw,min}}} & \text{if } \mathbf{s} \text{ and } \mathbf{Fr} \text{ are acceptable,} \\ 1 + \sum_{i=1}^{N} \left\{ \max\left(0 - \mathbf{s}_i / 0.305, 0\right) + \max\left(\mathbf{s}_i / 0.305 - 1, 0\right) + \max(\mathbf{Fr}_i - 1, 0) \right\} & \text{otherwise} \end{cases} \tag{1}$$

where $\mathbf{x}$ is a parameter sample in a normalized hypercube search space $[0, 1]^D$, $D = 2 \times n$ is the problem dimension, n is the number of cross sections to optimize, $\mathbf{g}(\cdot)$ is a mapping function $\mathbf{g} \colon \mathbf{x} \rightarrow \mathbf{e}$ that maps a parameter sample $\mathbf{x}$ to encroachment limits $\mathbf{e}$, which the HEC-RAS model takes as its input parameters to compute $\mathbf{s}$ and $\mathbf{Fr}$, $\mathbf{M}(\cdot)$ is the HEC-RAS model, $f(\cdot)$ is the objective function, A_{fw} is the floodway surface area, $A_{\text{fw,min}}$ and $A_{\text{fw,max}}$ are the minimum and maximum surface areas of the floodway, respectively as shown in Figure 4, $\mathbf{s}$ and $\mathbf{Fr}$ are the surcharge and Froude number vectors, respectively, for all cross sections, s_i and $\mathbf{Fr}_i$ are the surcharge and Froude number, respectively, for cross section i, and N is the total number of cross sections in the model including n cross sections to optimize and those used as boundary conditions.

If the surcharges and Froude numbers of all cross sections (s and $\mathbf{Fr}$, respectively) are within acceptable ranges, the objective function evaluates the ratio of the surface area difference between the current and minimum floodways (A_{fw} and $A_{fw,min}$, respectively) to the surface area difference between the maximum and minimum floodways ($A_{fw,max}$ and $A_{fw,min}$, respectively). Provided that all the hydraulic conditions are satisfied, the values of this ratio are 0 and 1 when the floodway is at its minimum and maximum widths, respectively. If one or more of the hydraulic conditions are violated, the objective function is set to unity and extra penalties are added based on how severely the surcharge and Froude number are deviating from their respective allowable limits. The penalty functions are designed so that the magnitude of deviation from the intended surcharge range or Froude number range is directly correlated to the magnitude of penalty added to the objective function. Any penalty will force the objective function to take on a larger value, which is an undesirable trait in a minimization problem. Now, floodway optimization is defined as a mathematical equation that can be evaluated objectively by computer code. The main goal of the proposed approach is to minimize the objective function $f(\mathbf{M}(\mathbf{g}(\mathbf{x})))$ by optimizing the variable vector $\mathbf{x}$ that represents encroachment limits $\mathbf{e}$, using a heuristic algorithm.

To better explain how the objective function works, Table 1 shows example outputs from a simple hypothetical model and their objective function values in the Equation (1) column as well as Yu's objective function values in the Equation (2) column. This hypothetical model consists of two cross sections ($N = 2$). Its minimum and maximum floodway areas are 50 km^2 and 100 km^2, respectively. Trial 1 evaluates the minimum floodway area, but the surcharges and Froude numbers for both cross sections violated hydraulic conditions (i.e., $s_i > 0.305$ m and $\mathbf{Fr}_i \geq 1$ for $i = 1, 2$). Since there are hydraulic violations, the penalty case of Equation (1) is used to calculate the objective function value in the Equation (1) column. That is, $f(\mathbf{M}(\mathbf{g}(\mathbf{x}))) = 1 + \max(0 - 0.366/0.305, 0) + \max(0.366/0.305 - 1, 0) + \max(1.1 - 1, 0) + \max(0 - 0.397/0.305, 0) + \max(0.397/0.305 - 1, 0) + \max(1.2 - 1, 0) = 1.8$. Similarly, trials 2–3 have some violations and use the same equation to evaluate the objective function. Trial 4 does not violate any hydraulic conditions, so the acceptable case of Equation (1) is used to calculate the objective function value. In this case, $f(\mathbf{M}(\mathbf{g}(\mathbf{x}))) = \frac{80-50}{100-50} = 0.6$. Similarly, trials 5–6 do not have any violations and use the same floodway area ratio as their objective function values. Assuming that there are no more trial simulations, trial 4 would be the best floodway model because it minimizes the floodway area while satisfying all hydraulic requirements. In actual optimization runs, trials will evolve based on a heuristic algorithm introduced in Section 2.3.

Table 1. Example model outputs and their objective function values. For all trials, $N = 2$, $A_{fw,min} = 50$ km^2, and $A_{fw,max} = 100$ km^2.

Trial	A_{fw}	s_1 (m)	s_2 (m)	$\mathbf{Fr}_1$	$\mathbf{Fr}_2$	Violations	Equation (1)	Equation (2)
1	50	0.366	0.397	1.1	1.2	4	1.80	0.50
2	60	0.336	0.366	1.0	0.8	3	1.30	0.30
3	70	0.310	0.320	1.1	0.9	1	1.17	0.07
4	80	0.295	0.295	0.9	0.7	0	0.60	0.07
5	90	0.244	0.214	0.8	0.6	0	0.80	0.50
6	100	0.305	0.305	0.9	0.8	0	1.00	0.00

For comparison, Yu's objective function [35] is defined as:

$$f(\mathbf{M}(\mathbf{g}(\mathbf{x}))) = \sum_{i=1}^{N} \left| \frac{s_i}{0.305} - 1 \right|. \tag{2}$$

This lump-sum way of integrating absolute differences can be problematic because bad performance in some cross sections with a high surcharge can be compensated for in other cross sections with a low surcharge. This compensation prohibits Equation (2) from differentiating good trials from bad ones. For example, both cross sections in trial 3 violated the surcharge requirement (i.e., $s_i > 0.305\,\text{m}$ for $i = 1, 2$) while the two cross sections in trial 4 did not (i.e., $s_i \leq 0.305\,\text{m}$ for $i = 1, 2$). However, Equation (2) evaluates to 0.07 for both trials. As shown above, this objective function cannot adequately rank different trials for better optimization because information about individual cross sections gets lost. At the same time, since this objective function evaluates surcharges closer to the maximum allowed limit (i.e., $0.305\,\text{m}$) favorably, trials with lower surcharges are penalized even if they are actually more desirable. For example, trial 6 with both surcharges at the maximum $0.305\,\text{m}$ evaluates to 0.00 while trial 5 with lower surcharges evaluates to 0.50. In the end, populations in a GA will evolve towards the $0.305\,\text{m}$ limit and the surcharge requirement can easily be violated when the surcharge is too close to the allowed limit.

2.3. Isolated-Speciation-Based Particle Swarm Optimization

Isolated-Speciation-based Particle Swarm Optimization (ISPSO) [37] is a multi-modal heuristic optimization algorithm based on collective intelligence of individual particles in a swarm. In ISPSO, parameter samples are referred to as particles, which are collectively called a swarm. They fly around the parameter space and form multiple species based on spatial proximity. Individual particles keep track of their experience and share information with neighbors in the same species. Their velocities and next positions are determined by combining their private experience and neighbors' experience. In this way, particles in one species converge to a local solution. Since there are multiple species in the search space, particles are able to find multiple local solutions, possibly including the global solution as well. More details about how this optimization algorithm works, mathematical examples, and a practical engineering problem can be found in [37].

Unlike gradient-based techniques, heuristic optimization algorithms do not depend on nor require local slope evaluation of the objective function surface. Since the HEC-RAS model transforms model inputs and parameters to model outputs non-linearly, the objective function surface may not be smooth and can be very complicate. Because of this complicate nature of the objective function surface, it becomes important to avoid dependency on the landscape gradient of the objective function surface to prevent solution finding algorithms from falling into inferior local pits. Particles in ISPSO are able to find multiple solutions in different regions of the search space without getting trapped into such inferior pits. For this reason, ISPSO has successfully been applied to stochastic rainfall generation [40–43], storm tracking [44], uncertainty analysis [45,46], and climate change studies [47,48]. The current version of ISPSO is implemented in the R language [49], which was also used to evaluate the objective function and run the HEC-RAS program.

2.4. Automated Floodway Optimizer for HEC-RAS

The Automated Floodway Optimizer for HEC-RAS (AFORAS) is a tool that automatically optimizes the floodway in a HEC-RAS model using ISPSO. The system is unique in that the modeler need not make manual adjustment trying different combinations of encroachment limits until an acceptable solution is found. For fully automating the optimization procedure, the Application Programming Interface (API) of HEC-RAS was used to interface the HEC-RAS program and ISPSO's R code. A command-line program called the Command-Line Interface for HEC-RAS (CLIRAS) was developed to control the HEC-RAS program from an R environment. CLIRAS can change HEC-RAS plans, read cross section information such as river stations, bank stations, encroachment stations, flood extents, etc., update encroachment stations, and, finally, execute the HEC-RAS program. These features of CLIRAS are essential in controlling the HEC-RAS program without manual user interventions and for the ISPSO R code to be able to execute the HEC-RAS model for a specified number of times non-interactively. ISPSO executes CLIRAS internally to update and run the HEC-RAS model and extract results from it to evaluate the objective function.

AFORAS integrates the ISPSO R code, CLIRAS, and HEC-RAS as shown in Algorithm 1. The maximum number of iterations $\text{iter}_{\max}$ tells ISPSO the total number of iterations to perform for optimization. The maximum number of HEC-RAS model runs is defined by $\text{iter}_{\max}$ times the swarm size S ($\text{iter}_{\max} \times S$). The HEC-RAS model is represented by $\mathbf{M}(\cdot)$ and requires that two plans be defined: (1) 100-year floodplain and (2) floodway. The boundary conditions specify how the downstream or upstream end of the floodway should tie into adjacent existing floodways. There are four possible boundary conditions: (1) No existing floodways at the upstream and downstream ends of the study reach (BC = None), (2) floodway only at the downstream end (BC = DS), (3) floodway only at the upstream end (BC = US), and (4) floodways at both ends (BC = Both). The boundary conditions fix the encroachment limits at the most upstream, downstream or both cross sections, and therefore the problem dimension can be determined based on the number of cross sections and the number of boundary conditions. For example, when there are no upstream or downstream floodways to tie into (BC = None), all cross sections should be optimized, whereas the number of cross sections to optimize reduces by either 1 or 2 if one boundary condition (either BC = DS or BC = US) or two boundary conditions (BC = Both) are specified, respectively. The number of cross sections to optimize is indicated by n and the problem dimension is the total number of left and right encroachment limits on those n cross sections, which is $D = 2 \times n$. The recommended swarm size of $S = 10 + \left\lfloor 2\sqrt{D} \right\rfloor$ [50] was used.

Algorithm 1 Pseudocode for automated floodway optimization for HEC-RAS.

Require: $\text{iter}_{\max}$ ▷ Maximum number of iterations
Require: $\mathbf{M}(\cdot)$ ▷ HEC-RAS model with 100-year and floodway plans and profiles
Require: $\text{BC} \in \{\text{None, DS, US, Both}\}$ ▷ Boundary conditions for the encroachment limits
    Extract cross section information from $\mathbf{M}(\cdot)$
    $N \leftarrow$ Number of cross sections
    $n \leftarrow N-$ Number of boundary conditions
    $D \leftarrow 2 \times n$ ▷ Problem dimension
    $S \leftarrow 10 + \left\lfloor 2\sqrt{D} \right\rfloor$ ▷ Swarm size
    $A_{\text{fw,min}}, A_{\text{fw,max}} \leftarrow$ Minimum and maximum possible areas of the floodway
    $\mathbf{X} \in [0,1]^{S \times D} \leftarrow S$ number of D-tuples randomly sampled from $[0,1]^D$ ▷ Initial population
    Let $\mathbf{g}\colon [0,1]^D \to \mathbb{R}^D$ that maps particles to encroachment limits
    $\text{iter} \leftarrow 1$
    repeat   ▷ ISPSO loop
        for $i \leftarrow 1, \ldots, S$ **do**
            $\mathbf{X}_i \leftarrow$ Row i from $\mathbf{X}$ ▷ i^{th} trial encroachment limits or particle i in ISPSO
            Simulate $\mathbf{M}(\mathbf{g}(\mathbf{X}_i))$ using CLIRAS   ▷ Execute the HEC-RAS program
            Evaluate $f(\mathbf{M}(\mathbf{g}(\mathbf{X}_i)))$ ▷ Equation (1)
            if $i = 1$ or $f(\mathbf{M}(\mathbf{g}(\mathbf{X}_i))) < f(\mathbf{M}(\mathbf{g}(\mathbf{X}_{\text{best}})))$ **then**   ▷ If $\mathbf{X}_i$ is better than $\mathbf{X}_{\text{best}}$
                $\mathbf{X}_{\text{best}} \leftarrow \mathbf{X}_i$ ▷ Store the best encroachment limits from the current iteration
            end if
        end for
        if $\text{iter} = 1$ or $f(\mathbf{M}(\mathbf{g}(\mathbf{X}_{\text{best}}))) < f(\mathbf{M}(\mathbf{g}(\mathbf{x}_{\text{best}})))$ **then**   ▷ If $\mathbf{X}_{\text{best}}$ is better than $\mathbf{x}_{\text{best}}$
            $\mathbf{x}_{\text{best}} \leftarrow \mathbf{X}_{\text{best}}$ ▷ Store the best encroachment limits so far
        end if
        Evolve $\mathbf{X}$ using ISPSO   ▷ Evolution of the swarm in ISPSO
            $\text{iter} \leftarrow \text{iter} + 1$
    until $\text{iter} = \text{iter}_{\max}$ or other conditions are satisfied
    Optimized encroachment limits $\leftarrow \mathbf{g}(\mathbf{x}_{\text{best}})$ ▷ Found the best encroachment limits

---

The valid ranges of the floodway encroachment limits vary from cross section to cross section as shown in Figure 4 (dot–diamond). As can be seen in Figures 1 and 4, even for the same cross section, the left and right encroachment limits can have different scales. Because our objective is to find the best feasible left and right encroachment limits, the problem space is constructed from D encroachment limits. However, different scales in different encroachment limits highly skew the problem space, which can negatively affect the performance of optimization. To address this issue, the problem space is normalized to $[0, 1]^D$ using a mapping function $\mathbf{g}(\cdot)$. The mapping function $\mathbf{g}(\cdot)$ is defined such that it transforms a particle (i.e., a trial set of normalized encroachment limits) in a hypercube search space $[0, 1]^D$ back to a D-tuple of encroachment limits $\mathbf{e}$, which is a direct input to the HEC-RAS model $\mathbf{M}(\cdot)$. The particle at the lower limits of all dimensions (i.e., $x_i = 0$ for $1 \leq i \leq D$) defines the minimum possible floodway (dashed polygon in Figure 4) determined by the bank stations (dots in Figure 4) while the particle at the upper limits (i.e., $x_i = 1$ for $1 \leq i \leq D$) defines the maximum possible floodway (unfilled solid polygon in Figure 4) determined by the 100-year floodplain extents (diamonds in Figure 4). In other words, bank stations are mapped to $x_i = 0$ while 100-year floodplain extents are mapped to $x_i = 1$. In this way, the floodway is guaranteed to be wider than the main channel such that the conveyance of flow is not highly obstructed. At the same time, the floodway cannot be wider than the 100-year floodplain. Floodway areas including the minimum and maximum areas $A_{\text{fw,min}}$ and $A_{\text{fw,max}}$, respectively, are calculated by straightening the reach and calculating the area of the polygon defined by the bank stations for $A_{\text{fw,min}}$ or the 100-year floodplain extents for $A_{\text{fw,max}}$. These minimum and maximum floodway areas $A_{\text{fw,min}}$ and $A_{\text{fw,max}}$, respectively, are used in the objective function in Equation (1) to assess the fitness of the floodway defined by trial encroachment limits $\mathbf{g}(\mathbf{X}_i)$. Once the problem is defined, ISPSO initializes the swarm and starts evolving the particles by evaluating trial encroachment limits. The final solution $\mathbf{e}_{\text{best}} = \mathbf{g}(\mathbf{x}_{\text{best}})$ is a set of optimized left and right encroachment limits, which then becomes input for the HEC-RAS floodway model.

2.5. Numerical Experiments

Three different approaches of producing the floodway were compared: (1) automated optimization by AFORAS, (2) manual determination by the authors, and (3) the default floodway in the model as a reference. For fair comparisons, the model was not modified except that floodway encroachment limits were updated. To update floodway encroachment limits using Algorithm 1, the mapping function $\mathbf{g}(\cdot)$ is defined to normalize encroachment limits in the HEC-RAS model into $[0, 1]^D$. To define the mapping function $\mathbf{g}(\cdot)$, the bank stations and 100-year flood extents were extracted and normalized to create a unit hypercube search space $[0, 1]^D$ where D is two times the number of cross sections to be optimized. After each iteration of evolution in Algorithm 1, particles in the problem space $[0, 1]^D$ are transformed back to actual floodway encroachment limits, which are in turn input into the HEC-RAS model for evaluating the objective function in Equation (1).

The number of cross sections to be optimized varies depending on the boundary condition. Four encroachment boundary conditions as defined in Section 2.4 were tested: (1) BC = None, (2) BC = DS, (3) BC = US, and (4) BC = Both. The upstream-most and downstream-most encroachment limits of the default floodway in the reference model were used as those encroachment boundaries that an optimized floodway is to be tied into. Only in this way, the default model need not be updated at all and can be used as a reference model for the other two approaches.

Numerical experiments were conducted using the floodway model to observe how AFORAS performs in different boundary conditions. A total of 30 independent runs of each boundary condition were performed. Based on Clerc's suggestion [50] and the number of dimensions in this problem, the recommended swarm size of 18 was used for the boundary condition BC = Both while 19 was used for the other three boundary conditions. For each AFORAS run, total 1000 iterations were performed resulting in total 18,000 and 19,000 model runs for the boundary condition BC = Both and the other boundary conditions, respectively.

3. Results and Discussion

3.1. Comparison of Different Approaches

The first AROFAS run for each boundary condition was used to compare the floodways between the three approaches. Table 2 shows the problem dimensions and the objective function values of the final floodways for the three approaches and four boundary conditions. The objective function value for the HEC-RAS floodway remains the same because its floodway was used as a reference and was not independently optimized. Since floodway optimization is a minimization problem, a floodway with the least objective function value is preferred. As can be seen in Table 2, among all the AFORAS floodway runs, the case with BC = Both yielded the worst objective function value (0.338), it but still performed better than the best floodway from the other two approaches (0.342). The numbers inside parentheses show that AFORAS reduced the objective function value by 1–29% from the manual method and by 12–40% from the HEC-RAS method. In terms of the floodway area, these numbers indicate the reduction of the floodway area outside the channel ($A_{fw} - A_{fw,min}$), not the total floodway area (A_{fw}). On average for all eight cases (i.e., manual and HEC-RAS cases), the reduction of the floodway area outside the channel was 20%, which can significantly increase encroachment areas for development.

Table 2. Objective function values for the test cases with different boundary conditions. Since Automated Floodway Optimizer for HEC-RAS (AFORAS) solves a minimization problem, lower values represent better models. The numbers inside parentheses indicate what percent of the objective function value could be improved by running AFORAS.

BC	Problem Dimension	AFORAS	Manual	HEC-RAS
None	24	0.270	0.348 (29%)	0.379 (40%)
DS	22	0.278	0.345 (24%)	0.379 (36%)
US	22	0.333	0.347 (4%)	0.379 (14%)
Both	20	0.338	0.342 (1%)	0.379 (12%)

Despite high problem dimensions from 20 to 24, AFORAS performed reasonably well and even outperformed the manual optimization. An interesting observation is that the performance improved as the problem dimension increased with more cross sections to optimize. When either the upstream-most or downstream-most cross section of the floodway is specified as a boundary condition, the fixed width of the encroachments in those cross sections prevents the floodway area from being further reduced beyond its boundary condition limits.

Figure 5 shows the final floodway for all cases with different boundary conditions. Vertical lines at encroachment station 0 represent the straightened river while negative and positive stations represent the left and right floodway encroachment limits on both sides of the river. For the case with BC = None, at the downstream-most cross section at river station 9.66 km, the floodway is narrower than those of the cases that have a restriction to the same cross section as a boundary condition (i.e., BC = DS and BC = Both). Similarly, for the case with BC = None, at the upstream-most cross section at river station 8.05 km, the floodway is narrower than those of the cases with BC = DS and BC = Both. Since these boundary conditions affect the objective function, lower objective function values in the BC = None case do not necessarily mean that AFORAS performs better when there are no boundary conditions specified. However, this result shows that AFORAS was able to take full advantage of cross sections that are not constrained to help reduce the footprint of the floodway and performed consistently better than the other two approaches.

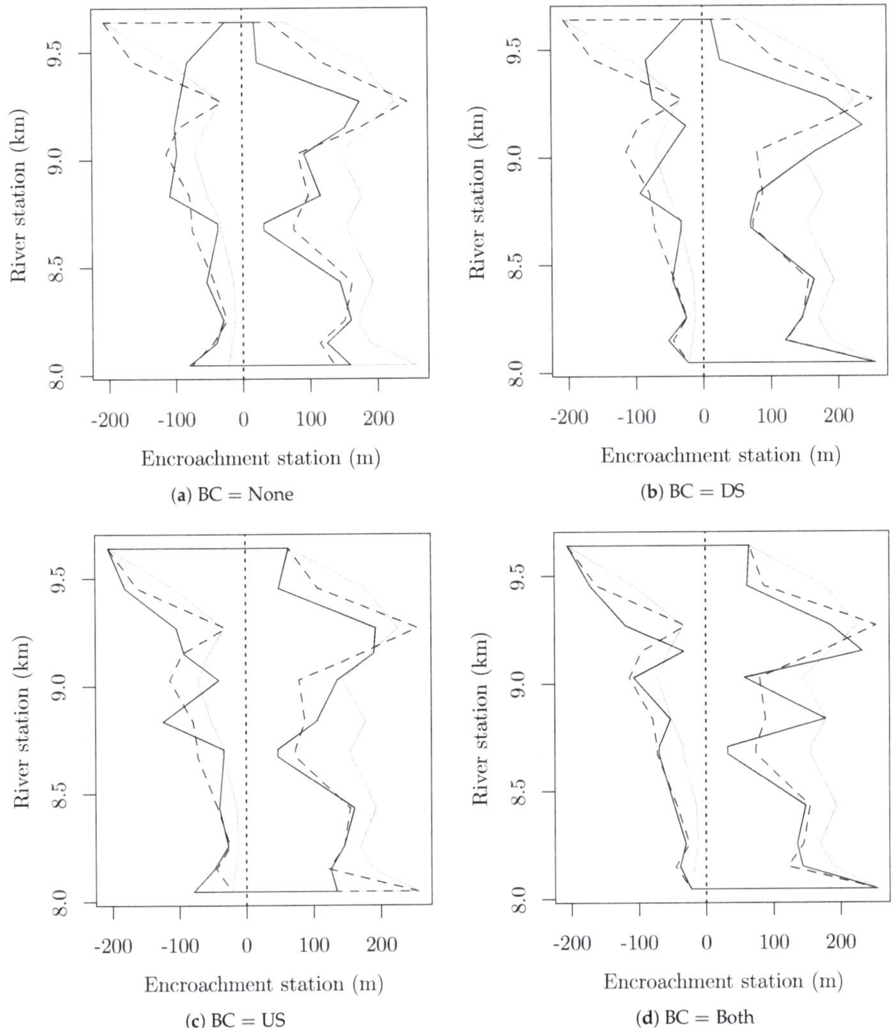

Figure 5. Graphical views of the left and right encroachment limits for the four different boundary conditions. Vertical lines at encroachment station 0 represents the river line. Negative and positive stations represent the left and right encroachment limits, respectively. ········ Straightened river, ———— Automated Floodway Optimizer for HEC-RAS (AFORAS) floodway, – – – – Manual floodway, and ——— Hydrologic Engineering Center's River Analysis System (HEC-RAS) floodway.

3.2. Sensitivity of Encroachment Limits to the Boundary Condition

In optimization, it is beneficial to visualize the landscape of the objective function surface to be able to understand a problem. However, since most of floodway optimization problems deal with more than one cross section, their problem dimensions are usually higher than two, which makes it difficult to visualize the objective function surface without projecting it onto a much smaller-dimensional space. Figures 6 and 7 show two-dimensional projections of 10,000 particles with grayscaled objective function values satisfying hydraulic requirements for BC = None and BC = Both, respectively. These plots show how sensitive particles were to the performance of optimization. The more spread out particles are along one encroachment limit, that encroachment limit was less sensitive to the performance because particles did not converge to a narrower band of the dimension. For example, in Figure 6e,f, particles tend to spread out almost randomly across the entire projected space. This widespread presence of particles means that they did not have any preference over specific encroachment limits in these cross sections at 8.67 km and 8.71 km during optimization and they were very insensitive to the performance compared to those in the other cross sections. Ideally, AFORAS should be able to take advantage of this insensitivity of particles to the performance of floodway encroachments in these two cross sections. The result of BC = None in Figure 5a (the very first run for this boundary condition) shows this behavior where AFORAS decided to take very narrow floodway encroachments as the final solution at these two cross sections. Since a bridge is located between these cross sections, ineffective flow areas were established around the bridge structure to model embankment areas. The areas immediately before and after the structure prevent the water from flowing freely downstream and result in ineffective flow areas. As a result, floodway encroachments within these ineffective flow areas could not affect the flood elevation significantly because those areas blocked by the floodway encroachments were already ineffective.

As shown in Figure 6a–d,g–k, in other cross sections except the upstream-most cross section, particles spread out more along the left encroachment axis than along the right encroachment axis. The narrow distribution of particles in the right encroachment axis means that the right encroachment limit was more sensitive to the performance and played a bigger role than the left encroachment limit during optimization. This result can be explained by the relative width of possible encroachment areas on the left overbank versus those on the right overbank. As shown in Figure 2, the portions of cross sections on the left side of the river line (i.e., left overbanks) are much narrower than those on the right side (i.e., right overbanks), especially at the downstream area. Because the encroachment widths on both overbanks are normalized to create a unit hypercube search space, wide-spread particles across the left encroachment do not necessarily mean that the left overbank area is wider. Also, for the same reason, significant movements of particles along the left encroachment dimension only slightly affect actual changes in the encroachment width, effectively making the left encroachment insensitive to the performance. However, since the left overbank gradually expands as it traverses upstream, this effect started slowly diminishing from 8.84 km.

The upstream-most cross section in Figure 6l showed a similar pattern to those around the bridge structure at 8.67 km and 8.71 km. Its insensitivity to the performance is due to the fact that HEC-RAS computes the flood elevation from downstream to upstream. Since the flood elevation within the floodway encroachment at the upstream-most cross section does not affect any downstream flood elevations, encroachment limits at this cross section may be placed anywhere as long as the hydraulic criteria are satisfied. For AFORAS, this insensitivity means that it should pick the narrowest possible floodway, which can be seen in the result in Figure 5. What it also means is that the problem with BC = None can be simplified by constraining the encroachment limits in the upstream-most cross section to the bank stations and solving the problem as if it had the boundary condition BC = US. In this way, the problem dimension of BC = None can be reduced by two and AFORAS should be able to achieve better performance and a faster convergence rate.

Two-dimensional plots for the boundary conditions BC = DS and BC = US are not presented in this paper because of space limitations and a prohibitively large number of data points. However, cross sections from 8.26 km to 9.27 km in these boundary conditions showed similar patterns to Figure 6c–j or Figure 7c–j. Cross sections at 8.05 km and 8.15 km in the boundary condition BC = US behaved similarly to Figure 6a,b, respectively (i.e., both cases do not constrain the downstream-most cross section). Also, cross sections at 9.45 km and 9.64 km in the boundary condition BC = DS behaved similarly to Figure 6k,l, respectively (i.e., both cases do not constrain the upstream-most cross section).

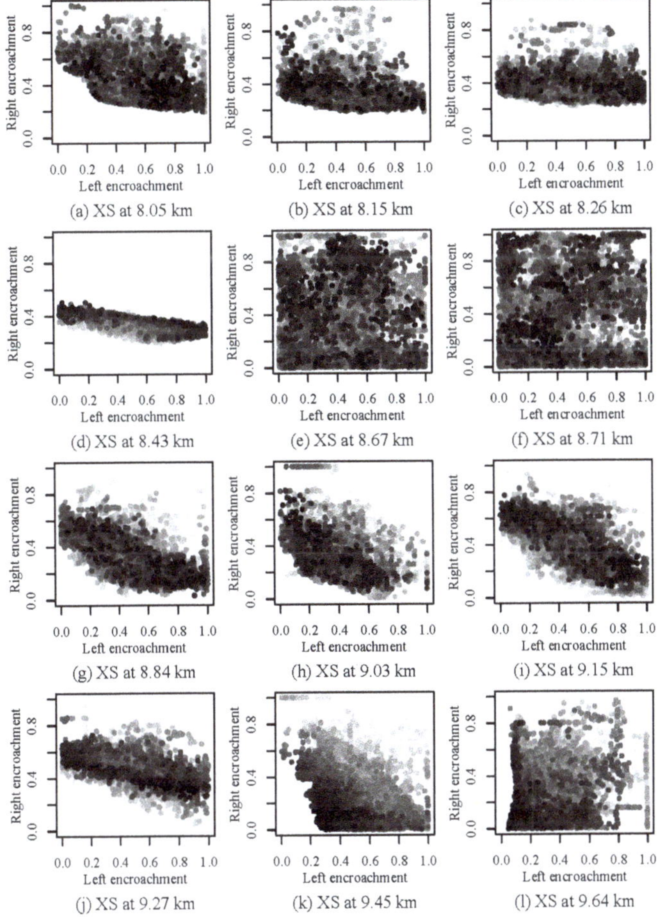

Figure 6. Two-dimensional projections of all particles $\mathbf{X}_i \in [0,1]^D$ for $1 \leq i \leq S$ satisfying the three hydraulic criteria from all 30 AFORAS runs for BC = None. Since the total number of particles meeting hydraulic requirements was excessively large—130,395 out of 570,000 (30 AFORAS runs × 19,000 model runs/AFORAS run)—10,000 particles were sampled to construct each subplot, which represents one cross section (XS). Particles that perform better are plotted darker in front of those that perform worse and are in a lighter gray.

When the downstream-most cross section is constrained (i.e., BC = DS or BC = Both), the right encroachment became even more sensitive to the performance at 8.15 km, just upstream of the downstream boundary cross section, and its encroachment plot for BC = DS is very similar to Figure 7b. This neighbor cross section was highly affected by the downstream boundary condition because of the backward calculation of HEC-RAS. When the upstream-most cross section is constrained (i.e., BC = US or BC = Both), AFORAS was able to achieve faster convergence by avoiding unnecessary trial-and-error sampling in the upstream-most cross section, which does not affect any other cross sections at all. The encroachment plot for 9.45 km, just downstream of the upstream boundary condition, became much narrower than Figure 6k with most particles clustering around the diagonal line from top-left to bottom-right as shown in Figure 7k. Table 3 summarizes these observations of two-dimensional projection subplots of all particles from 30 AFORAS runs.

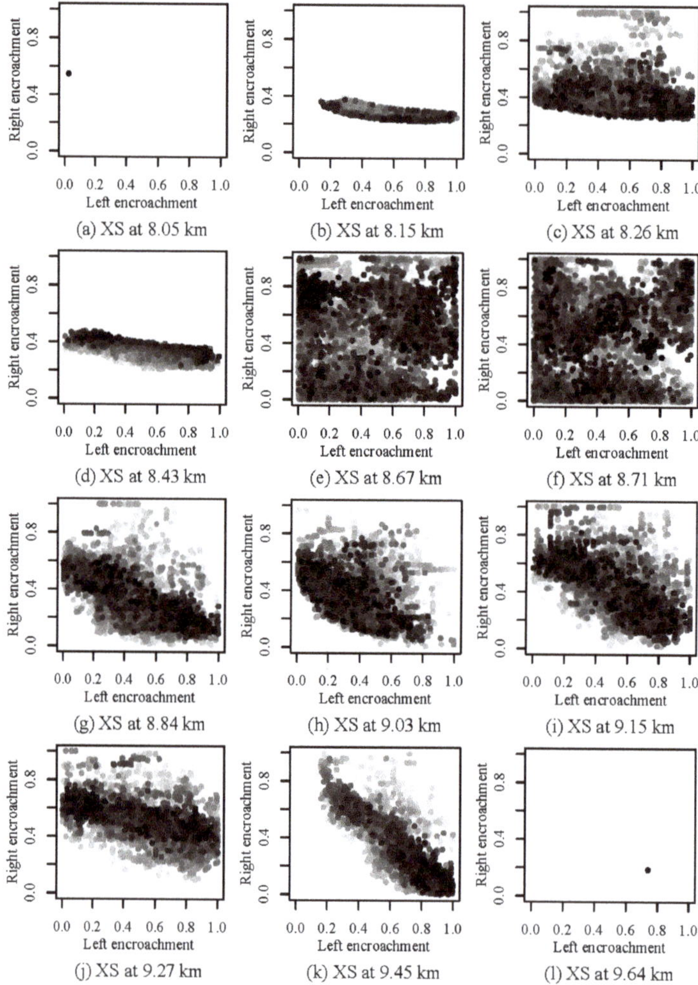

Figure 7. Two-dimensional projections of all particles $\mathbf{X}_i \in [0,1]^D$ for $1 \leq i \leq S$ satisfying the three hydraulic criteria from all 30 AFORAS runs for BC = Both. Since the total number of particles meeting hydraulic requirements was excessively large—142,937 out of 540,000 (30 AFORAS runs $\times$ 18,000 model runs/AFORAS run)—10,000 particles were sampled to construct each subplot, which represents one cross section (XS). Particles that perform better are plotted darker in front of those that perform worse and are in a lighter gray.

Table 3. Summary of two-dimensional projection subplots of all particles by Figures 6 and 7. A $\sim$ symbol indicates a similar pattern to the subplots given on the right. Figures 6c–j and 7c–j, respectively, have a similar pattern.

BC	XS 8.05 km	XS 8.15 km	XS 8.26 km–9.27 km	XS 9.45 km	XS 9.64 km
None	5a	5b	5c–5j	5k	5l
DS	6a	~6b	~5c–5j	~5k	~5l
US	~5a	~5b	~5c–5j	~6k	6l
Both	6a	6b	6c–6j	6k	6l

3.3. Optimization Performance

The convergence lines of all 30 AFORAS runs for different boundary conditions are presented in Figure 8 as 30 gray lines with the average performance as a black line. All the runs converged exponentially to the final values of the objective function. The minimum and maximum of the final values are 0.265 and 0.280, respectively, and their mean and standard deviation are 0.272 and 0.004, respectively. The standard deviation is approximately 1.5% of the mean, which indicates that the performance of AFORAS is very robust and reliable. The black line shows the mean of the 30 cumulative minimum values of the objective function. The average performance converges very quickly until 5000 model runs and experiences gradual improvement until about 15,000 model runs for BC = None and BC = DS, and 10,000 model runs for BC = US and BC = Both, after which the convergence rate slowed down significantly. Figure 8 shows that AFORAS achieved a faster convergence rate for the cases where the upstream-most cross section is constrained as a boundary condition. However, their objective function values are higher or worse than those for the other two cases. The higher objective function values for BC = US and BC = Both is because the upstream-most cross section used as a boundary condition is fairly wide compared to the bank stations, which act as the lower limits in that cross section for the other cases. Particles in BC = None and BC = DS where the upstream-most cross section was also optimized, found a very narrow floodway width at this cross section as the final solution. On the other hand, this cross section was fixed at a much larger width as a boundary condition for BC = US and BC = Both and produced worse objective function values. Since the two sets of boundary conditions have different search spaces, the higher objective function values do not necessarily mean poor performances.

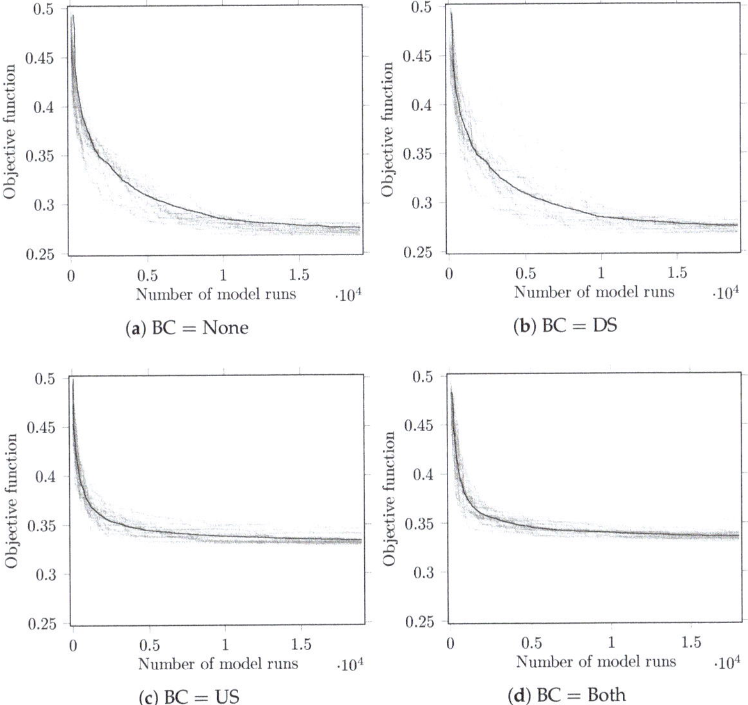

Figure 8. Cumulative minimum values of the objective function vs. the number of model runs. The gray lines and black line represent 30 runs of AFORAS and the mean of those runs, respectively.

3.4. AFORAS as a Tool for Floodway Optimization

To the authors' knowledge, Yu's study [35] was the first attempt to determine the floodway encroachment limits using HEC-RAS and a GA, but his approach did not consider the floodway area and subcritical flow state. Also, as we compared our objective function to his in Section 2.2, the lump-sum way of integrating information across all cross sections leads to information loss and a lack of differentiation between good and bad models. AFORAS takes a different approach for floodway determination by taking into account the flow state and minimizing the floodway area. At the same time, its objective function is formulated to separate out favorable models from those with hydraulic violations.

AFORAS integrates ISPSO and HEC-RAS for reach-wide optimization of the floodway area. The derivative- free search of ISPSO and a mathematical representation of floodway area optimality were integrated as AFORAS in a way that HEC-RAS can be executed automatically by computer code without user interventions. AFORAS was successful in seeking and improving the floodway encroachment limits for the case study. As discussed above, AFORAS was able to identify those cross sections that are insensitive to its performance and almost fully encroached the floodway at these locations so that the floodway area was kept to the minimum. Overall, AFORAS performed consistently better than the manual approach and the reference floodway, and showed reliable and consistent performance across different boundary conditions. It is also interesting to see that ISPSO, as a heuristic algorithm, is able to reliably solve high dimensional problems in this study. As summarized in Table 2, the problem dimension for this study varied from 20 to 24. Different problem dimensions result in different landscapes of the objective function, but, as shown in Figures 6 and 7, two-dimensional projections onto individual cross sections of the objective function surface exhibit similar patterns depending on which cross section was used as a boundary condition. Despite the differences in the surface of the objective function and its final values, convergence was achieved for all cases and the solutions found by AFORAS minimized the floodway area while keeping the surcharge and flow state within the allowable limits. These observations suggest that AFORAS can be a suitable tool for reach-wide floodway optimization.

While AFORAS can be a good candidate for reach-wide floodway optimization, it is only limited to handle reaches with subcritical flows. The proposed objective function within AFORAS has to be modified to accommodate cases with mixed or supercritical flows, which are not as common as subcritical flows already handled by AFORAS. More tests are needed to observe the performance of ISPSO given a different objective function that can handle varying flow states. Also, when calculating the floodway area, AFORAS currently assumes that the river line is straight and the floodway encroachment limits vary linearly between consecutive cross sections. While this simplification of geometries may be a reasonable assumption for optimization, a more realistic evaluation of the floodway area can be attained by incorporating the natural curvature of the river line. Finally, a sensitivity analysis needs to be performed to see how the numbers of particles and iterations for an ISPSO run affect the performance of AFORAS and find the right balance between computing time and the performance. Better a priori estimation of the total number of model runs may help reduce computing time. Future work and research on AFORAS will include addressing these limitations and running more HEC-RAS floodway models with different geometries and structures.

4. Conclusions

Since the floodway is an essential part of hydrologic and hydraulic studies of riverine flooding, in the United States, FEMA requires one to be determined for developed communities using their approved computer programs, one of which is HEC-RAS. HEC-RAS has widely been used for flood risk and floodway regulation studies by many researchers. Because the floodway encroachment area is often used for human activities, it is a local government's interest to expand this area by minimizing the floodway footprint area that conveys the flood water without affecting the water surface elevation too much. Our objective is to minimize the floodway area while maintaining the surcharge and subcritical flow state reach-wide, both of which are required by FEMA. The authors' literature review has revealed that very little work has been done in terms of floodway optimization. A recent attempt to determine the floodway encroachment limits using HEC-RAS and a GA did not consider the floodway area and subcritical flow state, which most of the streams in the United States exhibit. The proposed objective function takes into account the floodway area, surcharge, and subcritical flow state to make sure that the final optimized floodway not only meets FEMA's hydraulic requirements but also maximizes floodway encroachment areas for human activities in a reach-wide manner. By integrating the objective function and a heuristic algorithm called ISPSO, we proposed a floodway area optimization tool named AFORAS for reach-wide optimization of the floodway using HEC-RAS. We used a readily and freely available floodway model from the HEC-RAS 4.1.0 installation for a case study so that other researchers can replicate our results if they decide. Comparisons of the AFORAS, manual, and HEC-RAS approaches showed 1–40% improvements in the objective function value by AFORAS. AFORAS consistently provided superior results for all the boundary conditions. We also conducted a sensitivity analysis of encroachment limits to the boundary condition and a convergence test by running AFORAS 30 times for four different boundary conditions. Both left and right encroachment limits were insensitive to the performance in cross sections adjacent to a bridge structure while these encroachment limits exhibited different level of sensitivity to the performance in other cross sections. Because of the bridge opening and ineffective areas, encroaching these cross sections could not affect the flood elevation much and did not help improve the objective function compared to the other cross sections. The surface of the objective function may vary significantly for different HEC-RAS models or even for different combinations of the boundary conditions in the same floodway model. In this regard, it is advantageous for AFORAS to employ ISPSO over gradient-based optimization techniques because of the capability of ISPSO to solve high dimensional problems without requiring the derivative of the objective function. Limitations in the current AFORAS method include the lack of support for mixed and supercritical flows in the objective function, and the linear approximation of the river geometry and floodway. Also, since the total number of HEC-RAS runs has to be specified a prior, a quantitative analysis would be beneficial to reduce computing time by estimating how many model runs are required in advance. Addressing these limitations and recommending the required number of model runs a prior will be left for research in the near future.

Author Contributions: Conceptualization, T.M.Y.; Methodology, H.C. and T.M.Y.; Software, H.C.; Validation, T.M.Y., H.C. and J.H.; Formal Analysis, T.M.Y. and H.C.; Investigation, T.M.Y.; Resources, J.H.; Data Curation, H.C.; Writing—Original Draft Preparation, H.C., T.M.Y. and J.H.; Writing—Review & Editing, H.C., T.M.Y. and J.H.; Visualization, H.C.; Supervision, T.M.Y.; Project Administration, T.M.Y.; Funding Acquisition, N/A.

Funding: This research received no external funding.

Conflicts of Interest: The authors declare no conflict of interest.

References

1. FEMA. Guidance for Flood Risk Analysis and Mapping: General Hydrologic Considerations. 2018. Available online: https://www.fema.gov/media-library-data/1525201756728-390e667d3d0958347b8374f6321bd488/General_Hydrologic_Guidance_Feb_2018.pdf (accessed on 18 April 2018).
2. FEMA. Guidance for Flood Risk Analysis and Mapping: Hydraulics—One-Dimensional Analysis. 2016. Available online: https://www.fema.gov/media-library-data/1484864685338-42d21ccf2d87c2aac95ea1d7ab6798eb/Hydraulics_OneDimensionalAnalyses_Nov_2016.pdf (accessed on 28 December 2016).
3. Brunner, G.W. *HEC-RAS River Analysis System User's Manual Version 4.1*; US Army Corps of Engineers, Institute for Water Resources, Hydrologic Engineering Center: Davis, CA, USA, 2010.
4. Rastislav, F.; Martina, Z. The HEC-RAS model of regulated stream for purposes of flood risk reduction. *Sel. Sci. Pap. J. Civ. Eng.* **2016**, *11*, 59–70. [CrossRef]
5. ShahiriParsa, A.; Noori, M.; Heydari, M.; Rashidi, M. Floodplain zoning simulation by using HEC-RAS and CCHE2D models in the Sungai Maka River. *Air Soil Water Res.* **2016**, *2016*, 55–62. [CrossRef]

6. Golshan, M.; Jahanshahi, A.; Afzali, A. Flood hazard zoning using HEC-RAS in GIS environment and impact of Manning roughness coefficient changes on flood zones in semi-arid climate. *Desert* **2016**, *21*, 25–34.

7. Balogun, O.S.; Ganiyu, H.O. Development of inundation map for hypothetical Asa dam break using HEC-RAS and ArcGIS. *Arid Zone J. Eng. Technol. Environ.* **2017**, *13*, 831–839.

8. Ben Khalfallah, C.; Saidi, S. Spatiotemporal floodplain mapping and prediction using HEC-RAS-GIS tools: Case of the Mejerda River, Tunisia. *J. Afr. Earth Sci.* **2018**, *142*, 44–51. [CrossRef]

9. Goodell, C. *Breaking the HEC-RAS Code: A User's Guide to Automating HEC-RAS*; h2ls: Portland, OR, USA, 2014.

10. FEMA. Flood Insurance Study Guidelines and Specifications for Study Contractors. 2002. Available online: http://www.fema.gov/media-library-data/20130726-1546-20490-8681/frm_scg.doc (accessed on 24 December 2015).

11. Franz, D.D.; Melching, C.S. *Full Equations Utilities (FEQUTL) Model for the Approximation of Hydraulic Characteristics of Open Channels and Control Structures During Unsteady Flow*; Technical Report Water-Resources Investigations Report 97–4037; U.S. Department of the Interior, U.S. Geological Survey: Reston, Virginia, USA, 1997.

12. Howells, L.; McLuckie, D.; Collings, G.; Lawson, N. Defining the floodway—Can one size fit all? In Proceedings of the 44th Annual Floodplain Management Australia Conference, Coffs Harbour, Australia, 12–14 May 2004.

13. Thomas, C.R.; Honour, W.; Golasweski, R. Procedures for floodway definition: Is there a uniform approach? In Proceedings of the 50th Annual Floodplain Management Australia Conference, Gosford, Australia, 23–26 February 2010.

14. Selvanathan, S.; Dymond, R.L. FloodwayGIS: An ArcGIS visualization environment to remodel a floodway. *Trans. GIS* **2010**, *14*, 671–688. [CrossRef]

15. Esri. *ArcGIS Desktop: Release 10*; Esri: Redlands, CA, USA, 2011.

16. Rigby, T. Floodplain development manual (NSW) A document in need of review. In Proceedings of the 47th Annual Floodplain Management Australia Conference, Gunnedah, Australia, 27 February–1 March 2007.

17. Thomas, C.R.; Golaszewski, R. Refinement of procedures for determining floodway extent. In Proceedings of the 52nd Annual Floodplain Management Australia Conference, Batemans Bay, Australia, 21–24 February 2012.

18. Szemis, J.M.; Maier, H.R.; Dandy, G.C. A framework for using ant colony optimization to schedule environmental flow management alternatives for rivers, wetlands, and floodplains. *Water Resour. Res.* **2012**, *48*. [CrossRef]

19. Bogárdi, I.; Balogh, E. Flood system operation along levee-protected rivers. *J. Water Resour. Plan. Manag.* **2013**, *140*, 04014014. [CrossRef]

20. Luke, A.; Kaplan, B.; Neal, J.; Lant, J.; Sanders, B.; Bates, P.; Alsdorf, D. Hydraulic modeling of the 2011 New Madrid Floodway activation: A case study on floodway activation controls. *Nat. Hazards* **2015**, *77*, 1863–1887. [CrossRef]

21. Lund, J.R. Floodplain planning with risk-based optimization. *J. Water Resour. Plan. Manag.* **2002**, *128*, 202–207. [CrossRef]

22. Shafiei, M.; Bozorg-Haddad, O.; Afshar, A. GA in optimizing Ajichai flood levee's encroachment. In Proceedings of the 6th World Scientific and Engineering Academy and Society International Conference on Evolutionary Computing, World Scientific and Engineering Academy and Society, Lisbon, Portugal, 16–18 June 2005.

23. Mori, K.; Perrings, C. Optimal management of the flood risks of floodplain development. *Sci. Total Environ.* **2012**, *431*, 109–121. [CrossRef] [PubMed]

24. Yazdi, J.; Neyshabouri, S.S. A simulation-based optimization model for flood management on a watershed scale. *Water Resour. Manag.* **2012**, *26*, 4569–4586. [CrossRef]

25. Lu, H.W.; He, L.; Du, P.; Zhang, Y.M. An inexact sequential response planning approach for optimizing combinations of multiple floodplain management policies. *Pol. J. Environ. Stud.* **2014**, *23*, 1245–1253.

26. Porse, E. Risk-based zoning for urban floodplains. *Water Sci. Technol.* **2014**, *70*, 1755–1763. [CrossRef] [PubMed]

27. Woodward, M.; Gouldby, B.; Kapelan, Z.; Hames, D. Multiobjective optimization for improved management of flood risk. *JW Resour. Plan. Manag.* **2014**, *140*, 201–215. [CrossRef]

28. Woodward, M.; Kapelan, Z.; Gouldby, B. Adaptive flood risk management under climate change uncertainty using real ooption and optimization. *Risk Anal.* **2014**, *34*, 75–92. [CrossRef] [PubMed]

29. Lopez-Llompart, P.; Kondolf, G.M. Encroachments in floodways of the Mississippi River and Tributaries Project. *Nat. Hazards* **2016**, *81*, 513–542. [CrossRef]

30. Kondolf, G.M.; Lopez-Llompart, P. National-local land-use conflicts in floodways of the Mississippi River system. *AIMS Environ. Sci.* **2018**, *5*, 47–63. [CrossRef]

31. Czigáni, S.; Pirkhoffer, E.; Halmai, A.; Lóczy, D. Application of MIKE 21 in a multi-purpose floodway zoning along the lower Hungarian Drava section. *Revista de Geomorfologie* **2016**, *18*, 5–18. [CrossRef]

32. Dorigo, M.; Maniezzo, V.; Colorni, A. The ant system: Optimization by a colony of cooperating agents. *IEEE Trans. Syst. Man Cybern. Part B Cybern.* **1996**, *26*, 1–13. [CrossRef] [PubMed]

33. Goldberg, D.E. *Genetic Algorithms in Search, Optimization, and Machine Learning*; Addison-Wesley: Boston, MA, USA, 1989.

34. DHI. MIKE 21 Flow Model FM. Parallelisation Using GPU. Benchmarking Report. 2014. Available online: http://www.mikebydhi.com/~/media/Microsite_MIKEbyDHI/Publications/PDF/GPU_Benchmarking.ashx (accessed on 28 September 2018).

35. Yu, L. Using Genetic Algorithms to Calculate Floodway Stations with HEC-RAS. Master's Thesis, University of Dayton, Dayton, OH, USA, 2015.

36. Froehlich, D.C. A fair way to a floodway: Optimal delineation of floodplain encroachment limits. In Proceedings of the 1999 Water Resources Engineering Conference, Seattle, WA, USA, 8–12 August 1999.

37. Cho, H.; Kim, D.; Olivera, F.; Guikema, S.D. Enhanced speciation in particle swarm optimization for multi-modal problems. *Eur. J. Oper. Res.* **2011**, *213*, 15–23. [CrossRef]

38. Chanson, H. Development of the Bélanger equation and backwater equation by Jean-Baptiste Bélanger (1828). *J. Hydraul. Eng.* **2009**, *135*, 159–163. [CrossRef]

39. FEMA. Guidance for Flood Risk Analysis and Mapping: Floodway Analysis and Mapping. 2016. Available online: https://www.fema.gov/media-library-data/1484864412580-de4aa11166f23a8f06e6121925a5b543/Floodway_Analysis_and_Mapping_Nov_2016.pdf (accessed on 28 December 2016).

40. Kim, D.; Olivera, F.; Cho, H. Effect of the inter-annual variability of rainfall statistics on stochastically generated rainfall time series: Part 1: Impact on peak and extreme rainfall values. *Stoch. Environ. Res. Risk Assess.* **2013**, *27*, 1601–1610. [CrossRef]

41. Kim, D.; Olivera, F.; Cho, H.; Lee, S.O. Effect of the inter-annual variability of rainfall statistics on stochastically generated rainfall time series: Part 2: Impact on watershed response variables. *Stoch. Environ. Res. Risk Assess.* **2013**, *27*, 1611–1619. [CrossRef]

42. Kim, D.; Olivera, F.; Cho, H.; Socolofsky, S. Regionalization of the modified Bartlett-Lewis rectangular pulse stochastic rainfall model. *Terr. Atmos. Ocean. Sci.* **2013**, *24*. [CrossRef]

43. Kim, D.; Cho, H.; Onof, C.; Choi, M. Let-It-Rain: A web application for stochastic point rainfall generation at ungaged basins and its applicability in runoff and flood modeling. *Stoch. Environ. Res. Risk Assess.* **2016**, *31*, 1023–1043. [CrossRef]

44. Cho, H.; Kim, D.; Lee, K.; Lee, J.; Lee, D. Development and application of a storm identification algorithm that conceptualizes storms by elliptical shape. *J. Korean Soc. Hazard Mitig.* **2013**, *13*, 325–335. [CrossRef]

45. Cho, H.; Olivera, F. Application of multimodal optimization for uncertainty estimation of computationally expensive hydrologic models. *J. Water Resour. Plan. Manag.* **2014**, *140*, 313–321. [CrossRef]

46. Cho, H.; Kim, D.; Lee, K. Efficient uncertainty analysis of TOPMODEL using particle swarm optimization. *J. Korean Water Resour. Assoc.* **2014**, *47*, 285–295. [CrossRef]

47. Heo, J.; Yu, J.; Giardino, J.R.; Cho, H. Impacts of climate and land-cover changes on water resources in a humid subtropical watershed: A case study from East Texas, USA. *Water Environ. J.* **2015**, *29*, 51–60. [CrossRef]

48. Heo, J.; Yu, J.; Giardino, J.R.; Cho, H. Water resources response to climate and land-cover changes in a semi-arid watershed, New Mexico, USA. *Terr. Atmos. Ocean. Sci.* **2015**, *26*. [CrossRef]

49. R Development Core Team. *R: A Language and Environment for Statistical Computing*; R Foundation for Statistical Computing: Vienna, Austria, 2006; ISBN 3-900051-07-0. Available online: http://www.r-project.org (accessed on 9 November 2015).

50. Clerc, M. Standard Particle Swarm Optimization. 2012. Available online: http://clerc.maurice.free.fr/pso/SPSO_descriptions.pdf (accessed on 24 December 2013).

Article

Intelligent Storage Location Allocation with Multiple Objectives for Flood Control Materials

Wei Wang [1,2], Jing Yang [3], Li Huang [4,*], David Proverbs [5] and Jianbin Wei [6]

[1] Key Laboratory of Coastal Disaster and Defence of Ministry of Education, Hohai University, Nanjing 210098, China
[2] College of Harbor, Coastal and Offshore Engineering, Hohai University, Nanjing 210098, China
[3] Business School, Sichuan University, Chengdu 610065, China
[4] School of Public Administration, Hohai University, Nanjing 210098, China
[5] Computing, Engineering and Built Environment at Birmingham City University, Birmingham B4 7XG, UK
[6] Jiangsu Provincial Hydraulic and Flood Control Material Reserve Centre, Nanjing 210000, China
* Correspondence: lily8214@126.com; Tel.: +86-25-83787354

Received: 16 June 2019; Accepted: 18 July 2019; Published: 25 July 2019

Abstract: Intelligent storage is an important element of intelligent logistics and a key development trend in modern warehousing and logistics. Based on the characteristics of flood control materials and their intelligent storage, this study established a flood control material storage location allocation model reflecting the multiple objectives of retrieval efficiency and shelf stability and used a weighting method to transform a multi-objective optimization problem into a single-objective optimization problem. We then used the facilities and equipment planning and storage location allocation in the intelligent storage area for provincial flood control materials at the Zhenjiang warehouse of the Jiangsu water conservancy and flood control material reserve center as a case study. Empirical analysis was performed and used the genetic algorithm and Matrix Laboratory (MATLAB) software to optimize the storage location allocation of provincial flood prevention supplies at this warehouse, and it achieved effective results.

Keywords: flood control materials; intelligent warehousing; location allocation; multi-objective optimization

---

1. Introduction

While water resources play an important role in human society, the world suffers from all types of natural disasters [1]. In many countries, floods are the most likely natural disaster, and compared with other natural disasters, they are easily predicted and prevented [2]. Therefore, it is very important for flood disaster management departments to formulate scientific and comprehensive plans for flood control. Consequently, government agencies spend massive amounts of money and manpower on flood control and rescue.

Floods are easily predicted, which means that preparations can be made to prevent flooding from actually happening. Developed countries such as England, the United States, Holland, Denmark, Germany, France, Belgium, and Austria have invested significant time and effort into flood control and disaster mitigation research since the 1960s and have achieved significant results in meteorology, hydrology, water conservancy, water quality, topography, the influence of social and economic activities on the prevention of flood damages, the runoff-conflux model of floods and its prediction and analysis, and other fields. During the process, the Wallingford National Institute of England, the Delft Hydraulics Institute, the Danish Hydraulic Institute, (DHI, Hesholm, Denmark), the United States Environmental Protection Agency, (EPA, Washington, DC, USA) and Army Corps of Engineers, (ACE, Washington, DC, USA), and other research institutes have stood out [3,4]. In the 1970s, the United States proposed flood control that used non-structural measures, which realized flood management through legislation, flood forecasting, flood dispatching, flood detention, flood insurance, floodplain management, and soil and water conservation, developed flood contingency plans and other methods, standardized people's precautions and preparations against heavy rain and flooding, and provided guidance according to actual circumstances so that the losses from floods could be reduced and better social and economic benefits could be achieved. Compared with structural measures, non-structural measures were no longer the focus of flood control; instead, they emphasized the timely and scientific implementation of flood control command and dispatch through the collection, analysis, and processing of relevant flood information. This was done in order to improve the potential capability of the existing flood-fighting structural measures and standardize people's actions against floods and the developmental activities within the floodplain, thus realizing flood control and disaster mitigation [5]. Non-structural measures were important supplements to structural measures.

Among these non-structural measures, flood control material and emergency logistics were important for fighting floods. Flood control material is one of the three major elements of flood control and rescue, including material for preparation, response, and mitigation phases, and the reserve management of flood control materials is the key link in carrying out flood control work and plays an irreplaceable role in national flood control and rescue work. However, flood emergency preparedness lacks logistical insights [6,7], and relevant scholars have already conducted some primary studies. Garrido, Lamas, and Pino (2015) put forward a flood logistics model. The model attempts to optimize emergency supply inventories and vehicle availability [8]. Leeuw, Vis, and Jonkman (2012) developed an emergency logistics framework that supports preventing catastrophic breaches of flood defenses during extreme situations [9]. Alem, Clark, and Moreno (2016) developed a new two-stage stochastic network flow model to determine how to rapidly supply humanitarian aid to victims of floods [10].

Intelligent storage is an important element of intelligent logistics and a key development trend in modern warehousing and logistics. The effectiveness of intelligent storage is chiefly influenced by the storage location allocation, and optimal storage location allocation can enhance the storage space utilization rate, shorten the storage and retrieval distances and times, accelerate the turnover of goods, ensure inventory stability, and increase the operating performance of an intelligent storage system.

The key to effective storage is optimizing the storage location allocation strategies, and scholars have performed in-depth, systematic studies of storage location allocation. Roodbergen et al. analyzed storage location strategies [11] and found that the storage location allocation methods of commercial intelligent storage systems chiefly consist of the following types: fixed storage location assignment [12–15], random storage location assignment [16–18], class-based storage location assignment [19,20], random class-based storage location assignment [21], and shared storage location assignment. Most of these studies assume that a warehouse is initially empty, which is clearly not in accordance with the needs of real projects. Furthermore, even when the studies in the literature consider a warehouse with a non-empty initial state, they perform simulation experiments only involving a single batch of goods entering the warehouse. However, available empty storage locations will be abundant when a single batch of goods is put into storage, and there will thus be considerable freedom when selecting an optimal storage location. Consequently, it is possible that only sub-optimal locations will be left for the next batches of goods entering storage, and the arrangement of the storage locations as a whole will be irrational. Lee et al. proposed similarity coefficients to cluster goods and then assigned goods to storage locations in accordance with the clustering results [15]. As for algorithms, most studies have adopted intelligent optimization algorithms, such as the genetic algorithm [22], the simulated annealing algorithm [23], the tabu search method [24], the data mining-based algorithm [25], and other algorithms [26,27], which can greatly shorten the computing time and enable optimal solutions to be found after relatively few iterations.

The foregoing review reveals that most scholars have constructed intelligent storage systems via a systems engineering approach and synergistically applied technologies, including message identification, communications technology, automatic control, and intelligent algorithms, in their intelligent storage systems. In particular, many scholars have used mathematical modelling to construct mathematical models of intelligent storage systems and then used simulation experiments to validate the models [28–30]. Furthermore, operation optimization methods are most commonly used in intelligent storage analysis and decision-making [31], and the development of commercial intelligent storage location allocation methods has provided a foundation for research and development (R&D) related to intelligent storage for emergency logistics.

Because emergency supplies tend to be infrequently used, intelligent storage systems are seldom used for emergency supplies, and therefore, there has been little research on storage location allocation in intelligent emergency logistics [32–36]. The emergency supply reserve warehouses of some power agencies have used the Internet of Things to construct intelligent storage systems for their emergency power supplies [37], which has enabled substantial increases in storage automation and storage and retrieval performance, reduced injuries and supply losses, and enhanced the emergency response capabilities.

There has been relatively little research on the application of intelligent storage in emergency logistics, and no research on key intelligent storage technologies for emergency disaster relief supplies has yet been published. Furthermore, because intelligent storage systems have not been used for flood control, there have been few studies concerning this aspect. However, the research on the key intelligent storage technologies that are used in commercial logistics and particularly the research on the storage strategies that are used in intelligent storage systems are already quite mature. However, because of the characteristics of the storage and management of flood control materials, including limited types of supplies, large quantities, low batch numbers, large quantities in a batch, the need for quick retrieval in the event of an emergency, and strong constraints on response times, the findings of the research on the application of intelligent storage systems to commercial logistics are not fully applicable to flood control materials. Therefore, there is an urgent need for research on the practical application of technologies for the intelligent storage of emergency flood control materials.

Based on the characteristics of flood control materials and their intelligent storage, this study established a flood control material storage location allocation model with the multiple objectives of retrieval efficiency and shelf stability and used a weighting method to transform a multi-objective optimization problem into a single-objective optimization problem. The study then optimized the storage location allocation of provincial flood prevention supplies using MATLAB. The optimized allocation can comprehensively improve the support ability, utilization efficiency, technical level, and management level of the flood control material reserve system, reduce the losses and hazards caused by emergencies, and achieve remarkable social and economic benefits.

2. Analysis of the Storage Strategies for the Intelligent Storage of Flood Control Materials

After summarizing the commonly used storage strategies (including fixed location storage, random location storage, class-based storage, class-based random storage, shared storage, and item-location coupling storage) for location assignment, and considering the attributes of flood control materials, we decided to adopt the class-based fixed location storage [1].

Storage strategies constitute the major principles of flood control material storage area planning and must be naturally combined with storage location allocation principles in order to determine a storage operating model. To scientifically and rationally implement the storage of flood control materials, the storage location allocation for such materials must adhere to the following principles:

(1) Place the most frequently used materials near exits and the less frequently used materials farther away from exits. Because flood control materials do not require speedy storage but require extremely quick retrieval, this paper chiefly considers retrieval efficiency and does not consider storage input efficiency.

(2) Place heavier materials on lower levels and lighter materials on higher levels.

(3) Place bulkier materials on lower levels and more compact materials on higher levels.

(4)   Store materials along lanes and ensure that the blockage of individual lanes does not affect the retrieval efficiency.

(5)   Accelerate the turnover by storing and retrieving materials to and from proximate locations.

(6)   With regard to the relationships between materials, store the same types of materials close together. Because closely related materials are typically used together in emergency response actions, store them in adjacent locations as much as possible.

3. Intelligent Storage Location Allocation Model for Flood Control Materials

3.1. Construction of a Multi-Objective Optimization Model

The types of the flood control materials include motor oil, life jackets, tents, boats, portable lights, etc., and the material specifications cover the lengths, widths, heights, and importance of the materials. Here, in order to simplify the problem, a certain number of single-class materials will be grouped as a single standard pallet group based on their size. The combination is the standard pallet group of the material and the length, width, height, and weight of the standard pallet group of the various materials satisfy the shelf restrictions.

The paper takes single-row shelves as the research target, and we establish a multi-layer shelf coordinate system in the optimization module with the starting position of the work vehicle at the exit table as the origin $O(0,0)$, the X-axis is the longitudinal axis direction of the shelf, the Y-axis is the vertical direction of the shelf, each column and each layer of the shelf is one unit in length in the X-axis and Y-axis directions, and the position of the materials is determined as $P(i,j)$.

The main variables and parameters are defined as follows.

We assume that a certain set of shelves in a warehouse has n levels and m rows, and the position of the materials is determined as $P(i,j)$ $(0 \leq i \leq m, 0 \leq j \leq n)$.

The flood control materials include r total categories; W_k is the material weight (single standard pallet group weight) of the type-k material $(0 \leq k \leq r)$; t_{ij} is the time that is needed for a forklift to transport the goods at the storage location at the ith row and jth level; v_x and v_y are, respectively, the horizontal and vertical operating speeds of the shuttle vehicles or forklifts; L and H are respectively, the length and height of a storage location; p_k is the calling frequency of the materials $(0 \leq k \leq r)$, the total number of times the materials are used within a certain period of time, which is equivalent to the number of times that the materials are retrieved divided by the time.

We define the decision variable as x_{ijk}. When the type-k material $(0 \leq k \leq r)$ is stored in $P(i,j)$, $x_{ijk} = 1$; otherwise, $x_{ijk} = 0$, where $0 \leq i \leq m, 0 \leq j \leq n, 0 \leq k \leq r$.

Based on the classification of materials, different materials should be stored in different warehouse areas, and storage location allocation must be performed in different warehouse areas. Based on storage location allocation principles, this study constructed the following model.

(1)   In accordance with the principle of lighter materials on top and heavier materials on the bottom, assuming that a certain set of shelves in a warehouse has n levels and m rows, where the level closest to the floor is the first level and the row closest to the exit is the first row, the goal of storage location allocation optimization is to minimize the sum S of the products of the weights of the materials on pallets and the levels on which the materials are located. The first objective function of the shortest optimization objective function is as follows:

$$minS = \sum_{i=0}^{m} \sum_{j=0}^{n} \sum_{k=1}^{r} W_k x_{ijk}(i-1) \tag{1}$$

(2)   In accordance with the principles of close access and quick turnover, minimizing the sum T of the transport times of the materials in each storage location and minimizing the sum of the products of the usage frequency of each material and the forklift operating time when retrieving the material. The second objective function that optimizes the shelf stability is as follows:

$$minT = \sum_{i=0}^{m} \sum_{j=0}^{n} \sum_{k=1}^{r} t_{ij} x_{ijk} p_k \tag{2}$$

Based on the weighting method, the weights α and β (where $0 \le \alpha \le 1$, $0 \le \beta \le 1$, and $\alpha + \beta = 1$) of the fusion model are determined according to the importance of the shortest delivery time and shelf stability. The two objective functions for the shortest delivery time and shelf stability are merged, and the final warehouse optimization multi-objective model is established as follows:

$$minh = \alpha \sum_{i=0}^{m} \sum_{j=0}^{n} \sum_{k=1}^{r} W_k x_{ijk}(i-1) + \beta \sum_{i=0}^{m} \sum_{j=0}^{n} \sum_{k=1}^{r} t_{ij} x_{ijk} p_k, \; s.t. \begin{cases} \sum_k x_{ijk} = 1 \\ x_{ijk} = 0 \text{ or } 1 \end{cases} \tag{3}$$

3.2. Simplifying the Storage Location Allocation Model

Storage location allocation is a composite multi-objective optimization problem, and multi-objective optimization problems are typically solved in two ways. One way uses a weighting method, a maximum method, a constraint method, or goal programming to quantitatively address the multiple objectives and obtain a unique feasible solution. The second way uses a multi-objective optimization algorithm based on artificial intelligence, such as a multi-objective genetic algorithm, an ant colony optimization algorithm, or a simulated annealing algorithm, to perform the optimization.

In this study, considering the characteristics of a flood control materials warehouse, a weighting method was used to transform the multi-objective problem into a single-objective problem. In view of the equal importance of the two objectives of the shortest delivery time and shelf stability, in this study, the two objectives were both assigned weights of 0.5, which resulted in the following objective function:

$$minh = 0.5 \sum_{i=0}^{m} \sum_{j=0}^{n} \sum_{k=1}^{r} W_k x_{ijk}(i-1) + 0.5 \sum_{i=0}^{m} \sum_{j=0}^{n} \sum_{k=1}^{r} t_{ij} x_{ijk} p_k \tag{4}$$

Each side is multiplied by 2 to yield the final optimization model for location allocation as follows:

$$minH = \sum_{i=0}^{m} \sum_{j=0}^{n} \sum_{k=1}^{r} W_k x_{ijk}(i-1) + \sum_{i=0}^{m} \sum_{j=0}^{n} \sum_{k=1}^{r} t_{ij} x_{ijk} p_k, \; s.t. \begin{cases} \sum_k x_{ijk} = 1 \\ x_{ijk} = 0 \text{ or } 1 \end{cases} \tag{5}$$

3.3. Determining the Parameters

3.3.1. Facility and Equipment Status and Their Parameters

In accordance with the distances of shelves from the exit and their lifting heights, an optimal storage location is a location in an intelligent storage system that is at a height within 20% of the storage area that is closest to the floor and within the 20% of the storage area that is closest to the exit. These storage areas have the characteristics of easy storage, short pathways, and low mechanical operating losses. An intelligent storage system is chiefly composed of a material storage and transport system and a warehouse management system. Here, the material storage and transport system comprises shelves, the storage and retrieval entrance/exit, and warehousing equipment.

(1) Shelves: Multi-level shelves are closely spaced, and individual shelves may function independently. To facilitate this research, this study assumed a single set of shelves with a shelf height of H and length of L. Because the shelves had m rows and n levels, there were $n \times m$ storage locations, and the storage locations had identical specifications, namely, a height of h and a length of l. Thus, $H = n \times h$, and $L = m \times l$.

(2) Storage and retrieval entrance/exit: The material storage and retrieval system included a shuttle vehicle system and forklift. One storage and retrieval entrance/exit was located outside of each

set of shelves. Because vehicles can typically drive directly into flood control material warehouses for loading or unloading, this study considered only the transport of materials from shelves to a storage and retrieval entrance/exit.

(3) Warehousing transport equipment: Only one shuttle vehicle was used for single pick-up actions and was responsible for serving one set of shelves. The shuttle vehicle was located at a fixed initial position at the storage and retrieval entrance/exit in the beginning, and the time that is needed for the shuttle vehicle to leave its initial position, complete the placement (retrieval) of goods and return to its initial position was defined as the operation time. The warehousing equipment includes a shuttle car and a forklift. The shuttle car is responsible for the horizontal work. The forklift is responsible for the vertical work and other forklift operations. The working speed of the vehicle includes the maximum idle speed of the shuttle v_{x1} and the maximum speed of the shuttle load $vx2$. The vehicle's horizontal acceleration is denoted as a_x, the forklift's vertical speed is denoted as v_y, and the forklift's speed is denoted as v_f. Here, we can refer to the forklift's basic operating parameters to calculate the value or use our experience to set the value.

The shuttle's movement consists of its horizontal movement, the shelves, and the use of a forklift to move vertically up and down the shelves. The speed of a loaded shuttle vehicle is different from that of an empty shuttle vehicle. Its linear acceleration rate when starting is identical to the linear deceleration rate when stopping. After the shuttle vehicle reaches its maximum speed, it maintains that speed during its operations. See Table 1 for the warehouse shelving parameters.

Table 1. Warehouse shelf parameters.

Shuttle Vehicle Parameters		Shelf Parameters	Entrance/Exit Platform
Horizontal velocity of the unloaded shuttle vehicle v_{x1}	1.2 m/s		
Horizontal velocity of the loaded shuttle vehicle v_{x2}	0.6 m/s	Storage location length: 1.4 m, total of 10 rows	
Horizontal acceleration a_x	0.3 m/s²		Same height as the first level of shelves
Forklift vertical velocity v_y	0.3 m/s	Storage location height: 1.5 m, total of three levels	
Grasping velocity v_f	0.5 m/s	Storage location width: 1.2 m	
Grasping operation time t_f	5 s		

3.3.2. Determining the Storage Location Operation Time

In the coordinate system encompassing multiple levels of shelves, the X-axis represents the length of the shelves, and the Y-axis represents the height of the shelves. Goods were stored within storage locations. The time that is needed for the transport system consisting of shuttles and forklifts to move to individual storage locations on the fixed shelves was calculated using kinematics. We constructed a time-minimization model to obtain the amount of time that is needed to access goods at each storage location.

Storage locations were designated using two-dimensional coordinates, where x indicated the row coordinate and y indicated the level coordinate. The origin $O(0, 0)$ was set as the shuttle's initial location on the entrance/exit platform, and the shuttle's movement from O to the storage location $P(i, j)$ and back again completed an operating cycle (see Figure 1).

The horizontal movement distance is calculated as follows:

$$Lx = i \times l \tag{6}$$

The vertical operating distance is calculated as follows:

$$Hy = (j - 1) \times h \tag{7}$$

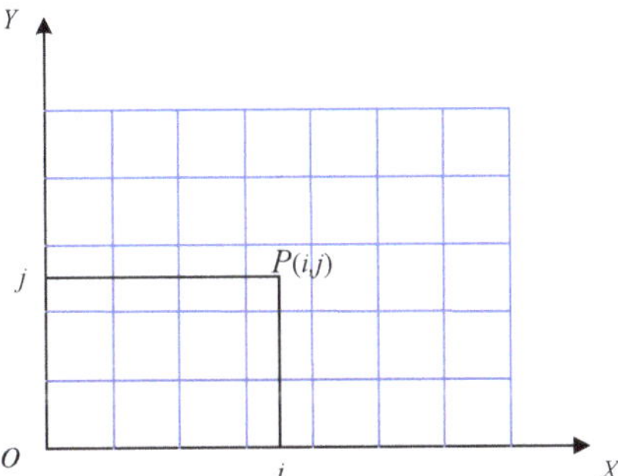

Figure 1. Operating pathway coordinate system.

The horizontal operating times were designated as t_{x1} and t_{x2}, where t_{x1} is the time for the loaded shuttle vehicle to reach the storage location, and t_{x2} is the return time of the loaded shuttle vehicle. The vertical operating time was designated as t_y, and the time-minimization model assumed that the shuttle vehicle accelerated uniformly until reaching the maximum velocity v_{y1} and then continued to move at its maximum velocity. The shuttle vehicle decelerated at a uniform rate after approaching its target storage location. Here, it was also necessary to consider the situations where the shuttle vehicle did not reach its maximum velocity when travelling a short route and where it travelled a sufficiently long route to reach its maximum velocity.

When the horizontal movement distance was too short, the shuttle vehicle could not reach the maximum velocity v_{x1} before it had to decelerate uniformly from its original velocity. When the movement distance was sufficiently long, the shuttle vehicle accelerated until reaching the maximum velocity v_{x1}, then moved at a uniform velocity and finally decelerated at a uniform rate. While the shuttle vehicle moved in the same manner when returning, it was loaded and therefore could not reach the maximum velocity v_{x1} of the unloaded condition. At this time, the maximum velocity that it reached was v_{x2}.

In the vertical direction, the forklift lifted the shuttle vehicle from the entrance/exit platform at a constant velocity of v_y, and the operation time was calculated using the data parameters in Table 1 as follows:

$$t_{x1} = \begin{cases} 2 \times \sqrt{\frac{L_X}{a_x}}, & L_X \le \frac{v_{x1}^2}{a_x} \\ \frac{v_{x1}}{a_x} + \frac{L_x}{v_{x1}}, & L_X > \frac{v_{x1}^2}{a_x} \end{cases} \quad (x = 1, 2, \ldots, m) \tag{8}$$

$$t_{x2} = \begin{cases} 2 \times \sqrt{\frac{L_X}{a_x}}, & L_X \le \frac{v_{x2}^2}{a_x} \\ \frac{v_{x2}}{a_x} + \frac{L_x}{v_{x2}}, & L_X > \frac{v_{x2}^2}{a_x} \end{cases} \quad (x = 1, 2, \cdots, m) \tag{9}$$

$$t_y = \frac{H_y}{v_y}, \quad (y = 1, 2, n) \tag{10}$$

When operating, the shuttle vehicle first moved vertically to the level of the target storage location and then moved horizontally to the appropriate location. The total time needed by the shuttle vehicle for a single operation was therefore calculated as follows:

$$t_{ij} = t_{x1} + t_{x2} + t_y + t_f \tag{11}$$

Taking shelves with three levels and 10 rows as an example, the operation time for the shuttle vehicle to reach each storage location is as shown in Table 2.

Table 2. Storage location operating time.

Level \ Row	1	2	3	4	5	6	7	8	9	10
1	13.7	17.8	21.5	25.0	28.5	32.0	35.5	39.0	42.5	46.0
2	18.7	22.8	26.5	30.0	33.5	37.0	40.5	44.5	47.5	51.0
3	23.7	27.8	31.5	35.0	38.5	42.0	45.5	48.5	52.5	56.0

4. Empirical Analysis

This study empirically analyzed the storage location allocation in the intelligent storage area for flood control materials at the Zhenjiang warehouse of the Jiangsu water conservancy and flood control material reserve center [1].

4.1. Case Warehouse

To emphasize the issues in intelligent storage location allocation with multiple objectives in Chinese flood control material reserve management, a case study was conducted.

The Jiangsu Provincial Hydraulic and Flood Control Material Reserve Centre (HFCMRC) Zhenjiang Warehouse is the only central- and provincial-level flood-fighting material warehouse in Jiangsu Province. It is one of the most representative flood-fighting material reserve warehouses in China. For this reason, this study chose the HFCMRC Zhenjiang Warehouse as the case example.

Interviews, document analyses, and observations were used for the data collection in this case study. A series of face-to-face semi-structured interviews with managers and staff members from the government sector (flood control and food control materials) and the business sector (food control materials) were conducted from September 2017 to August 2018. The inventory, invocation, and warehousing data of Jiangsu provincial flood control materials were provided by HFCMRC and the HFCMRC Zhenjiang Warehouse.

4.2. Retrieval of Materials in Storage

In accordance with the types and quantities of materials that were used at the Jiangsu provincial water conservancy and flood control material reserve center and in line with the warehouse's size and shelving arrangement, we selected five types of supplies: motor oil, life jackets, tents, outboard motors, and powerful handheld flashlights. These supplies are most frequently used, have regular shapes, and can be suitably placed on multi-level shelves. See Table 3 for the specific quantities of these items in storage.

Table 3. Types, specifications, and quantities of materials in storage.

Material	Specifications (Boxes/Bags)	Total Quantity in Centralized Storage
Motor oil	0.4 m × 0.3 m × 0.3 m	200 boxes
Life jackets	0.75 m × 0.45 m × 0.55 m	WYC86-5 type, 22,000 pieces, 91-YB type, 10,000 pieces, total of 1600 boxes
Tents	0.4 m × 0.7 m × 0.2 m	150 pieces
	Poles: length: 2 m, diameter: 0.15 m	-
Outboard motors	0.5 m × 0.7 m × 1.2 m (25 HP)	60 pieces
	0.4 m × 0.7 m × 1.0 m (18 HP)	100 pieces
	0.5 m × 0.8 m × 1.2 m (40 HP)	100 pieces
Handheld flashlights	0.56 m × 0.44 m × 0.35 m	200 pieces, approximately 16 boxes

4.3. Determining of the Intelligent Storage Area Location, Size, and Dimensions

According to the general construction plan of the Zhenjiang warehouse of the Jiangsu water conservancy and flood control material reserve center, the warehouse's storage room has a minimum length and width of 30 m and 18 m, respectively. Because the location of the warehouse has not yet been determined, the warehouse's dimensions were set as 30 m × 18 m, which provided a total of 540 m². This ensured that the design would be applicable to any storage area in the warehouse.

4.4. Facility Layout and Equipment Types of Intelligent Storage Area

(1) Selection of shelves

There are limited types of stored flood control materials. They have large batch quantities, are often heavy and bulky, are not easily to manually carry, are not frequently used, and must be accessed quickly when needed. In view of these characteristics, we considered the use of pallet racks, drive-in racks, shuttle racks, and cantilever racks.

The specifications of the intelligent shelf storage locations in this study were preliminarily set as 1.5 m × 1.1 m × 1.5 m. Table 4 lists the advantages and disadvantage of the various types of shelves and the other equipment required.

Table 4. Shelf types.

Shelf Type	Advantage	Disadvantage	Price
Pallet racks	Economical, convenient to set up and take down	Low storage density, requires many lanes, usually 3–5 levels are used, the height of the shelves must be restricted and must generally be less than 10 m	Low price
Drive-in racks	High storage density	Forklifts must enter shelves, accessing goods may easily damage shelves	Low price
Shuttle racks	High utilization rate, high operating performance, flexible working approach, high safety coefficient	Stored products must be uniform, not suitable for random storage, high initial investment	Relatively high price
Intelligent access warehouse shelves	High utilization rate, quick access	High price, difficult to access goods after power outage or malfunction	High price

From the above types of shelves, pallet racks and drive-in racks must be equipped with forklift shuttle racks, and these warehouses must possess shuttles and forklifts. Intelligent access warehouse shelves also must be equipped with forklifts. Among these types of shelves, the order of space utilization of warehouses from large to small is as follows: intelligent access warehouse shelves, shuttle racks, drive-in racks, and pallet racks. The total valuation of shelves in descending order is as follows: intelligent access warehouse shelves, shuttle racks, drive-in racks, and pallet racks. The price of a shuttle is 100,000 yuan, and a 500 square meter warehouse needs to be equipped with two to three sets. The shelves can be customized. Here, they are specified to be 1.5 m × 1.1 m × 1.5 m. The prices of each type of shelf are different, and the price of pallet racks is the lowest. A 500 square meter warehouse needs three to four shuttles. Considering the low utilization rate of the flood control warehouse, we can rent forklifts, which can save costs and avoid idle assets. The price of a forklift is approximately 500,000 yuan. If the decision is made to purchase a forklift, a Linde forklift or Zhejiang Nori forklift is recommended. Furthermore, electric forklifts are economical and environmentally friendly and require narrower lanes compared with diesel forklifts, thereby making electric forklifts more preferable. Their price is between 60,000 yuan and 80,000 yuan.

(2) Selection of pallets

Different countries have different pallet specifications. The most common pallet specification in China is 1200 mm × 1000 mm, which is also one of the most common pallet specifications in Europe. These pallets are low-price, flat, wooden pallets with good durability, which makes them well suited to flood control materials, and they have a price of approximately Renminbi (RMB) 35–60 each.

When pallets are used to store materials, attention should be paid to the reasonableness of the materials that are stored on the pallets. The materials should cover at least 80% of the pallet area, the height of the center of gravity of the stored materials should not exceed two-thirds of the pallet width, and the height of the materials above the pallet should not exceed 1200 mm. In this study, the various types of materials, their quantities, and their stacking arrangements are given in Table 5. The materials in this table were placed on shuttle racks, and generators and towed water pumps were stored on the floor. If more materials are added in the future, they can be placed on shuttle racks and placed at an upper level. Figures 2–6 present the schematic diagrams of the arrangements of the materials.

Table 5. Material types, quantities, and stacking arrangement.

Name		Quantity	Volume Stacked on Pallet	Number of Pallets Needed	Pallet Utilization Rate
Motor oil		200 boxes	1.2 m × 0.9 m × 0.8 m	9	90%
Life jackets		1600 boxes	1.1 m × 0.9 m × 0.75 m	400	82.5%
Handheld flashlights		16 boxes	1.12 m × 0.88 m × 0.7 m	2	82.1%
Tents		150 pieces	0.4 m × 0.7 m × 0.2 m	25	70%
			Poles: length: 2 m, diameter: 0.15 m	15	-
Outboard motors	25 HP *	60 pieces	1.2 m × 1.0 m × 0.7 m	30	100%
	18 HP	100 pieces	1.2 m × 1.0 m × 0.8 m	33	100%
	40 HP	100 pieces	1.2 m × 1.0 m × 0.8 m	50	100%

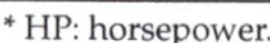

* HP: horsepower.

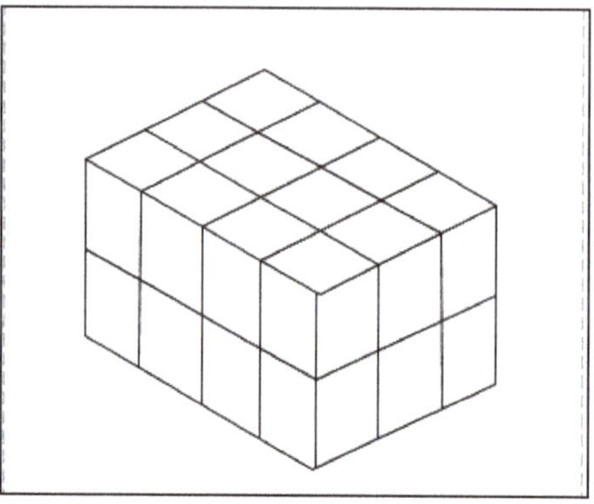

Figure 2. Motor oil stacking model.

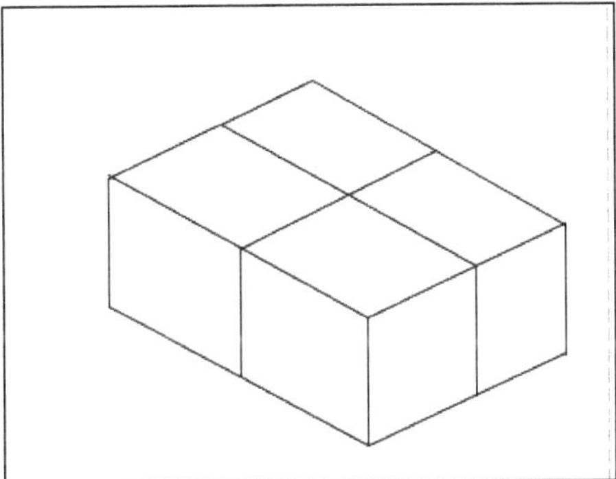

Figure 3. Life jacket stacking model.

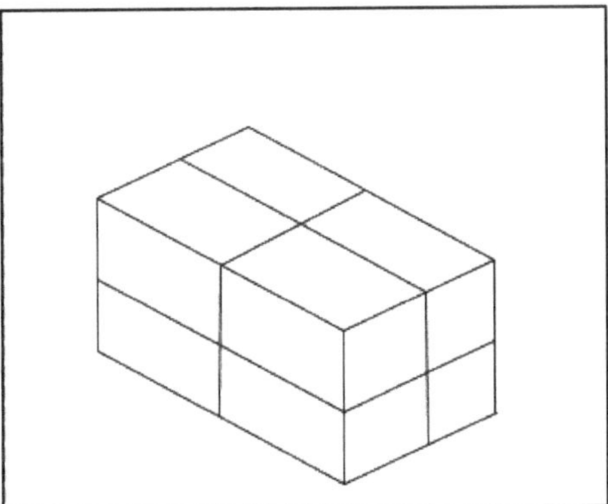

Figure 4. Flashlight stacking model.

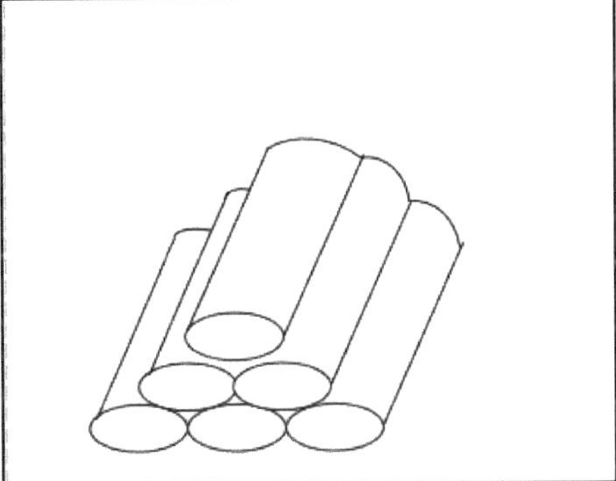

Figure 5. Tent stacking model.

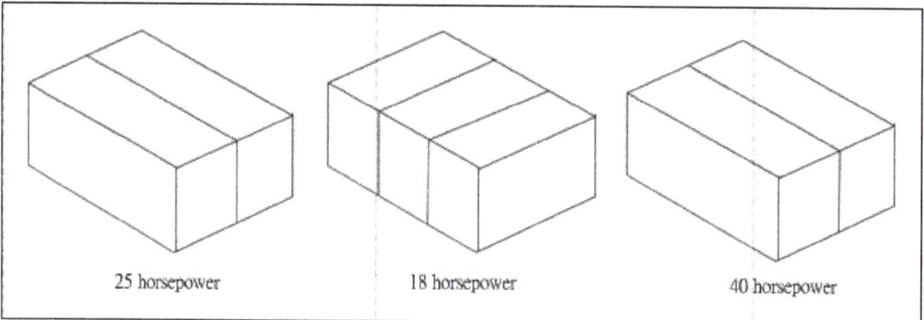

Figure 6. Outboard motor stacking model.

4.5. Storage Location Allocation in the Intelligent Storage Area

We first arranged the warehouse's internal layout (see Figure 7) in accordance with the existing intelligent warehouse area and material storage needs.

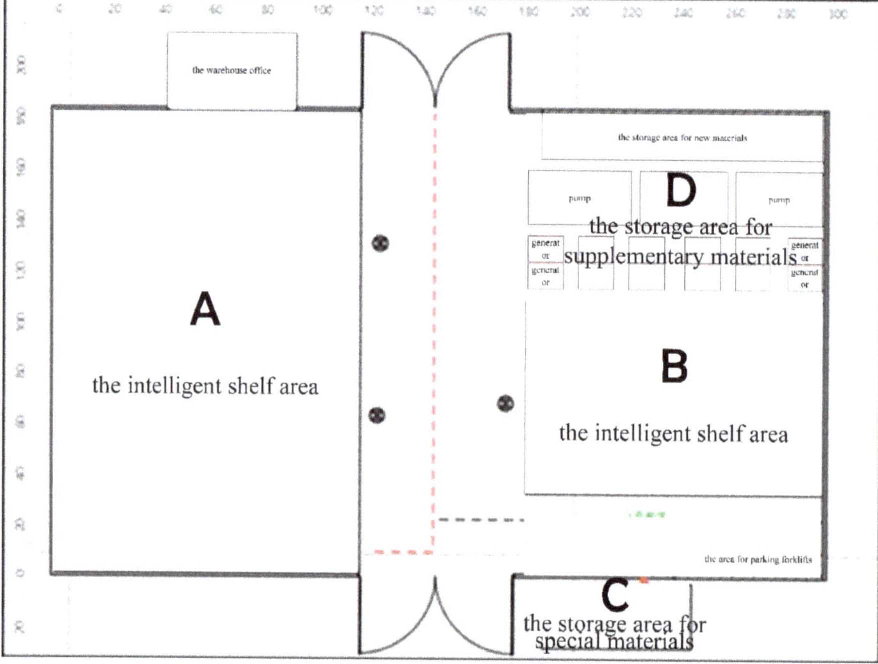

Figure 7. Plan layout.

Compartment A and compartment B are both intelligent shelf areas, and compartment C is for storing special materials, such as tent poles and very small amounts of materials. It is equipped with ordinary shelves and cantilevered shelves. Compartment D is the pending area, which can be used as the storage area for supplementary materials, such as generators, pumps, and other supplies. Compartment A is approximately 11 × 18 = 198 square meters, compartment B is approximately 12.1 × 7.5 = 84 square meters, compartment C is approximately 12 × 1 = 12 square meters, and compartment D is approximately 12 × 6.5 = 78 square meters. The blank areas are the lanes. The middle lane is 6 meters wide, and the other lane is approximately 3 meters wide. The upper part is the entrance, and the bottom is the exit. Temporary sorting areas can be established on both sides of the entrance and exit. If a one-time delivery is sufficient, the upper entrance can be used as a temporary exit to improve the distributional efficiency of flood control materials. In addition, if the warehouse is subject to realistic conditions, the entrances and exits can also be combined together, and the exit can be used as the entrance.

Based on our storage location allocation model and the flood control material warehouse's material use records, we assigned outboard motors, life jackets, tents, motor oil, and flashlights to one category, and generators and towed water pumps to another category. The materials that are suitable for storage in an intelligent warehouse were roughly divided into three areas: Area I contained outboard motors, life jackets, and tents; Area II contained motor oil and flashlights; and Area III contained generators, towed water pumps, and space for other materials that might be stored in the intelligent warehouse in the future. The materials in Area I were the most frequently used and had high inventory levels, the materials in Area II were frequently used but had low inventory levels, and the materials in Area III were moderately used, had low inventory levels, and were very heavy, bulky, and difficult to move.

Therefore, we arranged the storage area in accordance with the material types and quantities and the number of storage locations on the shelves. Type I materials were placed in areas A and B, with life jackets placed in Area A and outboard motors and tents placed in Area B. Type II materials consisted of tents, tent poles, and flashlights. Because of the close relationship between the tents and tent poles, they were placed in Area C. Type III materials consisting of generators and towed water pumps were placed sequentially on the floor in Area D.

After the materials had been placed in these sub-areas in accordance with the storage location allocation principles, the materials could be quickly and precisely located and retrieved from the warehouse, which increased the efficiency and facilitated their inspection, inventory, and maintenance.

Depending on their form, the shelves were classified into two main types. Type 1 consisted of pallet racks, which required many lanes and had a relatively low overall spatial utilization rate. Type 2 consisted of close-packed shelves, including drive-in racks, shuttle racks, and an intelligent 3-D storage area. These shelves required fewer lanes and used relatively little space.

The storage locations on the pallet racks are shown in Figure 8. A total of 90 storage locations were situated on each level in areas A and B, and five levels could be used for the storage of existing materials.

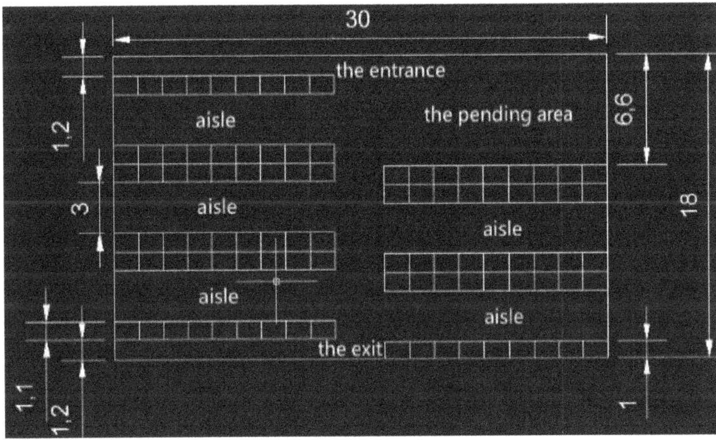

Figure 8. Pallet-type shelf storage locations.

The type 2 close-packed shelves, including drive-in racks, shuttle racks, and intelligent 3-D storage area shelves, could be arranged in the following manner. Objects could be placed horizontally on the shelves, and the main aisles could be used to ensure that the life jackets could be stored and removed via separate pathways. There were 176 storage locations on each level and four levels of shelves, which resulted in a total of 704 storage locations. See Figure 9 for a plan diagram of the storage locations.

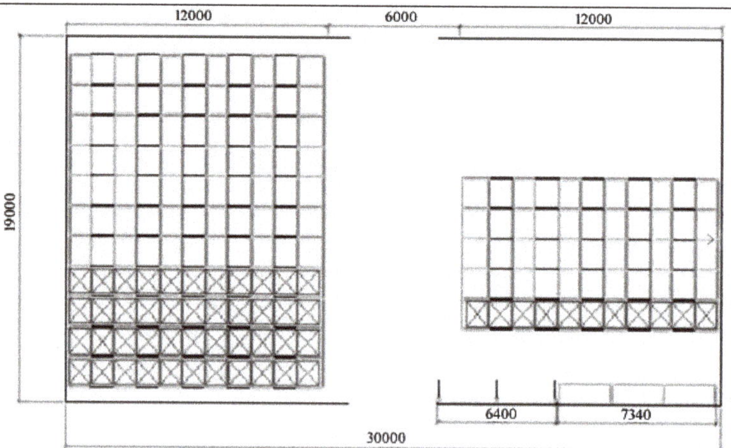

Figure 9. Close-packed shelves.

Area C, which was used for the storage of tents and poles, had an overall volume of 5 m × 1.2 m × 4 mm and a floor area of approximately 12 m². The tent poles were placed on the shelves in bundles of 10 poles. The storage locations were at a vertical distance of 0.5 m from each other, their total height was 3 m, and there were 12 storage locations. The size of the storage locations in the ordinary shelves located on the right was 1.4 m × 1.0 m × 1.5 m. Each level had five storage locations, and there were four levels. These shelves were used for the placement of small items, such as motor oil and flashlights that required convenient access. Area D contained 78 m² of empty space that could be used for the storage of additional materials or to meet other needs in the future.

See Table 6 for the material parameters at the Zhenjiang warehouse.

Table 6. Material parameters.

Type of Material	Frequency of Storage and Retrieval (Storage and Retrieval Frequency/Year)	Weight (Kg/Pallet)	Number of Occupied Storage Locations
Life jackets	0.016	32	400
Tents	0.011	180	25
Outboard motors	0.011	220	113
Motor oil	0.011	120	9
Handheld flashlights	0.011	32	2

Our storage location optimization model was programmed using the MATLAB software and yielded an *H*-function variation curve when run (see Figure 10).

The following conclusions can be derived from the curve in Figure 10. As the number of iterations increases, the objective function value *H* steadily decreases and reaches a relatively constant value after the number of iterations reaches 170. This indicates that the objective function has basically achieved convergence. The final value was 1.44×10^5.

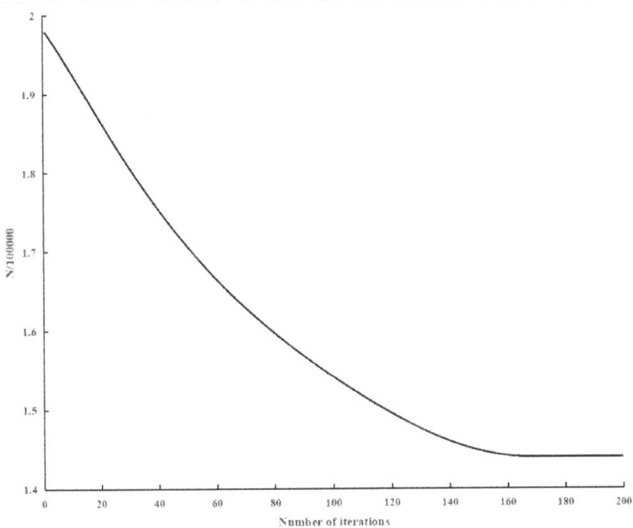

Figure 10. Optimizing process.

The solution that was obtained above optimized the model using the genetic algorithm, and MATLAB yielded the storage location allocation results for the close-packed shelves in different areas at the Zhenjiang intelligent warehouse (see Table 7).

Table 7. Simplified storage location allocation results.

Materials	Storage Location Allocation
Life jackets	Area A, rows 1–11, columns 1–11, levels 1–4
Tents	Area B, rows 1–11, columns 1–3, level 3
Outboard motors	Area B, rows 1–11, columns 1–5, levels 1–2
Motor oil	Area C, level 2
Handheld flashlights	Area C, level 3

5. Conclusions

This paper introduces the intelligent storage of flood control materials in detail, including the storage principles and allocation strategy, and establishes a multi-objective optimization model and a simplified method for the allocation of flood control materials. Based on the optimization strategy obtained from MATLAB, this paper empirically analyses the distribution of storage positions for the provincial flood control materials in the Zhenjiang branch warehouse of Jiangsu Province and obtains some good results. It can provide referential value and recommendations for effectively improving the storage efficiency and management of flood control materials and reducing losses from floods.

Finally, we make the following related contributions.

(1)   We scientifically assess the management of flood control materials. A flood control material classification management system is established. It is based on the attributes, the quantity, the occupied capital, and the frequency of the use of the flood control materials, and manages each material following the principles of scientific classification and a rational layout. It can be used to continuously improve the reserves and management of flood control materials and emergency material supplies.

(2)   We draw on the strengths of the information management of flood control material intelligent warehousing. The flood control material intelligent warehouse can guarantee the full-scale information management and control of water conservation and flood control material storage, while saving the maximum time, manpower, and material costs. In addition, it improves the storage efficiency and management of water conservation and flood control materials, and hence has bright application prospects. Fewer water conservation and flood control materials will be discharged from the warehouse in smaller batches and on larger scales within shorter response times using the Intelligent Warehouse Management System. Here, with the automation technology, the capacity of the flood control material intelligent warehouse can be more than doubled, and the warehouse capacity and utilization can be reinforced to accomplish higher land use efficiency. In addition, the construction of the Intelligent Warehouse Management System provides important support and guarantees for the application of water conservation, flood control, and emergency rescue big data.

(3)   Given that flood control materials are delivered with few varieties in larger quantities and smaller batches and on larger scales, the system is subject to strong out-of-stock timeless and fast response times. The previous studies on intelligent warehouse systems for commercial logistics were not applicable to flood control material storage. Future research should focus on the technical research and practical application of flood control and emergency materials. The key to storage efficiency is optimizing the material allocation and the scheduling of the storage equipment. Further research on the allocation of flood control materials and the scheduling of storage equipment based on intelligent warehousing is urgently needed.

Of course, different types of intelligent warehousing technology and equipment have different characteristics. When applied to other intelligent warehousing technology and equipment, the research in this paper must adjust the model accordingly. In future research, we will further enhance the generality of the model to solve these problems. In addition, we also hope that more emergency logistics, such as drought-proof materials and wind-proof materials, will be included to optimize the location allocation of intelligent storage of integrated emergency materials.

Author Contributions: All authors contributed equally to this work. In particular, W.W. developed the original idea for the study and designed the methodology. W.W. and L.H. revised the manuscript. J.Y. drafted the manuscript. J.Y. and L.H. performed the case study and collected the data. J.W. performed the investigation. D.P. helped analyze the results and further reviewed the literature. All authors have read and approved the final manuscript.

Funding: This work was funded by the Humanities and Social Sciences of Ministry of Education Planning Fund (No. 18YJAZH092); the Jiangsu Water Conservancy Science and Technology Project (Nos. 2017059 and 2018071); Key Laboratory of Coastal Disaster and Defence of Ministry of Education, Hohai University (Nos. 201913); the Fundamental Research Funds for the Central Universities (Nos. 2019B20414); and Sichuan University (No. SKSYL201819).

Acknowledgments: The authors would like to thank all the HFCMRC staff and especially Jianbin Wei, Yi Han and Qingli Hua for their help with the data collection.

Conflicts of Interest: The authors declare no conflict of interest.

References

1.   Wang, W.; Huang, L.; Le, W.W.; Zhang, Y.; Ji, P.; Proverbs, D. Research on Class-Based Storage Strategies for Flood Control Materials Based on Grey Clustering. *Water* **2018**, *10*, 1506. [CrossRef]
2.   Chang, M.S.; Tseng, Y.L.; Chen, J.W. A scenario planning approach for the flood emergency logistics preparation problem under uncertainty. *Transp. Res. Part E* **2007**, *43*, 737–754. [CrossRef]
3.   Kates, R.W. *Hazard and Choice Perception in Flood Plain Management*; Department of Geo Research Paper No. 78; University of Chicago Press: Chicago, IL, USA, 1962.
4.   Saddagh, M.H.; Abedini, M.J. Enhancing MIKE11 updating kernel and evaluation its performance using numerical experiments. *J. Hydrol. Eng.* **2012**, *17*, 252–261. [CrossRef]

5.   Interagency Floodplain Management Review Committee (IFMRC). *Sharing the Challenge: Floodplain Management into the 21st Century*; US Government Printing Office: Washington, DC, USA, 1994.

6.   Wang, W.; Huang, L.; Liang, X.D. On Simulation-based Reliability of Complex Emergency Logistics Network in Post-accident Rescue. *Int. J. Environ. Res. Public Health* **2018**, *15*, 79. [CrossRef] [PubMed]

7.   Wang, W.; Huang, L.; Guo, Z.H. Optimization of Emergency Material Dispatch from Multiple Depot Locations to Multiple Disaster Sites. *Sustainability* **2017**, *9*, 1978. [CrossRef]

8.   Garrido, R.A.; Lamas, P.; Pino, F.J. A stochastic programming approach for floods emergency logistics. *Transp. Res. Part E Logist. Transp. Rev.* **2015**, *75*, 18–31. [CrossRef]

9.   Leeuw, S.; Vis, I.F.A.; Jonkman, S.N. Exploring logistics aspects of flood emergency measures. *J. Conting. Crisis Manag.* **2012**, *20*, 166–179. [CrossRef]

10.   Alem, D.; Clark, A.; Moreno, A. Stochastic network models for logistics planning in disaster relief. *Eur. J. Oper. Res.* **2016**, *255*, 187–206. [CrossRef]

11.   Roodbergen, K.J.; Vis, I.F.A. A survey of literature on automated storage and retrieval systems. *Eur. J. Oper. Res.* **2009**, *194*, 343–362. [CrossRef]

12.   Chan, F.T.S.; Chan, H.K. Improving the productivity of order picking of a manual-pick and multi-level rack distribution warehouse through the implementation of class-based storage. *Expert Syst. Appl.* **2011**, *38*, 2686–2700. [CrossRef]

13.   Kuo, P.; Krishnamurthy, A.; Malmborg, C.J. Design models for unit load storage and retrieval systems using autonomous vehicle technology and resource conserving storage and dwell point policies. *Appl. Math. Model.* **2007**, *31*, 2332–2346. [CrossRef]

14.   Hausman, W.H.; Schwarz, L.B.; Graves, S.C. Optimal Storage Assignment in Automatic Warehousing Systems. *Manag. Sci.* **1976**, *22*, 629–638. [CrossRef]

15.   Lee, M.K. A storage assignment policy in a man-on-board automated storage/retrieval system. *Int. J. Prod. Res.* **1992**, *30*, 2281–2292. [CrossRef]

16.   Larco, J.A.; Koster, R.D.; Roodbergen, K.J.; Dul, J. Managing warehouse efficiency and worker discomfort through enhanced storage assignment decisions. *Int. J. Prod. Res.* **2016**, *55*, 1–16. [CrossRef]

17.   Tsamis, N.; Giannikas, V.; Mcfarlane, D.; Lu, W.; Strachan, J. Adaptive Storage Location Assignment for Warehouses Using Intelligent Products. *Nat. Commun.* **2015**, *6*, 271–279.

18.   Battini, D.; Calzavara, M.; Persona, A.; Sgarbossa, F. Order picking system design: The storage assignment and travel distance estimation (SA&TDE) joint method. *Int. J. Prod. Res.* **2015**, *53*, 1077–1093.

19.   Hsieh, S.; Tsai, K.C. A BOM oriented class-based storage assignment in an automated storage/retrieval system. *Int. J. Adv. Manuf. Technol.* **2001**, *17*, 683–691. [CrossRef]

20.   Manzini, R.; Accorsi, R.; Gamberi, M.; Penazzi, S. Modeling class-based storage assignment over life cycle picking patterns. *Int. J. Prod. Econ.* **2015**, *170*, 790–800. [CrossRef]

21.   Thonemann, U.W.; Brandeau, M.L. Optimal storage assignment policies for automated storage and retrieval systems with stochastic demands. *Manag. Sci.* **1998**, *44*, 142–148. [CrossRef]

22.   Pan, C.H.; Shih, P.H.; Wu, M.H.; Lin, J.H. A storage assignment heuristic method based on genetic algorithm for a pick-and-pass warehousing system. *Comput. Ind. Eng.* **2015**, *81*, 1–13. [CrossRef]

23.   Atmaca, E.; Ozturk, A. Defining order picking policy: A storage assignment model and a simulated annealing solution in AS/RS systems. *Appl. Math. Model.* **2013**, *37*, 5069–5079. [CrossRef]

24.   Xie, J.; Mei, Y.; Ernst, A.T.; Li, X.; Song, A. A Bi-Level Optimization Model for Grouping Constrained Storage Location Assignment Problems. *IEEE Trans. Cybern.* **2016**, *48*, 385–398. [CrossRef]

25.   Pang, K.W.; Chan, H.L. Data mining-based algorithm for storage location assignment in a randomised warehouse. *Int. J. Prod. Res.* **2017**, *55*, 4035–4052. [CrossRef]

26.   Zhang, P. The Research of Slotting Distribution and Flexsim Simulation Based on Automatic Storage and Retrieval System. Master's Thesis, Jilin University, Changchun, China, 2017.

27.   Li, J.; Moghaddam, M.; Nof, S.Y. Dynamic storage assignment with product affinity and ABC classification—A case study. *Int. J. Adv. Manuf. Technol.* **2016**, *84*, 1–16. [CrossRef]

28.   Choy, K.L.; Ho, G.T.S.; Lee, C.K.H. A RFID-based storage assignment system for enhancing the efficiency of order picking. *J. Intell. Manuf.* **2017**, *28*, 1–19. [CrossRef]

29.   Xiang, X.; Liu, C.; Miao, L. Storage assignment and order batching problem in Kiva mobile fulfilment system. *Eng. Optim.* **2018**, *50*, 1941–1962. [CrossRef]

30. Accorsi, R.; Baruffaldi, G.; Manzini, R. Picking efficiency and stock safety: A bi-objective storage assignment policy for temperature-sensitive products. *Comput. Ind. Eng.* **2017**, *115*, 240–252. [CrossRef]
31. Žulj, I.; Glock, C.H.; Grosse, E.H.; Schneider, M. Picker routing and storage-assignment strategies for precedence-constrained order picking. *Comput. Ind. Eng.* **2018**, *123*, 338–347. [CrossRef]
32. Özdamar, L.; Ekinci, E.; Küçükyazici, B. Emergency Logistics Planning in Natural Disasters: Models and Algorithms for Planning and Scheduling Problems. *Ann. Oper. Res.* **2004**, *129*, 217–245. [CrossRef]
33. Xiang, Y.; Zhuang, J. A medical resource allocation model for serving emergency victims with deteriorating health conditions. *Ann. Oper. Res.* **2016**, *236*, 177–196. [CrossRef]
34. Wu, X.; Cao, Y.; Xiao, Y.; Guo, J. Finding of urban rainstorm and waterlogging disasters based on microblogging data and the location-routing problem model of urban emergency logistics. *Ann. Oper. Res.* **2018**, 1–32. [CrossRef]
35. Chang, K.L.; Zhou, H.; Chen, G.J.; Chen, H.Q. Multiobjective location routing problem considering uncertain data after disasters. *Discret. Dyn. Nat. Soc.* **2017**, *3*, 1–7. [CrossRef]
36. Ahmadi, M.; Seifi, A.; Tootooni, B. A humanitarian logistics model for disaster relief operation considering network failure and standard relief time: A case study on San Francisco district. *Transp. Res. Part E* **2015**, *75*, 145–163. [CrossRef]
37. Liang, H.H.; Ding, G.M. *Research on Storage Technology of Metrology Automation System for Future Power Grid*; Guangdong Electric Power: Guangzhou, China, 2012; Volume 25, pp. 22–26.

Article

Guidelines for the Use of Unmanned Aerial Systems in Flood Emergency Response

Gloria Salmoral [1], Mónica Rivas Casado [1,*], Manoranjan Muthusamy [1], David Butler [2], Prathyush P. Menon [3] and Paul Leinster [1]

1   School of Water, Energy and Environment, Cranfield University, College Road, Cranfield MK430AL, UK; gloria.salmoral@cranfield.ac.uk (G.S.); Manoranjan.Muthusamy@cranfield.ac.uk (M.M.); Paul.Leinster@cranfield.ac.uk (P.L.)
2   Centre for Water Systems, College of Engineering, Mathematics and Physical Sciences, University of Exeter, Harrison Building, Streatham Campus, North Park Road, Exeter EX4 4QF, UK; D.Butler@exeter.ac.uk
3   Centre for Systems Dynamics and Control, College of Engineering, Mathematics and Physical Sciences, University of Exeter, Harrison Building, Streatham Campus, North Park Road, Exeter EX4 4QF, UK; P.M.Prathyush@exeter.ac.uk
*   Correspondence: m.rivas-casado@cranfield.ac.uk; Tel.: +44(0)1234750111 (ext. 4433)

Received: 15 December 2019; Accepted: 8 February 2020; Published: 13 February 2020

Abstract: There is increasing interest in using Unmanned Aircraft Systems (UAS) in flood risk management activities including in response to flood events. However, there is little evidence that they are used in a structured and strategic manner to best effect. An effective response to flooding is essential if lives are to be saved and suffering alleviated. This study evaluates how UAS can be used in the preparation for and response to flood emergencies and develops guidelines for their deployment before, during and after a flood event. A comprehensive literature review and interviews, with people with practical experience of flood risk management, compared the current organizational and operational structures for flood emergency response in both England and India, and developed a deployment analysis matrix of existing UAS applications. An online survey was carried out in England to assess how the technology could be further developed to meet flood emergency response needs. The deployment analysis matrix has the potential to be translated into an Indian context and other countries. Those organizations responsible for overseeing flood risk management activities including the response to flooding events will have to keep abreast of the rapid technological advances in UAS if they are to be used to best effect.

Keywords: drone applications; deployment time; monitoring; flood modelling; evacuation; rescue; resilience

---

1. Introduction

In recent decades, significant flood events have affected many countries around the world including those caused by Hurricane Katrina (Florida and Louisiana, August 2005), Hurricane Leslie (France, October 2018) and the 2018 monsoon season in India (Kerala, August 2018). The impacts of these flood events on people and communities are wide and varied and can be catastrophic. The Katrina floods resulted in over 1100 deaths in Louisiana [1], with estimated economic losses of $149 billion [2]; the 2018 floods in France resulted in 14 deaths and over $500 million of damage [3]; whilst the 2018 Kerala in India floods resulted in more than 400 deaths, displaced 1.8 million people and caused an estimated $3 billion worth of damage [4]. The last major flood event in England took place in winter 2015/2016, with December 2015 being the wettest month ever recorded in England. A total of 17,000 properties across the north of England were affected with named storms Desmond, Eva and Frank. The total economic damages were estimated to be £1.6 billion [5].

Effective and efficient flood emergency response has a key role in reducing the adverse impacts of flooding. Coordinating the response—including warning and informing prior to and during events, evacuation prior to the flooding, the rescue of people and the organization of volunteers [6]—has been a priority for governments in recent decades. Although there are clearly lessons to be learned from experiences in other countries, often the detailed arrangements need to be context specific.

In England, revised flood emergency operational arrangements were put in place building on the experience to date. These are outlined in the National Flood Emergency Framework [7]. The primary aims of the emergency response include the protection of human life, the alleviation of suffering and the restoration of disrupted services (e.g., water, electricity, transport). Within this framework and based on documented command and control protocols, decisions are taken at the lowest appropriate level with coordination at the highest necessary level.

In India, the Central Government has established the National Disaster Management Division within the Ministry of Home Affairs. This Division has introduced various initiatives and set up several organizations to deal with disasters, including floods. In 2016, a National Disaster Management Plan was published to provide the overall direction and national goals. Under the plan, the various ministries and departments at state and district levels have to develop their own specific management and response plans, and related operating procedures [8].

Various emergency response activities rely on the information provided by monitoring, models and multiple data sources. For example, in India hydrological and flood models are used by the Central Water Commission for modelling and forecasting purposes [8] to provide water level and river flow information to the authorities [9] with an online dissemination portal [10]. In England, there has been a continuing interest in developing flood models for fluvial, pluvial and sea flooding [11], with different data needs and outcomes [11,12], to help reduce the impacts on people, property and critical infrastructure [13]. From the response side, emergency responders request and collate a varied range of information, from aerial imagery to individual eyewitness reports, to support decision-making. Different information is required pre-, during- and post-events. For instance, during a flood event real-time or near real-time local information on how many people, buildings, and other infrastructure are at risk is required [14]; post-event aerial imagery can provide vital and detailed information about the extent of flooding and damage to properties [15].

In recent years, emergency responders have used Unmanned Aircraft Systems (UAS) to acquire core information pre-, during- and post-events [16–18]. UAS are small and light (less than 20 kg) remotely piloted aircraft generally equipped with a range of sensors for the collection of information. There are two main types of UAS platforms: fixed wing and vertical take-off and landing (VTOL). The former relies on wings that generate lift to fly, whereas VTLO rely on rotors. UAS can be equipped with different sensors [19] from cameras to warning systems. RGB cameras are able to provide high resolution imagery of up to 2 cm. Emergency responders in various countries have identified the added benefits of UAS in humanitarian responses in terms of the rapid assessment of damage, such as collapsed buildings or blocked roads and search and rescue operations [20,21]. In England for example, during winter 2015/2016, UAS were used by the Environment Agency (EA) to assess smaller scale flooding incidents in high detail; in particular, UAS were used to provide an up close and detailed live stream of an inaccessible river breach. This enabled an effective and efficient assessment of the area [22]. However, the rapid uptake and continuous development of the technology have resulted in the ad-hoc and opportunistic use of UAS over a strategic appraisal of how best to use them and for what purpose pre-, during- and post-events. Various types of UAS missions have been identified as being used in flood emergency responses (including strategic situation awareness, inspections, ground search, water search, debris/flood/damage estimation and tactical situation awareness) with an indication of the data products (e.g., images, videos, and orthomosaics) generated in the flights [18]). However, there is not yet a purpose-driven approach defining which UAS products would be of benefit at each stage of the preparation for and response to a flood event. The need of such logic-based decision support approach has been identified by multiple research and governmental organizations within the context of catastrophe response in India and England through two knowledge exchange workshops organized in Delhi (30 September 2018) and Bangalore (18 September 2019) within the engagement activities organized by the EPSRC research project 'Emergency flood planning and management using unmanned aerial systems' (www.efloodplan.net). To the authors' knowledge, a purpose-driven approach detailing how and when UAS with specific embedded sensors should be used to collect data to assist in flood event responses is not yet documented. It is envisaged that such an approach will be context-specific and influenced by the nature of the flood events that occur within a particular area, region or country, the data available from other sources, as well as the airspace regulatory framework for UAS use.

Based on these premises, the aim of this study is to develop purpose led guidelines for the efficient and effective deployment of UAS for flood risk management activities including emergency response pre-, during- and post-event phases. We demonstrate how a deployment analysis matrix can be designed and used to assist flood emergency response requirements in the context of catchment response and the nature of flood events for England and explore its potential to be translated into an Indian context. This will be achieved through the following four overarching objectives: (1) to map out the current role of existing organizations involved in emergency response in England and India; (2) to identify existing UAS applications within the components of a flood risk management system; (3) to determine context-specific requirements for UAS products to assist in flood risk management activities; and, (4) to develop an adaptive and transferable matrix analysis framework that can then be used as the basis for guidelines for the effective deployment of UAS for flood risk management activities leading to more resilient urban environments and including emergency response before, during and after a flood event.

2. Materials and Methods

2.1. Flood Emergency Response in England and India and the Potential Use of UAS

The institutional arrangements for flood emergency response and the current and potential applications of UAS technology were determined through a literature review and face-to-face interviews with key personnel with detailed understanding of the flood response arrangements in England and in India. England and India were selected for this study based on recent flood events occurring in these areas (Cockermouth, England, 2015 and Kerala, India, 2018). A total of 14 interviews were held in India (7) and England (7), including participants with experience in flood emergency response from the national and regional authorities, private sector and Non-governmental organizations (NGOs). A semi-structured questionnaire based on a set of twenty open questions was used (Supplementary Materials). This format enabled the interviews to be focused on the research objectives, but with the flexibility to evaluate responses and explore issues that emerged during interviews. The raw data products (i.e., those are provided without extra processing or internet connection) and derived products (i.e., produced by post-processing of the raw data products) produced by UAS applications, as well as the key factors affecting UAS deployment and flight plan configuration were also identified in the literature review. Additional information on the main applications and potential use of UAS was also obtained from a knowledge exchange workshop organized in India within the engagement process of the EPSRC research project "Emergency flood planning and management using unmanned aerial systems" (Delhi, 30 May 2018). The UAS applications were grouped into a set of five flood management components, which included flood warning, flood monitoring and flood risk assessment, evacuation route identification, damage assessment and rescue.

2.2. Development of an UAS Deployment Analysis Matrix

This study uses a 3×3 matrix to identify potential uses of UAS in flood risk management activities and as the basis of guidelines on the deployment of UAS for flood risk management activities. A number of factors were considered for the x and y axes of the matrix when considering UAS deployment and flight plan configurations. The three main factors identified were related to catchment size, flood source type, and phase of a flood event (Table 1). Catchment size influences the amount of data gathered [23] and the type of UAS that is required to provide the spatial coverage [24].

Catchment flood response was chosen as one of the key factors because this gives an indication of the time available to deploy an UAS and the use of particular applications and technologies in a given situation. The catchment flood response was determined based on the time between the start of a rainfall event and the potential for the flooding of properties. Based on climatic and catchment conditions in England, the flood response was considered to be 'slow' when flooding occurs more than 8 h after the rain event, 'medium' when flooding occurs between 3 and 8 h and 'fast' when the onset of flooding takes place in less than 3 h [25,26]. We also considered in the deployment analysis matrix the phase in which UAS will be deployed in the overall approach to flood risk management activities [27]: 'pre-event', 'during-event' and 'post-event' (Figure 1). Pre-event refers to activities such as flood modelling activities, the construction of flood risk reduction assets and the planning that will be needed to respond effectively to a defined flood magnitude (i.e., based on a return period). During-event starts as soon as the first flood warning is issued, whilst post-event refers to the recovery and clean-up phase when the water has receded and for example is no longer within people's houses or blocking transport routes.

Table 1. Key factors identified as relevant for the development of the Unmanned Aircraft Systems (UAS) deployment analysis matrix.

	Factor	Description
Catchment	Catchment size	Size of the catchment where the event has taken place [23,24]
	Catchment flood response	Catchment response to flooding after the rain event occurs [28,29]
Flood	Flood source type	Source of flooding classified as groundwater, pluvial and fluvial [30]
	Flood extent	The spatial coverage of the flood event and the associated emergency response coverage (e.g., single scene, regional coverage and national coverage) [31]
Response	Emergency response phases	Pre-event, during-event and post-event [27,32]

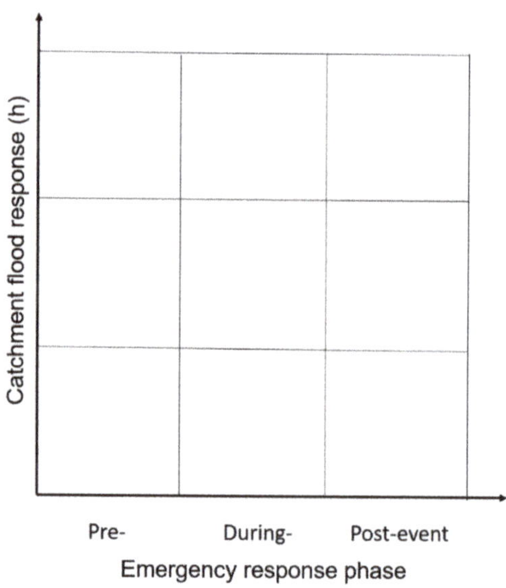

Figure 1. Format of the UAS deployment analysis matrix.

Each of the nine cells within the matrix were populated with the UAS applications identified from the literature review and via a second set of ten one-to-one interviews in England. The interviews targeted specialists in the use of UAS for monitoring, surveying and incident response within the EA, Cranfield University, University of Exeter and an independent expert in flood risk management who had extensive experience of emergency responses at a senior level. During each interview, responders were presented with a set of cards defining the UAS applications, the processing time required to obtain the UAS products and the accuracy or resolution of such products. Processing time, accuracy and resolution were defined based on values reported in the literature. Responders were able to allocate the UAS applications with a given processing time and accuracy within the context of a recent flood event in which they were involved. The data gathered in the matrix was analyzed to identify the consistency between participants in allocating an application to a particular matrix cell. Consistency between participants was assessed through direct comparison of the number of responses per application and cell.

A workshop organized in India (Bangalore, 18 September 2019) helped provide insights about the transferability of the designed deployment analysis to an Indian context. Discussions with experts on flood risk management activities including emergency response in India were held to determine the potential transferability of the matrix to other countries.

2.3. Technological Needs for the Use of UAS in Flood Emergency Response

To assess how the technology should be further developed to meet flood emergency response needs, an extended online survey was also carried out (See Supplementary Materials for details of the survey). The UAS applications—flood extent, flood depth and flow velocity—were selected as they are considered to be key for making decisions during a flood event [33,34]. For the three UAS applications, participants were asked about the current and desired time to process these specific geomatics products and the associated accuracy requirements. Accuracy refers to the expected error range in flood extent (m), flood depth (cm) and flow velocity (m s^{-1}) in the generated geomatics product. The survey was built in Qualtrics software and distributed to relevant experts and at two flood risk management related events: Oasis Conference (London, 18 June 2019) and Flood and Coast (Telford, UK, 20 June 2019). A total of 25 participants completed the survey. Data collected were compared using descriptive statistics to assess the current and desired accuracies and time to process each application. The information gathered enabled an assessment to be made as to whether there are any knowledge and technology gaps that need to be addressed to achieve a desired time and accuracy for a given application.

3. Results

3.1. Organizations Involved in Flood Emergency Response: England and India

Results from the literature review and one-to-one interviews highlighted that in England there are over 17 organizations involved in flood emergency response. This is similar to the number in India, where 16 key organizations were identified (see Supplementary Materials).

In England, the response to localized flooding is led by the local emergency responders without any significant involvement from central government [35]. For some flood events, local responders are supported by central government via Department for Environment, Food and Rural Affairs (Defra) as the designated Lead Government Department for responding to floods. Defra normally co-ordinates the cross-government response to lower level national flooding events (level 1) and manage it within the department. As the extent and impact of the flooding increases, it is likely that there will be increasing involvement by others in central government with the activation of the Cabinet Office Briefing Room (COBR), which brings together ministers, seniors government officials, representatives from national response agencies and organizations impacted by the flood event. Level 2 events (serious impact) are still coordinated by Defra but through COBR. More serious events (level 3—catastrophic) are fully escalated to central co-ordination by the Civil Contingencies Secretariat within COBR [7] (Figure 2). The Army and other military forces may be requested to help in a flood response [36].

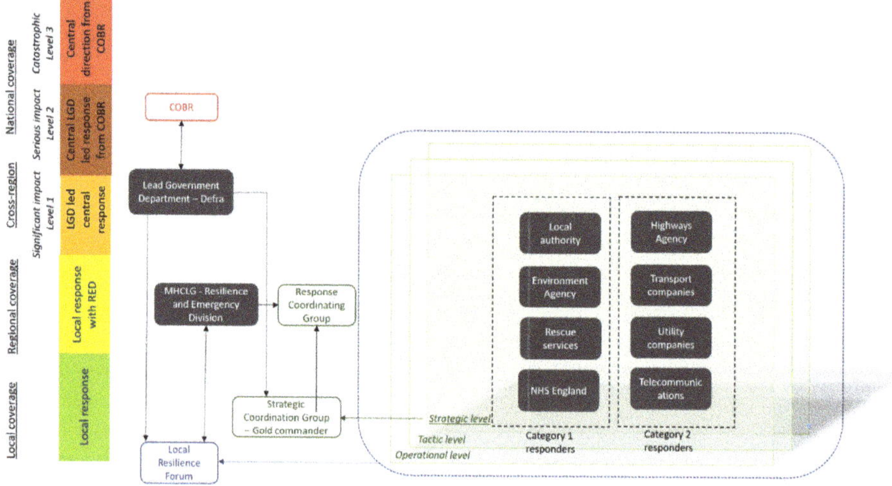

Figure 2. Schematic diagram of flood emergency response in England showing the main agencies and groups involved, the levels of emergency responses at the local level (operational, tactical, strategic), the categories of responding organizations (Category 1 and Category 2) and the likely government arrangements (from local response to central direction from COBR) which depends upon flood extent (local, regional, cross-region and national). Category 1 comprises the organizations that are at the core of the response to most emergencies, whereas Category 2 responders are co-operating bodies involved in incidents that affect their sector. The color schemes in the government arrangements reflect the increasing levels of emergency response (from green to red). COBR: Cabinet Office Briefing Rooms, Defra: Department for Environment, Food and Rural Affairs, MHCLG: Ministry of Housing, Communities and Local Government, RED: Resilience and Emergency Division, and LGD: Lead Government Department.

At present, there are a number of organizations that may use UAS during a flood event. These include the EA, the Fire and Rescue Service, the Police and insurance companies as well as private individuals. As an example, in England, the local or national incident responders may request the deployment of UAS to the specialist Geomatics Team of the EA to provide information related to flood damage and impacts [37]. The Geomatics Team will evaluate if UAS are the most appropriate means of obtaining the information. One of the interviewees informed us that arrangements are in place that allow the EA's Geomatics Team to deploy UAS in any part of England within six hours. The UAS images can be sent via a live feed to an EA incident room. The decision as to who will fly UAS in a particular situation is agreed locally event by event. To date, there is not an established approach to decide which organization will fly UAS for what purpose during and after an event. This can result in a duplication of effort or in important information not being gathered during a particular event.

India also has a tiered approach to flood emergency response (Figure 3). The national government develops policies and provides advice and assistance when there are major events, whilst the States are the responsible for carrying out risk assessments and planning and implementing mitigation measures [8]. At the district level, flood events are categorized into three levels of impact [38]: Level 1—there are sufficient resources and capacity to respond at the district level; Level 2—the impact is beyond existing capacities and support from State agencies is needed; and Level 3—the impact is beyond the existing capacities of district and state resources and support from national agencies is needed (Figure 3). If an emergency escalates beyond their capabilities, the local administration must seek assistance from the district administration or the State Government. If the State Government considers it necessary, it can seek central assistance [8]. The Ministry of Water Resources (river flooding) and Ministry of Urban Development (pluvial flooding) function under the overall guidance of the Ministry of Home Affairs [8,39] when responding to flood events.

In India the police, navy and army have permission to fly UAS for security and rescue reason. However, as in England, there is not an established system to deploy UAS in flood emergency response activities. For example, in the 2013 Uttarakhand floods, the National Disaster Response Force (NDRF) deployed UAS with technical support from research institutions [40,41]. The Indian security forces and the Indo-Tibetan Border Police also deployed UAS to assist in the relief efforts of the National Disaster Response Force by helping find survivors in remote locations [42] and in areas cut off by landslides [43]. During the interviews performed in Delhi, Indian participants highlighted the opportunities to use UAS at the district level by the District Collector, who is responsible for district-level responses to a flooding event.

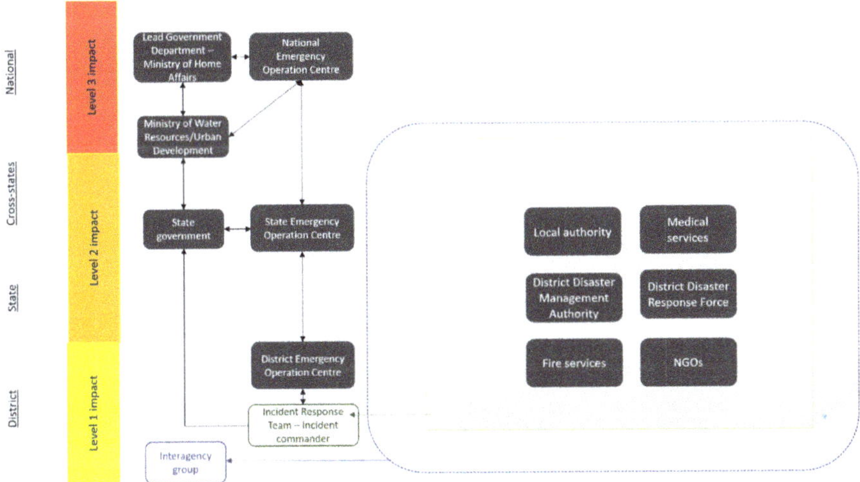

Figure 3. Schematic diagram of flood emergency response in India showing the main agencies and groups involved and the levels of emergency responses (level 1, level 2 and level 3) which depends upon the flood extent (district, state, cross-states, national).

3.2. The Potential Use of UAS in Flood Emergency Response

From the existing scientific literature, sixteen UAS applications that could be used in flood risk management activities were identified. The UAS applications can be assessed in terms of their use before an event in flood risk assessments, determining terrain elevations, flood extent modelling, identifying evacuation routes and flood warning. During an event to inform responders about actual flood extents, flood sources and routes, whether evacuation routes remain clear, identifying people in need of rescue and provision of emergency relief supply. Post events as part of damage and impact assessments. UAS raw products included high definition (HD) video, infrared imaging, Red-Green-Blue (RGB) imagery, RGB video, RGB video streaming and thermal imaging. A total of fourteen post-processing outcomes were identified: those derived from models (e.g., flood models, evacuation models), bespoke algorithms (e.g., image feature recognition), UAS-specific software (e.g., terrain elevation measurements) (Table 2).

Table 2. Detailed list of flood risk management components, with identified applications, UAS raw products and post-processed outcomes from the literature within the context of flood emergency response. The processing time for outcome generation and the accuracy of the outcome is also indicated. DEM, DTM and DSM stand for Digital Elevation Model, Digital Terrain Model and Digital Surface Model, respectively.

Flood Management Component	Applications	Examples	UAS Raw Product	Processed Outcome	Time	Details [3]
Flood warning	Evacuation warning	UAS with an embedded audible alarm to provide an alert about an upcoming flood and the need for evacuation (Mozambique).	N/A	N/A	Real-time	Flight path reach
Flood monitoring and flood risk assessment	Visualization of flood extent	Delineation of inundated areas by digitizing the boundaries at the contrasting land surface/water boundary [28].	RGB imagery RGB video	Flood extent, Ponding locations	Real time (<1 h)	0.2 m high resolution RGB imagery [4]
	Flood extent detection	Application of an algorithm to detect flood areas automatically [44,45].	RGB imagery	Orthoimage, flood extent	>48h [1]	0.2 m high resolution RGB imagery [4]
		A spectral difference index is generated from the RGB photos to map flood water extent [46].	RGB imagery	Orthoimage, flood extent	Real time (<1 h)	0.2 m high resolution in RGB imagery
	Modelling flood extent	A concept of transfer learning where a Convolutional Neural Network model is trained based on one dataset, transferred and used to classify another dataset to delineate flood extent [47]. A high accuracy terrain model combined with hydraulic calculations performed on transverse profiles produce the flood-prone areas at 1% and 5% exceedance probabilities of discharge [48]. A DEM mapped with UAS is used in hydrologic and hydraulic modelling to provide a flood hazard map [49].	RGB imagery	Orthoimage, DEM, flood extent	<48 h	1.5 cm ultra-high resolution in RGB imagery with 93% accuracy in flood extent
	Point measurement of flow velocity	A set of floating wireless sensors are deployed within the flood extent by multiple UAS to capture water velocity readings at multiple locations [50]. A combination of floating, infrared light-emitting particles and a programmable embedded colour vision sensor to simultaneously detect and track objects [51].	Water velocity readings. Infrared imaging	Water velocity	Real time (<1 h)	Estimated 0.15 m/s accuracy [5]
	Optical water velocity	Bespoke algorithms are used to track the movement of tracers [52], or texture [53–55] in the water surface in consecutive frames obtained from video footage.	RGB imagery RGB video streaming HD video	Orthoimage, water velocity / Water velocity	>48 h [1] / Real time (<1 h)	Estimated 0.5 m/s accuracy [5]
	Visual flood depth	Flood depth is estimated through the observation of wrack marks (post-event) [28] or via the observation of the water level (during event) against know height points (e.g., cars, bridges, feature buildings).	(Oblique) RGB imagery RGB video	Flood depth	Real time (<1 h)	High resolution RGB imagery
	Point measured of flood depth	UAS with ultrasonic sensors used to detect water level [56]. A pre-event DEM is required to estimate flood depth.	Water depth readings	Water level, point cloud, DEM, DTM, DSM	<48 h [2]	6 cm accuracy

Table 2. *Cont.*

Flood Management Component	Applications	Examples	UAS Raw Product	Processed Outcome	Time	Details [3]
	Flood source identification (fluvial, pluvial, groundwater)	Sources of flooding can be identified based on damaged caused within/outside the fluvial flood extent [15,57] or based on differences in water temperature [58].	RGB frames Thermal frames	Orthoimage, DEM, DTM, DSM	>48 h	1 m resolution (DEM)
Evacuation routes identification	Modelled evacuation route identification	Modelling of evacuation routes by using UAS as end devices of M2M architecture [59]. The input model needs DEM, DTM and/or DSM as well as hydrological/hydraulic input data.	RGB imagery RGB video	Map of evacuation routes, DSM, DTM, DEM, orthoimage	<24 h	20–60% tracking accuracy
	Surface changes and displacements of landslides	Surface changes and displacements of landslides [60,61].	RGB imagery RGB video	3D point clouds, DEM, orthoimage	<48 h [2]	1 cm ultra-high resolution (DEM), accuracy: 7.4 cm (horizontal) × 6.2 cm (vertical)
Damage assessment	Visual detection of affected properties, businesses, hospitals, schools	To collect information on damage to hospitals for patient rescue and for efficient allocation of resources [62].	RGB imagery	Location of affected properties	Real time (<1 h)	Resolution at building level
	Resistance and resilience measures identification	Identification of residential properties with resistance measures (i.e., flood aperture guards for doors and windows, flood resistant airbricks, and raised doors or steps leading to a property) [15].	RGB imagery	Orthoimage, DEM	>48 h	Resolution at building level
Rescue	Identification of safe shelter points	To identify where to best place NGO camps [21] and to identify land that could be safer to relocate families [20].	RGB imagery RGB video	Map with location of points	Real time (<1 h)	Resolution at building level
	Detection of stranded people	The use of UAS to locate stranded people [63,64] even at night [65,66] and specifically during floods [67].	RGB imagery RGB video Infrared imaging	N/A	Real time (<1 h)	Resolution at individual level
	Delivery of ad-hoc supplies	The use of UAS to deliver equipment or resources that guarantee the survival of stranded people e.g. to carry a radio to communicate [64], floating devices [65].	RGB imagery RGB video	N/A	Real time (<1 h)	Resolution at individual level

[1] A 36 h of photogrammetric processing is assumed [15] plus the time needed to apply the algorithm. [2] A 36 h of photogrammetric processing is assumed [15]. [3] When more than one study shows values of resolution and/or accuracy, it is indicated in the table the details of the lowest values for resolution and/or accuracy found. [4] Assumed the value of 0.2 m based on the real-time optical flood extent detection. [5] It is assumed that above 3 m/s, there are damages in the infrastructure [68], which is multiplied by the median of two coefficients of variation in flow velocity [52] to estimate the medium accuracy in m/s.

3.3. The UAS Deployment Analysis Matrix for England: Pre-, During- and Post-Event

Results (Figure 4) showed that pre-event for all catchment responses, the UAS applications were primarily concerned with digital elevation models for use in flood models, the condition of flood risk management assets, identification of safe shelter points and evacuation routes and providing warnings.

During the event applications providing information in real-time were prioritized. A combination of rapid visualization with high resolution of flood extent and flood depth were chosen. This can be provided with the current UAS technical capabilities. In fast response catchments, the participants' preference was for flow velocity with medium accuracy (instead of high accuracy). Additional time and costs are needed to achieve higher accuracies. Real-time applications relating to rescue activities (i.e., identification of safe shelter points, detection of stranded people and delivery of ad-hoc supplies) and damage assessment (i.e., visual detection of affected properties) were also identified as priorities. Participants stated that applications requiring more than 4 h of processing time to generate products are unlikely to be of use to responders in many flood events.

With the limitations of current UAS applications, the updating of evacuation routes was identified as being important only in catchments with a flood response longer than 12 h or where the duration of flooding in a faster responding catchment persists for more than 12 h. Applications that need more than 48 h of processing time, such as modelling flood extent and the identification of resilience and resistance measures, were still identified as being important as the data collected can be used subsequently to improve flood models and the response for future flood events.

During an event time is the priority, whereas after the event accuracy was most relevant. The focus in post event data collection was on the provision of more precise information for flood extent, flood depth and flood source so that flood impact can be estimated more accurately. After an event, there is also a continuing need for information that will assist with the rescue and recovery activities and the estimation of property level flood impacts.

3.4. Preferences in England for UAS Applications in Flood Emergency Response

The online survey evaluated the existing and desired processing time and accuracy in UAS applications for flood emergency response in England, as a way to determine whether technological development is needed to better inform emergency response. Results from the online survey indicated that 44% of participants currently have access to flood extent data within 12 h, with accuracies from 2 cm to 50 m. Only 3 of the 25 participants indicated they have access to flood extent data in less than 1 h with accuracies between 1 and 50 m. The preference of 52% of the participants was to have access to flood extent data within 0.5 h (i.e., near real time) with an accuracy of <10 m. There was also another significant group of participants (28%), who would seek to have access to flood extent data within 12 to 24 h with an accuracy of <10 m. Similarly, for flood depth and flow velocity respondents considered that having data more quickly with improved accuracy compared with the current products would be of benefit, with a desire for data to be available in <0.5 h with accuracies of 1 to 5 cm in flood depth and 0.1–0.5 m s^{-1} in flow velocity (Figure 5). There are current UAS technologies that are able to meet these requirements (Table 2).

Catchment flood response	Pre-event	During-event	Post-event
Fast	- Evacuation warning [<1h, flight path reach] — 7 - Flood source identification [>48h, 1m resolution] — 4 - Evacuation route identification [12-24h, 20-60% tracking accuracy] — 5 - Resistance and resilience measures identification [>48h, building level resolution] — 7 - Identification of safe shelter points [<1h, building level resolution] — 5	- Evacuation warning [<1h, flight path reach] — 7 - Flood extent [<1h, 0.2 m resolution] — 9 - Flow velocity [<1h, 0.15 m/s accuracy] — 6 - Flow velocity [<1h, 0.5 m/s accuracy] — 8 - Flood depth [<1h, high resolution] — 9 - Evacuation route identification [12-24h, 20-60% tracking accuracy] — 5 - Visual detection of affected properties [<1h, building level resolution] — 7 - Identification of safe shelter points [<1h, building level resolution] — 7 - Detection of stranded people [<1h, individual level resolution] — 6 - Delivery of ad-hoc supplies [<1h, building level resolution] — 8	- Flood extent [>48h, 0.2m resolution] — 5 - Flood depth [24-48h, 6 cm accuracy] — 4 - Flood source identification [>48h, 1 m resolution] — 7 - Surface changes and displacements of landslides [24-48h, 1 cm resolution] — 7 - Visual detection of affected properties [<1h, building level resolution] — 7 - Detection of stranded people [<1h, individual level resolution] — 4 - Delivery of ad-hoc supplies [<1h, building level resolution] — 5
Medium	- Evacuation warning [<1h, flight path reach] — 7 - Flood source identification [>48h, 1m resolution] — 4 - Evacuation route identification [12-24h, 20-60% tracking accuracy] — 5 - Resistance and resilience measures identification [>48h, building level resolution] — 6 - Identification of safe shelter points [<1h, building level resolution] — 5	- Evacuation warning [<1h, flight path reach] — 7 - Flood extent [<1h, 0.2 m resolution] — 9 - Flow velocity [<1h, 0.15 m/s accuracy] — 8 - Flow velocity [<1h, 0.5 m/s accuracy] — 7 - Flood depth [<1h, high resolution] — 9 - Flood source identification [>48h, 1 m resolution] — 6 - Evacuation route identification [12-24h, 20-60% tracking accuracy] — 7 - Visual detection of affected properties [<1h, building level resolution] — 8 - Identification of safe shelter points [<1h, building level resolution] — 8 - Detection of stranded people [<1h, individual level resolution] — 7 - Delivery of ad-hoc supplies [<1h, building level resolution] — 8	- Flood extent [>48h, 0.2m resolution] — 4 - Flood depth [24-48h, 6 cm accuracy] — 4 - Flood source identification [>48h, 1 m resolution] — 7 - Surface changes and displacements of landslides [24-48h, 1 cm resolution] — 8 - Visual detection of affected properties [<1h, building level resolution] — 7 - Detection of stranded people [<1h, individual level resolution] — 4 - Delivery of ad-hoc supplies [<1h, building level resolution] — 6
Slow	- Evacuation warning [<1h, flight path reach] — 6 - Flood source identification [>48h, 1 m resolution] — 4 - Evacuation route identification [12-24h, 20-60% tracking accuracy] — 5 - Resistance and resilience measures identification [>48h, building level resolution] — 7 - Identification of safe shelter points [<1h, building level resolution] — 5	- Evacuation warning [<1h, flight path reach] — 7 - Flood extent [<1h, 0.2 m resolution] — 8 - Flow velocity [<1h, 0.15 m/s accuracy] — 7 - Flow velocity [<1h, 0.5 m/s accuracy] — 6 - Flood depth [<1h, high resolution] — 9 - Flood source identification [>48h, 1m resolution] — 5 - Evacuation route identification [12-24h, 20-60% tracking accuracy] — 8 - Surface changes and displacements of landslides [24-48h, 1 cm resolution] — 4 - Visual detection of affected properties [<1h, building level resolution] — 7 - Identification of safe shelter points [<1h, building level resolution] — 7 - Detection of stranded people [<1h, individual level resolution] — 7 - Delivery of ad-hoc supplies [<1h, building level resolution] — 8	- Flood extent [>48h, 0.2m resolution] — 4 - Flood depth [24-48h, 6 cm accuracy] — 4 - Flood source identification [>48h, 1 m resolution] — 7 - Surface changes and displacements of landslides [24-48h, 1 cm resolution] — 8 - Visual detection of affected properties [<1h, building level resolution] — 7 - Detection of stranded people [<1h, individual level resolution] — 4 - Delivery of ad-hoc supplies [<1h, building level resolution] — 6

Emergency response

Figure 4. Respondents' preferences in England for each UAS application in relation to catchment response and UAS deployment at different emergency response phases (before, during, post). The numbers indicate the number of participants. There were 10 participants, and therefore the maximum score possible is 10. Only UAS applications with a score ≥ 4 are shown as a means of indicating the majority opinion.

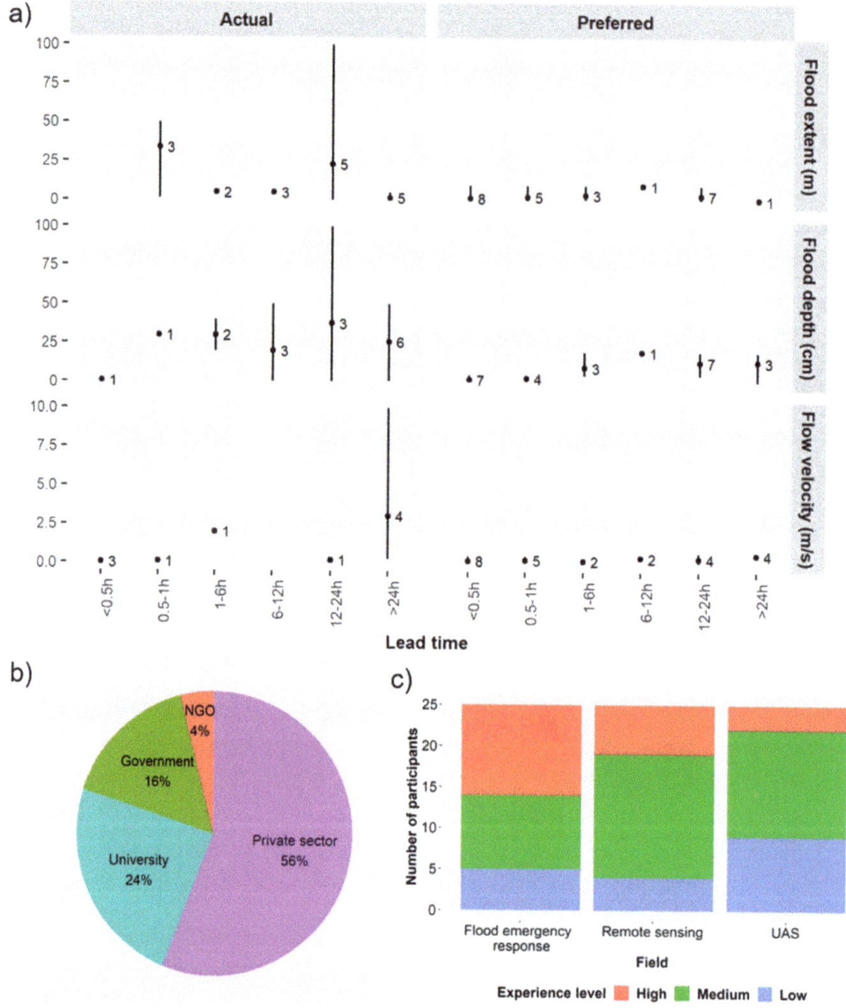

Figure 5. (**a**) Actual and preferred accuracy values and time needed to process flood extent, flood depth and flow velocity. The marker indicates the average of flood extent, flood depth and flow velocity, whereas the vertical lines shows the minimum and maximum values. The number of participants with a preference for a given combination are indicated against each measure. (**b**) Type of organizations that completed the survey. (**c**) Participants' experience level in flood emergency response, remote sensing and UAS. Results obtained from 25 participants.

For flood extent 13 participants considered time more important than accuracy when generating a product that will assist flood emergency response, whereas 12 participants thought accuracy was more important than time. Some participants stated that the most important factor for them was the trustworthiness of the data source. For flood depth (16) and flow velocity (17) participants were more interested in improved accuracy than in the time taken to obtain the data.

4. Discussion

4.1. A Purpose-Driven Approach to UAS Deployment in the Context of Flood Emergency Response

The operational use of UAS in flood emergency response is still limited [69]. A more systematic analysis of their application and capabilities in relation to their use in flood risk management including as part of an effective response to an event is, therefore, required if a purpose-driven approach to their deployment is to be realized.

During Storm Desmond in Cockermouth (Cumbria, England) in December 2015, more than 300 mm of rain fell over a 24 h period with an estimated <1% annual exceedance probability for both rainfall and river flows [70]. The Ministry for Housing, Communities and Local Government—Resilience and Emergencies Division (DCLG-RED), the Department for Environment, Food and Rural Affairs (Defra) and the Cabinet Office Briefing Room (COBR) were involved in coordinating the emergency response and supporting the local Cumbrian Strategic coordinating Group (SCG) [71]. UAS were used in Cockermouth during and after the storm to estimate the flood extent and identify impacted properties [72]. However, the use of UAS in Cockermouth could also have facilitated the identification of different types of flood sources (e.g., pluvial versus fluvial), as highlighted by [15]. In 2015 the range of applications UAS could be used for and how best to deploy them during flooding events had not been studied in a systematic way and, therefore, they were used in a reactive rather than strategically planned way.

Kerala is one of the Indian States that experiences the highest monsoon rainfall every year [73] and was affected by flooding in August 2018. The rain caused thousands of landslides in mountainous regions. Nearly 500 people died in the event [74]. Parts of the city of Cochin—the commercial capital of Kerala—were flooded, with a 90% increase in water cover and a water level rise of up to 5 m to 10 m [75]. As a result, major infrastructure assets including the airport, roads and railways had to be closed for safety reasons. The government issued evacuation orders and deployed the National Disaster Response Force teams within the area. During the emergency response over 223,000 people were evacuated to emergency relief camps [76]. UAS were used in Kerala to support rescue operations [77] and deliver aid [78]. There were also examples of people using UAS independently of the official response.

Although in both examples UAS were used in the response, the full capabilities were not necessarily exploited and the deployment of UAS was largely uncoordinated within the emergency response. The deployment analysis matrix developed here will enable those involved in flood risk management, including incident response, to take a structured approach to determining how best to use and deploy UAS within their specific context. The matrix-based approach will enable guidelines to be produced for the purpose-driven deployment of UAS within flood risk management activities including emergency response, as we discuss in the next section. This will help reduce duplication of effort and ensure the timely capture of important information that can be used to inform the current and future responses.

4.2. Guidelines for the Deployment of UAS within Flood Risk Management Activities Including Emergency Response

There are many benefits that can be derived from the use of UAS, to help reduce flood risk and the impacts on people, properties and the economy, if they are deployed in a structured and considered way that are currently not being fully utilized or exploited. The use of UAS has to be considered within the strategic planning for flood risk management activities including the response to flood events. This can build on experiences from the development of integrated flood forecasting, warning and response systems [79,80] and the use of real-time modelling to assist flood emergency response [81]. Our deployment matrix approach can be used as the basis for developing guidelines for the use of UAS within flood risk management before, during and post events. These guidelines are summarized in the following paragraphs.

Before a flood event:

- There is time to produce digital elevation and surface and terrain models which can then be used within flood models to estimate, for a given return period rainfall event, the likely extent, depth and velocity of flooding. The information produced can also be used to identify locations for temporary barriers, shelter points and evacuation routes.
- Flood models can be used to identify the likely fluvial, pluvial and coastal sources (e.g., [15]) and routes of flooding and can then enable potential evacuation routes to be identified (e.g., [82,83]).
- If flooding has been forecast for a particular area, UAS-based audio systems can be used to provide audible warnings to those at risk.

During a flood event:

- High levels of accuracy are often not a priority. The timeliness of the information being available to inform the response activities is paramount.
- Providing information in real-time is key as it enables effective prioritization of emergency response actions including: to identify where to deploy maintenance crews to deal with blockages and low spots in defences which are giving rise to unexpected sources of flooding; to identify whether the flooding is developing in line with modelled predictions in terms of extent and depth; to determine whether flooding is occurring in areas outside of those predicted by the models; to assess whether evacuation routes remain clear of flooding; to identify people and properties impacted by the flooding.
- During an event, typically time is limited to carry out new simulations [14]. During the event responders potentially identify gaps in the existing flood emergency plan (i.e., location of shelter points, evacuation routes, knowledge about resistance and resilience measures). However, in an event that exceeds the planned preparedness plan, it may be necessary to rerun the evacuation routes models.
- UAS can be used to determine the extent, depth and velocity of the flooding and the properties impacted. This information can be extremely valuable after the event to calibrate and refine the models and to determine whether additional flood risk reduction measures are required.
- Many organizations are involved in responding to flood events [81]. The use of UAS should be discussed well in advance of any need to deploy them including what information will be collected, how, by whom and for what purpose.
- UAS of the correct specification, including specified sensors and trained pilots will have to be available for deployment within an agreed standard of service.

After an event:

- UAS can be used to help determine the impact on people, properties and infrastructure, the flood extents and depths, and in identifying where best to deploy those still involved in rescue activities and in the recovery operations.
- UAS can be used to collect information that enable the calibration and validation of those models used pre-event for flood prediction purposes. Data collected at this stage will also look at features highlighting flood impact (e.g., properties affected). The EA with its strategic overview role for flood risk management in England would be well placed to develop guidelines for the use and deployment of UAS in this context and then to oversee their implementation.

4.3. Selecting the Correct UAS Platform

Multiple UAS platforms are available for use. The selection of the most appropriate platform for a particular application is a complex task. Factors that need to be considered include the capability of the gimbal to integrate the payload, weather conditions, the extent of the area to be flown, and the availability of pilots with the rights skills and regulatory permissions. UAS can be classified into vertical take-off-and-landing (VTOL), fixed wing and hybrids [84]. In Rivas et al. [85], the authors highlight that VTOL UAS are able to hover over a point and provide high resolution still imagery whereas fixed wing platforms enable wide area surveying [85,86].

The flooding of large areas, which will most likely occur in catchment areas with a slow response to floods, will require the use of fixed wing rather than VTOL platforms, as the former have longer endurance, although they are more difficult to fly and require specific training [87]. However, some fixed wing platforms do not have the capability of slowing down to speeds that enable them to collect high resolution imagery. Rivas Casado et al. [85] report 8 h to map the river channel of a 1.4 km reach, when using a rotor platform (Falcon 8 octocopter, ASCTEC, Krailling, Germany) whereas the same author reported a coverage of 142 ha within four hours in two flights undertaken with a Sirius Pro fixed wing [15]. Fixed wing platforms such as the Sirius Pro enable flights of up to 50 min at a cruising speed of 18 m s^{-1} [88]. In certain locations, VTOL UAS will also be required to overcome the limitations of terrain in terms of take-off and landing [19]. VTLOs can hover on site with high location accuracy and, therefore, take more detailed photographs at locations of interest. Hybrid models are able to combine the advantages of both fixed wing and VTOL platforms. The WingtraOne PPK VTO is a hybrid rotor and fixed wing platform able to provide Ground Sampling Distances of 0.7 cm/pixel and map 400 ha in a single flight [89]. The battery endurance is 55 min. The platform is able to fly under wind conditions of up to 45 km h^{-1} in cruise and up to 30 km h^{-1} for landing.

Battery endurance can also compromise performance. Overall, fixed wing platforms provide better battery endurance than VTOL platforms. Figure 6 and Table 3 show an alternative classification for UAS based on battery endurance and work range. Low-cost close range UAS include platforms with a range of generally up to 5 km and an endurance time of less than 45 min. Examples of such platforms include standard small VTOL platform such as the DJI Phantom 4 Pro (DJI Technology Co, Shenzhen, China) which can fly continuously for over 30 min [90] at a maximum speed of 20 m s^{-1}. Such platforms will be suitable to cover small catchments or specific areas of medium to large catchments. More expensive close-range platforms offer a working range of up to 50 km with battery endurance of up to 6 h. The mdMapper4-3000DµoG VHR VTLO (microdrones) is an example of such a platform able to capture RGB imagery at a ground sampling distance (GSD) of up to 0.6 cm/pixel [91]. The platform has an endurance of approximately 40 min when flying at an altitude of 120 m. The UAS can cover between 64 ha (80 mm lens, GSD = 0.6 cm/pixel) and 150 ha (35 mm lens, GSD = 1.3 cm/pixel) at a constant speed of 5 m s^{-1} within a single flight. A fixed wing example of a close-range platform is the Sirius Pro (Topcon Positioning System Inc., Livermore, CA, USA) which has a 50 min flight endurance and is able to operate under windy conditions (50 km h^{-1}) with gusts of up to 65 km h^{-1} [92]. Another example is the eBeeX which is able to gather RGB imagery at 1 cm/pixel and cover 220 ha in a single flight when flying at an altitude of 120 m. Short-range, medium-range and high-endurance platforms all require runways for their deployment and, although they provide a larger working range and endurance, are difficult to deploy in urban settings, especially during flood events.

In large flooded areas, there is likely to be a need to coordinate the deployment of multiple UAS within an affected area. The surveying of large areas will result in larger data sets and this will have a consequential impact on the time taken to generate products. More stable and perhaps heavier platforms, such as the microdrone md4-1000 [93], are needed for more extreme wind and rainfall conditions. Additional advances will enable the miniaturization of sensors, enhance the level of autonomy, increase battery life and the capability of flying in more extreme weather conditions.

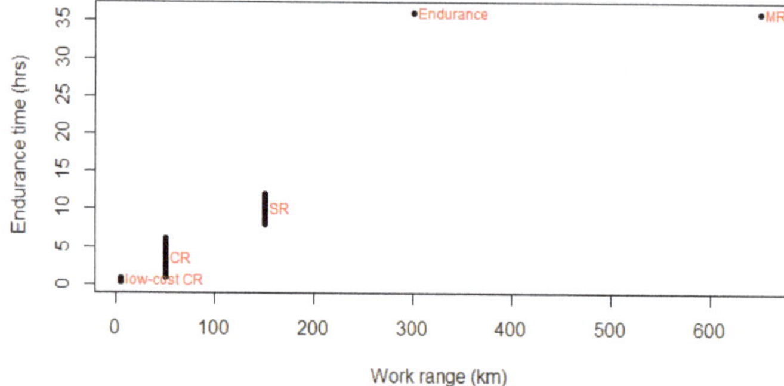

Figure 6. Simplified classification of UAS platforms based on their work range (km) and battery endurance time (h). The different classes include low-cost close-range (CR), close-range (CR), short-range (SR), medium-range (MR) and high-endurance (Endurance) platforms. A full description of these classes is provided in Table 3.

Table 3. Description of the simplified classification of UAS.

UAS Type	Range (km)	Endurance Time (h)	Remarks
Low cost close range	5	1/3—3/4	This type includes Micro and Nano air vehicles, low altitude flying with a maximum altitude of ca. 1000 m, no need for runways.
Close range	50	1—6	Need runways, altitude up to ca. 1500 m.
Short range	150	8—12	Need runways, altitude of a few thousand meters.
Mid-range	650	-	Need runways, altitude up to 9000 m.
Endurance	300	36	Need runways, altitude of 20,000 m or more.

4.4. The Deployment Decision Approach to an Indian Context

Our deployment analysis matrix approach for the use of UAS in flood risk management activities in England has the potential to be transferred to other countries (e.g., India) with different climatic, topographic and socioeconomic contexts. In India, environmental conditions can be extreme in terms of the intensity and extent of the rainfall. Transferability of the matrix will need to take into account the catchment response. In India, some catchments are of a larger scale than those in England and can be flooded for weeks with a need for recurrent monitoring of large areas. One of the challenges faced is the need to evacuate large numbers of people over extended areas in a short period [82]. The size of the areas affected will require the deployment of certain types of UAS (Section 4.3) for particular applications. In India, rural areas can also present access, travel time and maintenance challenges [82,94] with landslides limiting access to remote areas [95,96].

5. Conclusions

UAS are currently used in a largely ad hoc manner in flood risk management activities with practice differing significantly even within countries. Even so, their use has proved to be beneficial. However, if they were to be used in a purpose-driven and strategically coordinated way, they can provide more coherent and targeted information that will have added value for flood risk management activities, including during the response to events. The data and information produced by UAS can be used to improve flood risk management activities, structures, tools and approaches helping to reduce flooding and its associated impacts on people, properties, infrastructure and the economy.

We have identified a range of products that can be delivered by UAS and have developed an analysis matrix approach to help target their deployment. The UAS deployment matrix forms the basis for developing guidelines to assist those involved in flood risk management activities, including emergency responders, in developing a more strategic and targeted approach to the use of UAS before, during and after flood events. The approaches developed will need to be context specific including who will use what type of UAS and for what purpose before, during and after an event. The deployment matrix we have developed for England has the potential to be translated into an Indian context, and in other countries.

Further research should focus on exploring future technological developments of UAS platforms and sensors, their potential applications within a flood emergency response context and how these will feed into the existing deployment guidelines. Technological developments would be particularly helpful in the miniaturization of sensors, their integration on more stable UAS platforms and increased flight (i.e., battery) endurance. The fast pace of technological advances within the field of UAS requires a flexible and adaptive approach, which facilitates operational uptake as soon as advances are commercially available. The various organizations involved in the use of UAS in flood risk management will have to keep the deployment guidelines under review if they are to make the best use of the available and developing technologies to achieve flood management and resilience targets.

Supplementary Materials: The following are available online at http://www.mdpi.com/2073-4441/12/2/521/s1, Table S1: Organisations involved in the flood emergency response in England and India and their associated role.

Author Contributions: Conceptualization, G.S., P.L., M.R.C., M.M., D.B. and P.P.M.; methodology, G.S., M.R.C., P.L. and M.M.; validation, G.S.; formal analysis, G.S. and M.M.; investigation, G.S., M.R.C., P.L. and M.M.; resources, M.R.C., P.L., D.B. and P.P.M.; data curation, G.S. and M.M.; writing—original draft preparation, G.S., M.R.C. and P.L.; writing—review and editing, G.S., P.L., M.R.C., M.M., D.B. and P.P.M.; supervision, M.R.C. and P.L.; project administration, M.R.C.; funding acquisition, D.B., P.L., P.P.M. and M.R.C. All authors have read and agree to the published version of the manuscript.

Funding: This research was funded by EPSRC, EP/P02839X/1 "Emergency flood planning and management using unmanned aerial systems" and EPSRC EP/N010329/1 "BRIM: Building Resilience Into risk Management". Due to the ethically sensitivity nature of the research (interviews), supporting data cannot be made available.

Acknowledgments: The authors acknowledge the support from the following institutions and experts from England and India: Environment Agency, Central Water Commission, National Institute Disaster Management, International Federation of Red Cross, Goonj, Seeds, R.K. Dave, and Ruchi Saxena.

Conflicts of Interest: The authors declare no conflict of interest.

References

1. U.S. House of Representatives. *A Failure of Initiative. Final Report of the Selected Bipartisan Committee to Investigate the Preparation for and Response to Hurricane Katrina*; U.S. House of Representatives: Washington, DC, USA, 2006.
2. Hallegatte, S. An adaptive regional input-output model and its application to the assessment of the economic cost of Katrina. *Risk Anal.* **2008**, *28*, 779–799. [CrossRef] [PubMed]
3. *AON Global Catastrophe Recap: October 2018*; AON Empower Results: London, UK, 2018. Available online: http://thoughtleadership.aonbenfield.com/Documents/20181107-ab-analytics-if-oct-global-recap.pdf (accessed on 11 February 2020).
4. Venkataraman, A.; Suhasinini, R.; Abi-Habib, M. After Worst Kerala Floods in a Century, India Rejects Foreign Aid. Available online: https://www.nytimes.com/2018/08/23/world/asia/india-kerala-floods-aid-united-arab-emirates.html (accessed on 24 November 2019).
5. Environment Agency. *Managing Flood and Coastal Erosion Risks in England: 1 April 2011 to 31 March 2017*; Environment Agency: Bristol, UK, 2018. Available online: https://www.gov.uk/government/publications/flood-and-coastal-risk-management-national-report (accessed on 11 February 2020).
6. Paciarotti, C.; Cesaroni, A.; Bevilacqua, M. The management of spontaneous volunteers: A successful model from a flood emergency in Italy. *Int. J. Dis. Risk Reduct.* **2018**, *31*, 260–274. [CrossRef]
7. *Defra the National Flood Emergency Framework for England. December 2014*; Department for Environment, Food and Rural Affairs: London, UK, 2014.

8. *NDMA National Disaster Management Plan (NDMP)*; National Disaster Management Authority Government: New Delhi, India, 2016. Available online: https://ndma.gov.in/images/policyplan/dmplan/National%20Disaster%20Management%20Plan%20May%202016.pdf (accessed on 11 February 2020).

9. WMO; US NOAA; US NWS; HRC. Proceedings of the USAID/OFDA First Steering Committee Meeting (SCM 1) of The South Asia Flash Flood Guidance (SAsiaFFG) Project, New Delhi, India, 26–28 April 2016; p. 29.

10. CWC Flood Forecast. Central Water Commission. Available online: http://14.143.182.4/ffs/ (accessed on 28 October 2019).

11. Oleo, M. *Developing Data Standards for Flood Models*; Cranfield University: Cranfield, UK, 2018.

12. Neelz, S.; Pender, G. *Benchmarking the Latest Generation of 2D Hydraulic Modelling Packages*; Environment Agency: Bristol, UK, 2013; ISBN 0234-1026.

13. Evans, E.P.; Simm, J.D.; Thorne, C.R.; Arnell, N.W.; Ashley, R.M.; Hess, T.M.; Lane, S.N.; Morris, J.; Nicholls, R.J.; Penning-Rowsell, E.C.; et al. *An Update of the Foresight Future Flooding 2004 Qualitative Risk Analysis*; An independent review by Sir Michael Pitt; Cabinet Office: London, UK, 2008.

14. Ward, P.J.; Jongman, B.; Salamon, P.; Simpson, A.; Bates, P.; De Groeve, T.; Muis, S.; De Perez, E.C.; Rudari, R.; Trigg, M.A.; et al. Usefulness and limitations of global flood risk models. *Nat. Clim. Chang.* **2015**, *5*, 712–715. [CrossRef]

15. Rivas Casado, M.; Irvine, T.; Johnson, S.; Palma, M.; Leinster, P. The use of unmanned aerial vehicles to estimate direct tangible losses to residential properties from flood events: A case study of Cockermouth Following the Desmond Storm. *Remote Sens.* **2018**, *10*, 1548. [CrossRef]

16. *UN-OCHA Unmanned Aerial Vehicles in Humanitarian Response*; United Nations Office for the Coordination of Humanitarian Affairs: New York, NY, USA, 2014.

17. Kim, K.; Davidson, J. Unmanned Aircraft Systems Used for Disaster Management. *Transp. Res. Rec. J. Transp. Res. Board* **2015**, *2532*, 83–90. [CrossRef]

18. Fernandes, O.; Murphy, R.; Adams, J.; Merrick, D. Quantitative Data Analysis: CRASAR Small Unmanned Aerial Systems at Hurricane Harvey. In Proceedings of the 2018 IEEE International Symposium on Safety, Security, and Rescue Robotics (SSRR), Philadelphia, PA, USA, 6–8 August 2018; pp. 1–6.

19. Watts, A.C.; Ambrosia, V.G.; Hinkley, E.A. Unmanned aircraft systems in remote sensing and scientific research: Classification and considerations of use. *Remote Sens.* **2012**, *4*, 1671–1692. [CrossRef]

20. Reliefweb. Drones Used for Good-Relief Organisation Uses Drones to Map Typhoon Haiyan Recovery Efforts. Available online: https://reliefweb.int/report/philippines/drones-used-good-relief-organisation-uses-drones-map-typhoon-haiyan-recovery (accessed on 25 April 2019).

21. Santos, L.A. In the Philippines, Drones Provide Humanitarian Relief. Available online: https://www.devex.com/news/in-the-philippines-drones-provide-humanitarian-relief-82512 (accessed on 31 May 2019).

22. Mathew, A. Up, up and very Far away: Remote Sensing in Flood Response. Available online: https://defradigital.blog.gov.uk/2017/03/30/up-up-and-very-far-away-remote-sensing-in-flood-response/ (accessed on 18 April 2019).

23. Schumann, G.J.-P.; Muhlhausen, J.; Andreadis, K.M. Rapid Mapping of Small-Scale River-Floodplain Environments Using UAV SfM Supports Classical Theory. *Remote Sens.* **2019**, *11*, 982. [CrossRef]

24. DeBell, L.; Anderson, K.; Brazier, R.E.; King, N.; Jones, L. Water resource management at catchment scales using lightweight UAVs: Current capabilities and future perspectives. *J. Unmanned Veh. Syst.* **2016**, *4*, 7–30. [CrossRef]

25. Dale, M.; Dempsey, P.; Dent, J. *Defra/Environment Agency Flood and Coastal Defence R & D Programme*; Extreme Rainfall Event Recognition Phase 2 Work Package 5: Establishing a user requirement for a decision-support tool; R & D Technical Report-FD2208; 2005; pp. 1–20.

26. Kjeldsen, T.R. *Flood Estimation Handbook Supplementary Report No. 1 the Revitalised FSR/FEH Rainfall-Runoff Method*; Centre for Ecology and Hydrology: Wallingford, UK, 2007.

27. Erdelj, M.; Król, M.; Natalizio, E. Wireless Sensor Networks and Multi-UAV systems for natural disaster management. *Comput. Netw.* **2017**, *124*, 72–86. [CrossRef]

28. Diakakis, M.; Andreadakis, E.; Nikolopoulos, E.I.; Spyrou, N.I.; Gogou, M.E.; Deligiannakis, G.; Katsetsiadou, N.K.; Antoniadis, Z.; Melaki, M.; Georgakopoulos, A.; et al. An integrated approach of ground and aerial observations in flash flood disaster investigations. The case of the 2017 Mandra flash flood in Greece. *Int. J. Dis. Risk Reduct.* **2018**, *33*, 290–309. [CrossRef]

29. Marchi, L.; Borga, M.; Preciso, E.; Gaume, E. Characterisation of selected extreme flash floods in Europe and implications for flood risk management. *J. Hydrol.* **2010**, *394*, 118–133. [CrossRef]

30. Environment Agency Flooding in England: A National Assessment of Flood Risk. *Environ. Agency* **2009**, 5–32.

31. Smith, M.W.; Carrivick, J.L.; Hooke, J.; Kirkby, M.J. Reconstructing flash flood magnitudes using "Structure-from-Motion": A rapid assessment tool. *J. Hydrol.* **2014**, *519*, 1914–1927. [CrossRef]

32. Restas, A. Drone Applications for Supporting Disaster Management. *World J. Eng. Technol.* **2015**, *3*, 316–321. [CrossRef]

33. *Defra Flood Risks to People-Phase 2-FD2321/TR2 Guidance Document*; Department for Environment Food and Rural Affairs: London, UK, 2006.

34. Coles, D.; Yu, D.; Wilby, R.L.; Green, D.; Herring, Z. Beyond 'flood hotspots': Modelling emergency service accessibility during flooding in York, UK. *J. Hydrol.* **2017**, *546*, 419–436. [CrossRef]

35. *Defra Flood and Water Management Act 2010*; Department for Environment Food and Rural Affairs: London, UK, 2010.

36. Tooth, J.-P. Storms & Emergencies: How A UK Military Response is Decided. Available online: https://www.forces.net/news/storms-emergencies-how-uk-military-response-decided (accessed on 24 November 2019).

37. Defra Exercise Tempest Tests the Environment Agency flood Response ahead of Winter. Available online: https://www.gov.uk/government/news/exercise-tempest-tests-the-environment-agency-flood-response-ahead-of-winter (accessed on 28 October 2019).

38. District Disaster Management Authority Madhubani. *District Disaster Management Plan-Madhubani*; District Disaster Management Authority Madhubani: Madhubani, Bihar, 2013.

39. Tripathi, P. Flood Disaster in India: An Analysis of trend and Preparedness. *Interdiscip. J. Contemp. Res.* **2015**, *2*, 91–98.

40. Gusain, R. Drones Come to NDRF's Rescue in Locating Stranded People in Kedarnath. Available online: https://www.indiatoday.in/india/north/story/national-disaster-response-force-takes-help-of-drones-to-locate-stranded-people-in-kedarnath-169487-2013-07-08 (accessed on 24 September 2019).

41. Sethi, N. Drones Scan Flood-Hit Uttarakhand. Available online: https://www.livemint.com/Politics/ZDib5YWR1G2Mcuth1kbwyO/Drones-scan-floodhit-Uttarakhand.html (accessed on 22 November 2019).

42. ENS UAVs Look for Survivors as New Landslides, Rain Hamper Rescue. Available online: http://www.newindianexpress.com/nation/2013/jun/25/UAVs-look-for-survivors-as-new-landslides-rain-hamper-rescue-490250.html (accessed on 24 September 2019).

43. Hoyle, C. Pictures: How Netra UAVs Helped Indian Disaster Relief Effort. Available online: https://www.flightglobal.com/news/articles/pictures-how-netra-uavs-helped-indian-disaster-relief-effort-387984/ (accessed on 24 September 2019).

44. Popescu, D.; Ichim, L.; Caramihale, T. Flood areas detection based on UAV surveillance system. In Proceedings of the 2015 19th International Conference on System Theory, Control and Computing (ICSTCC), Cheile Gradistei, Romania, 14–16 October 2015; pp. 753–758.

45. Popescu, D.; Ichim, L.; Stoican, F. Unmanned aerial vehicle systems for remote estimation of flooded areas based on complex image processing. *Sensors* **2017**, *17*, 446. [CrossRef]

46. Zhang, Y.; Canisius, F.; Guindon, B.; Feng, B.; Zhang, Y.; Canisius, F.; Guindon, B.; Zhen, C.; Feng, B. Effectiveness of RGB imagery from diverse sources for real-time urban flood water mapping. In Proceedings of the SPIE 10793, Remote Sensing Technologies and Applications in Urban Environments III, Berlin, Germany, 9 October 2018.

47. Gebrehiwot, A.; Hashemi-Beni, L.; Thompson, G.; Kordjamshidi, P.; Langan, T.E. Deep convolutional neural network for flood extent mapping using unmanned aerial vehicles data. *Sensors* **2019**, *19*, 1486. [CrossRef]

48. Şerban, G.; Rus, I.; Vele, D.; Breţcan, P.; Alexe, M.; Petrea, D. Flood-prone area delimitation using UAV technology, in the areas hard-to-reach for classic aircrafts: Case study in the north-east of Apuseni Mountains, Transylvania. *Nat. Hazards* **2016**, *82*, 1817–1832. [CrossRef]

49. Heimhuber, V.; Hannemann, J.C.; Rieger, W. Flood risk management in remote and impoverished areas-a case study of Onaville, Haiti. *Water* **2015**, *7*, 3832–3860. [CrossRef]

50. Abdelkader, M.; Shaqura, M.; Claudel, C.G.; Gueaieb, W. A UAV based system for real time flash flood monitoring in desert environments using Lagrangian microsensors. In Proceedings of the 2013 International Conference on Unmanned Aircraft Systems (ICUAS), Atlanta, GA, USA, 28–31 May 2013; pp. 25–34.

51. Thumser, P.; Haas, C.; Tuhtan, J.A.; Fuentes-Pérez, J.F.; Toming, G. RAPTOR-UAV: Real-time particle tracking in rivers using an unmanned aerial vehicle. *Earth Surf. Process. Landf.* **2017**, *42*, 2439–2446. [CrossRef]

52. Tauro, F.; Petroselli, A.; Arcangeletti, E. Assessment of drone-based surface flow observations. *Hydrol. Process.* **2016**, *30*, 1114–1130. [CrossRef]

53. Le Coz, J.; Patalano, A.; Collins, D.; Guillén, N.F.; García, C.M.; Smart, G.M.; Bind, J.; Chiaverini, A.; Le Boursicaud, R.; Dramais, G.; et al. Crowdsourced data for flood hydrology: Feedback from recent citizen science projects in Argentina, France and New Zealand. *J. Hydrol.* **2016**, *541*, 766–777. [CrossRef]

54. Fujita, I.; Notoya, Y.; Tani, K.; Tateguchi, S. Efficient and accurate estimation of water surface velocity in STIV. *Environ. Fluid Mech.* **2018**, *19*, 1363–1378. [CrossRef]

55. Tsuji, I.; Tani, K.; Fujita, I.; Notoya, Y. Development of Aerial Space Time Volume Velocimetry for Measuring Surface Velocity Vector Distribution from UAV. In Proceedings of the E3S Web of Conferences- River Flow 2018 - Ninth International Conference on Fluvial Hydraulics, Lyon-Villeurbanne, France, 5–8 September 2018; Volume 40, p. 06011.

56. Srikudkao, B.; Khundate, T.; So-In, C.; Horkaew, P.; Phaudphut, C.; Rujirakul, K. Flood Warning and Management Schemes with Drone Emulator Using Ultrasonic and Image Processing. In *Recent Advances Technology 2015*; Unger, H., Meesad, P., Boonkrong, S., Eds.; Springer: Cham, Switzerland, 2015; pp. 107–116. ISBN 9783319190235.

57. Muthusamy, M.; Rivas Casado, M.; Salmoral, G.; Irvine, T.; Leinster, P. A Remote Sensing Based Integrated Approach to Quantify the Impact of Fluvial and Pluvial Flooding in an Urban Catchment. *Remote Sens.* **2019**, *11*, 577. [CrossRef]

58. Chiabrando, F.; Piras, M.; Aicardi, I.; Vigna, B.; Noardo, F.; Lingua, A.M. A methodology for acquisition and processing of thermal data acquired by UAVs: A test about subfluvial springs' investigations. *Geomatics, Nat. Hazards Risk* **2016**, *8*, 5–17.

59. Maher, A.; Inoue, M. Generating Evacuation Routes by Using Drone System and Image Analysis To Track Pedestrian and Scan the Area After Disaster Occurrence. In Proceedings of the 10th SEATUC (South East Asian Technical University Consortium) Conference, Tokyo, Japan, 22–24 February 2016.

60. Lucieer, A.; De Jong, S.M.; Turner, D. Mapping landslide displacements using Structure from Motion (SfM) and image correlation of multi-temporal UAV photography Mapping landslide displacements using Structure from Motion (SfM) and image correlation of multi-temporal UAV photography. *Prog. Phys. Geogr.* **2013**, *38*, 97–116. [CrossRef]

61. Scaioni, M.; Longoni, L.; Melillo, V.; Papini, M. Remote Sensing for Landslide Investigations: An Overview of Recent Achievements and Perspectives. *Remote Sens.* **2014**, *6*, 9600–9652. [CrossRef]

62. Ohshimo, S.; Sadamori, T.; Shime, N. The western Japan chaotic rainstorm disaster: A brief report from Hiroshima. *J. Intensive Care* **2018**, *6*, 4–6. [CrossRef]

63. Anand, J.; Ramachandran, U. *Role of Various Sectors in Demonstrating Resilience during Chennai Flood 2015*; Asian Cities Climate Change Network; Taru Leading Edge: Ahmedabad, New Delhi, India, 2016.

64. Decorah Newspapers. Decorah Fire Department Drone Aids in Allamakee County River Rescue. Available online: https://decorahnewspapers.com/Content/News/Lead-Stories/Article/Decorah-Fire-Department-drone-aids-in-Allamakee-County-river-rescue/2/13/43505 (accessed on 5 June 2019).

65. DJI. *More Lives Saved: A Year of Drone Rescues Around The World*; DJI: Shenzhen, China, 2018.

66. Watson, F. Drone Used to Help Emergency Crews for the First Time in the Ozarks. Available online: https://www.kspr.com/content/news/Drone-used-to-help-emergency-crews-for-the-first-time-in-the-Ozarks-478245203.html (accessed on 5 July 2019).

67. Sharma, S. *Flood-Survivors Detection Using IR Imagery on an Autonomous Drone*; Stanford University: Stanford, CA, USA, 2017; pp. 8–12.

68. Gallegos, H.A.; Schubert, J.E.; Sanders, B.F. Structural Damage Prediction in a High-Velocity Urban Dam-Break Flood: Field-Scale Assessment of Predictive Skill. *J. Eng. Mech.* **2012**, *138*, 1249–1262. [CrossRef]

69. Yu, M.; Yang, C.; Li, Y. Big Data in Natural Disaster Management: A Review. *Geosciences* **2018**, *5*, 165. [CrossRef]

70. McCall, I.; Evans, C.; Cockermouth, S. *19 Flood Investigation Report*; Environment Agency, Cumbria County Council: Penrith, UK, 2016.

71. Deeming, H. *Project: Understanding Extreme Events as Catalysts for Flood-Risk Management Policy Change: A Case Study of the Impact of "Storm Desmond" in Cumbria, UK*; HD Research: Bentham, UK, 2016; pp. 1–27.

72. Krol, C. Storm Desmond: Aerial Footage Shows Extent of Flooding Damage in Cumbria. Available online: https://www.telegraph.co.uk/news/weather/12036038/Drone-footage-shows-flooding-in-Carlisle.html (accessed on 2 October 2019).

73. Pal, I.; Al-Tabbaa, A. Trends in seasonal precipitation extremes-An indicator of "climate change" in Kerala, India. *J. Hydrol.* **2009**, *367*, 62–69. [CrossRef]

74. Padma, T.V. Kerala floods made worse by mining and dams. *Nature* **2018**, *561*, 13–14. [CrossRef] [PubMed]

75. Vishnu, C.L.; Sajinkumar, K.S.; Oommen, T.; Coffman, R.A.; Thrivikramji, K.P.; Rani, V.R.; Keerthy, S. Satellite-based assessment of the August 2018 flood in parts of Kerala, India. *Geomat. Nat. Hazards Risk* **2019**, *10*, 758–767. [CrossRef]

76. BBC. Kerala Floods: Troops Rush in to Help Rescue Efforts. Available online: https://www.bbc.co.uk/news/world-asia-india-45231222 (accessed on 2 October 2019).

77. Putrevu, S. Here's how these Drone Startups can help in Rescue Operations in Bihar Floods. Available online: https://yourstory.com/2019/09/bihar-floods-drone-startups (accessed on 2 October 2019).

78. Manorama. Drones to Deliver Aid to Marooned Tribal Colonies in Idukki. Available online: https://english.manoramaonline.com/news/kerala/2018/08/20/drones-for-idukki-rescue-works.html (accessed on 2 October 2019).

79. Parker, D.; Fordham, M. An evaluation of flood forecasting, warning and response systems in the European Union. *Water Resour. Manag.* **1996**, *10*, 279–302. [CrossRef]

80. Plessis, L. Du A review of effective flood forecasting, warning and response system for application in South Africa. *Water SA* **2002**, *28*, 129–138. [CrossRef]

81. Maidment, D.R. Conceptual Framework for the National Flood Interoperability Experiment. *J. Am. Water Resour. Assoc.* **2017**, *53*, 245–257. [CrossRef]

82. Gupta, A.; Sarda, N.L. Efficient Evacuation Planning for Large Cities. In *Database and Expert Systems Applications. 25th International Conference, DEXA 2014 Munich, Germany, September 1–4, 2014 Proceedings, Part I*; Decker, H., Lhotská, L., Link, S., Spies, M., Wagner, R.R., Eds.; Springer: Cham, Switzerland, 2016; ISBN 9783319444024.

83. Das, S. Evaluating climate change adaptation through evacuation decisions: A case study of cyclone management in India. *Clim. Chang.* **2019**, *152*, 291–305. [CrossRef]

84. Singhal, G.; Bansod, B.; Mathew, L. Unmanned Aerial Vehicle classification, Applications and challenges: A Review. *Preprint* **2018**. [CrossRef]

85. Rivas Casado, M.; Ballesteros Gonzalez, R.; Kriechbaumer, T.; Veal, A. Automated identification of river hydromorphological features using UAV high resolution aerial imagery. *Sensors* **2015**, *15*, 27969–27989. [CrossRef]

86. Zhang, C.; Kovacs, J.M. The application of small unmanned aerial systems for precision agriculture: A review. *Precis. Agric.* **2012**, *13*, 693–712. [CrossRef]

87. Chapman, A. Types of Drones: Multi-Rotor vs. Fixed-Wing vs. Single Rotor vs. Hybrid VTOL. Available online: https://www.auav.com.au/articles/drone-types/ (accessed on 24 November 2019).

88. Topcon. *Topcon Sirius Unmanned Aerial Solution*; Topcon: Tokyo, Japan, 2015.

89. Wingtra. *WingtraOne – Technical specifications*; Wingtra AG: Zürich, Switzerland, 2018. Available online: https://wingtra.com/wp-content/uploads/Wingtra-Technical-Specifications.pdf (accessed on 12 February 2020).

90. Flynt, J. 11 Best Long Flight Time Drones. Available online: https://3dinsider.com/long-flight-time-drones/ (accessed on 5 November 2019).

91. Microdrones. *Product Line Up: Fully Integrated Systems for Professionals*; Microdrones: Ney York, NY, USA, 2020. Available online: https://cdn.microdrones.com/fileadmin/web/_downloads/brochures/english/2020_Brochure_mdSOLUTIONS_LETTER_EN_DL.pdf (accessed on 12 February 2020).

92. Topcon. Sirius Pro Specifications. Available online: https://www.topconpositioning.com/mass-data-collection/aerial-mapping/sirius-pro#panel-product-specifications (accessed on 19 January 2020).

93. Fernández-Hernandez, J.; González-Aguilera, D.; Rodríguez-Gonzálvez, P.; Mancera-Taboada, J. Image-Based Modelling from Unmanned Aerial Vehicle (UAV) Photogrammetry: An Effective, Low-Cost Tool for Archaeological Applications. *Archaeometry* **2015**, *57*, 128–145. [CrossRef]
94. Jain, M.; Korzhenevych, A.; Hecht, R. Determinants of commuting patterns in a rural-urban megaregion of India. *Transp. Policy* **2018**, *68*, 98–106. [CrossRef]
95. Hearn, G.J.; Shilston, D.T. Terrain geohazards and sustainable engineering in Ladakh, India. *Q. J. Eng. Geol. Hydrogeol.* **2017**, *50*, 231–238. [CrossRef]
96. Hearn, G.J.; Shakya, N.M. Engineering challenges for sustainable road access in the Himalayas. *Q. J. Eng. Geol. Hydrogeol.* **2017**, *50*, 69–80. [CrossRef]

Article

Analysis of the Public Flood Risk Perception in a Flood-Prone City: The Case of Jingdezhen City in China

Zhiqiang Wang [1,2,3,*], Huimin Wang [1,3,*], Jing Huang [1,3], Jinle Kang [1,3] and Dawei Han [2]

[1] State Key Laboratory of Hydrology-Water Resources and Hydraulic Engineering, Hohai University, Nanjing 210098, China; j_huang@hhu.edu.cn (J.H.); kjlhhu@hhu.edu.cn (J.K.)

[2] Department of Civil Engineering, Water and Environmental Management Research Centre, University of Bristol, Bristol BS8 1TR, UK; d.han@bristol.ac.uk

[3] Institute of Management Science, Business School, Hohai University, Nanjing 211100, China

* Correspondence: zqwang@hhu.edu.cn (Z.W.); hmwang@hhu.edu.cn (H.W.); Tel.: +86-025-83787221 (H.W.)

Received: 13 October 2018; Accepted: 1 November 2018; Published: 4 November 2018

Abstract: Understanding and improving public flood risk perception is conducive to the implementation of effective flood risk management and disaster reduction policies. In the flood-prone city of Jingdezhen, flood disaster is one of the most destructive natural hazards to impact the society and economy. However, few studies have been attempted to focus on public flood risk perception in the small and medium-size city in China, like Jingdezhen. Therefore, the purpose of this study was to investigate the public flood risk perception in four districts of Jingdezhen and examine the related influencing factors. A questionnaire survey of 719 randomly sampled respondents was conducted in 16 subdistricts of Jingdezhen. Analysis of variance was conducted to identify the correlations between the impact factors and public flood risk perception. Then, the flood risk perception differences between different groups under the same impact factor were compared. The results indicated that the socio-demographic characteristics of the respondents (except occupation), flood experience, flood knowledge education, flood protection responsibility, and trust in government were strongly correlated with flood risk perception. The findings will help decision makers to develop effective flood risk communication strategies and flood risk reduction policies.

Keywords: flood risk perception; natural flood management; disaster mitigation; flood-prone city; questionnaire survey

---

1. Introduction

Natural disasters are a major threat to the social and economic structure and they can easily wipe out the wealth accumulated in the past. In the future, flood risk is projected to increase in many regions due to effects of climate change and an increased concentration of people and economic properties [1,2]. Besides, the natural disaster frequency appears to be increasing in recent years [3], and the threat to development and economic losses from flood disasters are increasing too. Although many efforts have been done to reduce the risk and damage from natural disasters, floods remain the most devastation natural hazard in the world (World Bank, 2012). In 2017, Emergency Events Database (EM-DAT) data showed 318 natural disasters in the world, affecting 122 countries. These disasters resulted in 9503 deaths, 96 million people affected, and \$314 billion in economic losses. Among them, nearly 60% of the population affected by the disaster in 2017 were affected by the flood. Similar to previous years, China was the most disaster-affected country, with 25 events (c): 15 floods/landslides and 6 storms. During 2006–2015, flood disasters in China killed 6641 people, affected about one half billion people, and caused more than US\$87.5 billion damage (https://www.cred.be).

In order to study and reduce the negative impact of flood disasters on society and economy, researchers began to pay attention to flood risk assessment and flood risk management. Many researchers studied the objective flood risk, such as flood occurrence probability, flood inundation, and economic loss based on risk perspective. Other researchers believe that the subjective factors of the individuals can influence the judgment of the objective flood disaster risk. One of the important factors is the individuals' flood risk perception, and it has become an important topic to policy makers that are concerned with flood risk management [4]. Generally, risk perception refers to people's beliefs, attitudes, judgements and feelings towards events, and researchers believe that flood risk perception is the direct cause of flood risk prevention awareness and response behaviors [5–7]. Studying people's risk perception level is conducive to the implementation of effective flood risk management and disaster reduction policies, which has very important practical significance [8,9]. This is because:

1. People's behaviors are influenced by their risk attitudes towards the event;
2. People with different characteristics have different attitudes toward the same kind of event, and this difference can be useful for improving flood risk control and management;
3. Existing flood control engineering measures can reduce the real flood risk, but human behaviors is irrational, and their understanding of things is not sufficient, which can easily lead to behavioral deviation. It is difficult to achieve the desired results by only using technical means to reduce the risk of flooding;
4. Residents are both victims of disasters and executors of flood disaster prevention and mitigation policies. Studying their flood risk perception is helpful to understand their attitudes towards policies and possible behaviors.

Actually, research on risk perception began in the 1940s, when Gilbert White published the human adjustments to floods in the United States [10]. White found that people's behaviors could be directly affected by their previous flood experience, which created a precedent for study on human dimensions of risk in a multi-hazard environment [11,12]. The key early paper about public risk perception was written in 1960s by Chauncey Starr. Starr explored the correlation between the social acceptance of technological risks and the perception of social benefits and justified social costs from these technologies [13,14]. This method of revealed preference influenced the subsequent research. In 1978, Fischhoff and Slovic first proposed the use of psychometrics to assess risk. They used a scaled questionnaire of expressed preferences to directly capture people's different perceptions of risk and benefits, responding to limitation of Starr's revealed preference [15]. In the flood disasters area, relevant studies are based on their research, using psychological experiments and social surveys to assess people's flood risk perceptions. These studies on flood risk perception are mainly:

1. Study the risk perception of different subjects. These subjects include ordinary people [16,17], rural households [2,18], farmers [19,20], students [21], and tourists [4].
2. Identify and analyze a number of impact factors of flood risk perception. These impact factors include perception about the cause of hazard [9,22–25], previous flood experience [9,26–29], respondent's demographics [27,28], geographic location [28,30], residential history [28,31], the perception of responsibilities [32–34], trust in local government or institutions [27,33], different information source [30,35,36], and the public knowledge of risk mitigation actions [28,33,35].
3. Study the application of flood risk perception in actual flood disaster reduction [27,33,37,38]. For example, Daniel, etc. found the households' choice to purchase flood insurance was positively and significantly correlated with risk preference data and subjective risk perception data [38]. Paul, etc. studied the flood risk perception and implications for flood risk management such as flood protection plans in the Netherlands [37].
4. Study different risk perception measurement paradigms and indicators of risk perception. Psychometric paradigm and heuristics were the influential and popular theoretical framework in risk perception research [15,39,40]. With regard to the indicators of risk perception, Miceli indicate that flood risk perception encompasses both cognitive (likelihood, knowledge, etc.) and

affective (feelings, perceived control, etc.) aspects. Most studies use a different set of items to measure the flood risk perception. However, Gotham [41] and Horney [42] measured the risk perception with only one item or question.

Although China is a country with a long history of flood disasters, few studies have assessed the flood risk perception and analyzed the influence factors of flood risk perception. Kellens reviewed 57 studies of flood risks perception in leading journals and found the study of flood risk perception is mainly in the Western world (more than 40 studies) [43]. Related research in China is still in its infancy, and the published literature is very limited. Besides, the current study areas were usually rural area or the area near the river, less concerned about the differences of flood risk perception in different regions of the same small and medium-sized and flood prone city. Given the lack of research of flood risk perception in China and the significance and important role of risk perception research in flood risk management, the aims of this study are

1.   to evaluate the public flood risk perception in Jingdezhen City and compare the flood risk perception differences between the different districts,
2.   identify the key impact factors of the flood risk perception of respondents,
3.   to examine the influence of the impact factors of the flood risk perception, and
4.   to discuss the recommendations for the flood risk reduction measures based on the public flood risk perception.

2. Materials and Methods

2.1. Study Area

The study was undertaken in the Jingdezhen City as it is seriously affected by floods almost every year, causing huge economic losses and wide impacts. At the same time, Jingdezhen City is a typical small and medium-sized city in China, with rapid social and economic development. The local economic structure is more fragile than the bigger cities, and the structure and function of the river network can be more easily damaged.

In addition, the Jingdezhen government is also actively developing Integrated Flood Risk Management Plan to reduce the flood risk and it is expected to provide reference for other small and medium-sized cities in China. The concept of the plan is fully realized by the transition from flood control to flood risk management, combining engineering and non-engineering measures to form a comprehensive flood control and disaster reduction system for cities and meeting the needs of the whole society for water security. Therefore, it is very practical to choose Jingdezhen City as a research area. This study is one part of the proposed Integrated Flood Risk Management in Jingdezhen City.

Jingdezhen is located in the northeast of Jiangxi Province, China, and it belongs to the transition zone between the extension of Huangshan Mountain, Huaiyu Mountain and Poyang Lake Plain (as shown in Figure 1). It lies between $116°57'–117°42'$ E longitude and $28°44'–29°56'$ N latitude. Jingdezhen City covers a total area of 5256.23 km^2 and governs two counties and two districts, namely Leping County (1982.76 km^2), Fuliang County (2580.84 km^2), the Changjiang District (391.83 km^2), and Zhushan District (30.80 km^2). The agricultural land, construction land, and unused land in Jingdezhen City are 4754.47 km^2, 355.9 km^2 and 147.8 km^2, respectively, accounting for 90.4%, 6.8%, and 2.8% of the total area of the city. The highest and lowest elevations in Jingdezhen are 1618 m and 20 m, respectively, with plains on the southern part having an average altitude of 200 m. This region is characterized by a subtropical monsoon climate, with abundant sunshine and rainfall. The annual average temperature is 17 °C and the annual average sunshine time is 2009.8 h. The annual precipitation is 1763.5 mm, and the distributions of precipitation are quite uneven, with about 46% of precipitation occurring in the rainy season (from April to June). In the past 10 years, the economy of Jingdezhen has grown at an annual rate of more than 8%. In 2017, the regional GDP reached 87.825 billion Yuan. At the end of 2017, the total resident population was 1.665 million, of which the urban resident population was 1.098 million.

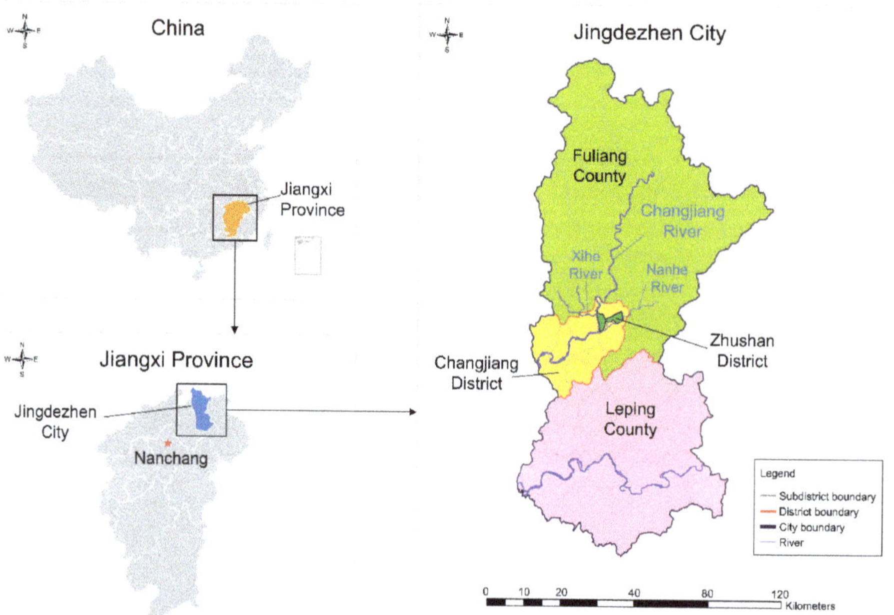

Figure 1. Study area: Jingdezhen City, Jiangxi Province, China.

Jingdezhen City is prone to floods. On the one hand, this is due to the extreme rainfall and flow during the flood season. Taking the flood in 2016 as an example, the city's 24-h average rainfall exceeded 200 mm with a return period of 20 years and the maximum flow of the Changjiang River was 7090 m^3/s. The flood return period of the main stream of the Changjiang River was 20 years, and some river sections were 50 years. On the other hand, Jingdezhen City lacks an effective flood control and drainage project. The existing flood control project can only defend floods with 5–8 years return period. The drainage site design standards are low as well, and some sites stopped working due to flooding. Besides, the length of the urban drainage pipe network is only 26.4 km and only 19% meets the drainage standard of 1-year return period flooding event. Meanwhile, Jingdezhen City has problems with its drainage pipe network, such as many bottleneck pipe sections, mild pipe network slopes, and insufficient pipe network outlet, resulting in regular flooding in Jingdezhen City.

Floods in Jingdezhen City

The climate of the Changjiang River Basin in Jingdezhen is significantly changeable, and the precipitation distribution is extremely uneven. The central city of Jingdezhen is located on both sides of the Changjiang River and its tributaries, the Nanhe River and the Xihe River. The terrain along the rivers are low, with an elevation of 24–31 m, which is threatened by floods. In 1955, 1996, 1998, 1999, 2010 to 2012, 2016, and 2017, the floods occurred in the Changjiang River, which caused serious flooding in Jingdezhen City, resulting in large economic losses and social impact.

According to historical records, in 1998, Jingdezhen City suffered a serious flood. The flooding time in the urban area reached 94 h, and the water depth in the urban low-lying area reached 10 m. The urban flooded area reached 31.4 square kilometers, accounting for 94.6% of the total urban area. Besides, the flood affected 354 thousand people and caused 3.23 billion Yuan damage. The flood in 2010 was also very serious, affected 32.52 thousand hectares of crops, with 666.8 thousand people affected and 1284 collapsed houses. The direct economic loss was about 2.96 billion Yuan. In 2016, from 4 pm on June 18 to 4 pm on the 19th, the rainfall in the Changjiang river basin reached 200.9 mm. The flood level of Dufengkeng Hydrological Station increased by 9.35 m in one day. The city has relocated a total of 119.7 thousand people, which was the largest relocation number of people since the past 10 years. And the direct economic losses amounted to 1.90 billion Yuan.

2.2. Sample Selection

All four districts were chosen as survey sampling sites because the floods affected all of them and the comparison was need latter. In the heavily flood-hit areas and those in city center, more subdistricts were selected, while fewer subdistricts were chosen in less affected areas. Changjiang District and Zhushan District are the urban centers of Jingdezhen City. The population is large and concentrated. The Changjiang River passes through these two areas and has many low-lying areas. There are more affected populations, so more questionnaires have been set up. Fuliang and Leping are far away from the city center, and their populations are relatively scattered. It is difficult to carry out all investigations in these two areas, thus fewer questionnaire surveys have been set up. As a result, five subdistricts were selected in Changjiang District, eight subdistricts were chosen in Zhushan District, one subdistrict was selected in Leping County, and two townships were selected in Fuliang County.

2.3. Questionnaire Design

This survey was part of the World Bank Project JIANGXI WUXIKOU INTEGRATED FLOOD MANAGEMENT PROJECT and it was approved by the Jingdezhen City Government. A semi-structured questionnaire was designed to investigate the public flood risk perception. In order to eliminate misunderstanding of the questionnaire, before the formal survey, some respondents with different educational levels were chosen to complete the questionnaire, and their feedbacks on the content of the questionnaire were collected. Then, based on these feedbacks, the project team changed all obscure and professional vocabulary into simple and easy to understand vocabulary. At the same time, the number of questions in the questionnaire was reduced too.

In the introduction part of the questionnaire, the purpose of the investigation and the relevant confidentiality principles were highlighted. This was used to inform respondents that the survey is anonymous and what data were collected. The main content of the questionnaire was divided into three main sections (Table 1). Each section had several items to measure. The first section included six items determining the most important sociodemographic factors of the respondents such as place of residence, gender, age, education level, occupation, and income per month [43]. The second section was other four important factors that could influence the public flood risk perception. The four impact factors were flood experience, flood knowledge education, flood protection responsibility, and trust in government. One impact factor comprised one item. The third section was the measurement of public flood risk perception. The impact and likelihood are most often employed variables to measure the flood risk perception as the flood risk is usually defined by the product of the likelihood of flood disaster with its consequences (impact) [43]. In this study, Jingdezhen City suffers from the flood almost every two years in history. Thus, the impact of the flood was more important than likelihood and only the flood impact was defined as the measurement of flood risk perception.

As mention above, some researchers include many other impact factors such as distance from the river, residence history, etc. But in this study, these factors were not included. This is because in this study, Jingdezhen City is a small and medium-sized city with a small migrant population and local people have lived here for a long time. In addition, not only people near the river are affected by flood, but people living in the low-lying city center are often affected. Likert scale technique was employed for the impact factors in Table 1.

Table 1. Definition of measurement and impact factors of public flood risk perception.

Section	Details	Item
Socio-demographic	Socio-demographic Characteristics of respondents	- District - Gender - Age - Education level - Occupation - Income per month
Other important impact factors	Respondents' flood experience	- The frequency of flood disaster experience
	Flood protection responsibility	- The responsibility of flood protection
	Flood knowledge education	- The extent of local flood knowledge education
	Trust in the government work	- Trust in the government on flood risk management
Flood risk perception	Respondents' perceived flood risk	- The impact of the flood disaster

2.4. Data Collection

The data used in this study came from a face-to-face questionnaire survey that was conducted from 8 July 2016 to 14 July 2016. Investigators who attended to the survey were Ph.D. students and MSc students. All of them had basic knowledge and background of natural disaster management. The project team invited four experts in the field of flood risk management to conducted four standardized training sessions for the investigators to make sure that the survey can be carried out smoothly. These trainings mainly focused on the introduction of project objectives and the basic skills for the investigation.

According to the sample selection, in total, 16 subdistricts were selected. Based on the distribution of subdistricts, investigators were divided into four groups of at least four investigators. In each group, there was a senior researcher who monitored the survey process, coordinated the questionnaires collection and checked the completeness and validity of collected questionnaires. Before each interview began, the purpose of the investigation and confidentiality principles were verbally explained by the investigators again. The participation of the respondents in this study was voluntary and consented, and enough time was given. Any confusions of the questions during the survey were explained by the investigators and the respondents had the rights to refuse to participate or withdraw from the survey at any time. In order to incentivize the public to participate the survey, every respondent was given a gift after they completed the questionnaire. At last, 900 questionnaires were distributed, 852 were collected, and the number of valid responses after the exclusion of incomplete questionnaires was 719 (the response rate was 84.4%).

2.5. Statistical Analysis

Firstly, descriptive statistics were applied to quantitatively describe and summarize the features of the socio-demographic characteristics of the respondents and the other four impact factors (flood experience, flood knowledge education, flood protection responsibility, trust in government). Secondly, analysis of variance (ANOVA) was conducted to examine the mean ranks of two or more independent variables with the null hypothesis of equality. This was to identify whether there existed correlations between the impact factors and public flood risk perception or not. Then, Post Hoc Tests was employed to find and compare the flood risk perception differences between different groups under the same impact factor among all respondents. Thus, the correlations were confirmed to be positive or negative. Besides, with regard to the gender variable, the mean difference comparison among gender using independent T test because only two groups exist, male and female.

All statistical analysis was carried out under a significance test value of 0.05 to confirm whether these impact factors affected the public flood risk perception. Impact factors with the significance value of less than 0.05 were considered to be significantly influential to public flood risk perception, as proposed in similar questionnaire-based studies [2,17,44]. The data was analyzed using the IBM SPSS Statistics software (Version 24.0.0.2, SPSS Inc., Chicago, IL, USA).

3. Results and Discussion

3.1. Preliminary Analysis

In this study, 719 valid survey questionnaires were analyzed. Table 2 summarized the response number for each area and the Socio-Demographic Variables of the respondents. The number of respondents in Changjiang District and Zhushan District were higher than in the other areas. Because the more subdistricts were selected, the more responses were collected. Table 3 summarized the distribution of other four important impact factors (flood experience, flood knowledge education, flood protection responsibility, and trust in government).

Table 2. The Socio-Demographic variables of the respondents (N = 719).

Variable	Jingdezhen City (Total)	Changjiang District	Zhushan District	Leping County	Fuliang County
Number of respondents n (%)	719 (100%)	244 (33.9%)	362 (50.3%)	52 (7.2%)	61 (8.5%)
Gender n (%)					
Male	334 (46.5%)	128 (52.5%)	157 (43.4%)	22 (42.3%)	27 (44.3%)
Female	385 (53.5%)	116 (47.5%)	205 (56.6%)	30 (57.7%)	34 (55.7%)
Age n (%)					
16–20 years	49 (6.8%)	15 (6.1%)	15 (4.1%)	15 (28.8%)	4 (6.6%)
21–35 years	272 (37.8%)	97 (39.8%)	121 (33.4%)	30 (57.7%)	24 (39.3%)
36–50 years	255 (35.5%)	83 (34.0%)	144 (39.8%)	4 (7.7%)	24 (39.3%)
51–70 years	129 (17.9%)	44 (18.0%)	74 (20.4%)	3 (5.8%)	8 (13.1%)
$\geq$ 71 years	14 (1.9%)	5 (2.0%)	8 (2.2%)	0 (0.0%)	1 (1.6%)
Education level n (%)					
Primary school and below	273 (38.0%)	91 (37.3%)	137 (37.8%)	18 (34.6%)	27 (44.3%)
Middle school	307 (42.7%)	104 (42.6%)	152 (42.0%)	29 (55.8%)	22 (36.1%)
High school	96 (13.4%)	29 (11.5%)	56 (15.5%)	4 (7.7%)	8 (13.1%)
Bachelor	41 (5.7%)	20 (8.2%)	16 (4.4%)	1 (1.9%)	4 (6.6%)
Master and above	2 (0.3%)	1 (0.4%)	1 (0.3%)	0 (0.0%)	0 (0.0%)
Occupation n (%)					
Work in company	141 (19.6%)	34 (13.9%)	73 (20.2%)	12 (23.1%)	22 (36.1%)
Work in government	45 (6.3%)	8 (3.3%)	33 (9.1%)	2 (3.8%)	2 (3.3%)
Self-employed person	305 (42.4%)	122 (50.0%)	128 (35.4%)	27 (51.9%)	28 (45.9%)
Student	37 (5.1%)	11 (4.5%)	24 (6.6%)	0 (0.0%)	2 (3.3%)
Retired person	78 (10.8%)	23 (9.4%)	48 (13.3%)	4 (7.7%)	3 (4.9%)
Others	113 (15.7%)	46 (18.9%)	56 (15.5%)	7 (13.5%)	4 (6.6%)
Income per month n (%)					
¥0–2000	324 (45.1%)	108 (44.3%)	173 (47.8%)	20 (38.5%)	23 (37.7%)
¥2001–5000	353 (49.1%)	122 (50.0%)	171 (47.2%)	28 (53.8%)	32 (52.5%)
¥5001–8000	36 (5.0%)	12 (4.9%)	15 (4.1%)	4 (7.7%)	5 (8.2%)
$\geq$¥8000	6 (0.8%)	2 (0.8%)	3 (0.8%)	0 (0.0%)	1 (1.6%)

Table 3. The distribution of the respondents regarding flood experience and trust in government (N = 719).

District	Jingdezhen City (Total)	Changjiang District	Zhushan District	Leping County	Fuliang County
Number of respondents *n* (%)	719 (100%)	244 (33.9%)	362 (50.3%)	52 (7.2%)	61 (8.5%)
Flood experience *n* (%)					
Less than once every two years	268 (37.3%)	81 (33.2%)	123 (34.0%)	37 (71.2%)	27 (44.3%)
Once every two years	109 (15.2%)	23 (9.4%)	78 (21.5%)	4 (7.7%)	4 (6.6%)
One or two times every year	211 (29.3%)	78 (32.0%)	104 (28.7%)	8 (15.4%)	21 (34.4%)
More than two times every year	131 (18.2%)	62 (25.4%)	57 (15.7%)	3 (5.8%)	9 (14.8%)
Flood knowledge education *n* (%)					
Never	189 (26.3%)	79 (32.4%)	63 (17.4%)	22 (42.3%)	25 (41.0%)
Few	188 (26.1%)	71 (29.1%)	84 (23.2%)	16 (30.8%)	17 (27.9%)
Medium	201 (28.0%)	61 (25.0%)	114 (31.5%)	10 (19.2%)	16 (26.2%)
Many	141 (19.6%)	33 (13.5%)	101 (27.9%)	4 (7.7%)	3 (4.9%)
Flood protection responsibility *n* (%)					
Government	348 (48.4%)	99 (40.6%)	196 (54.1%)	28 (53.8%)	25 (41.0%)
Flood management experts	177 (24.6%)	86 (35.2%)	62 (17.1%)	13 (25.0%)	16 (26.2%)
Company	12 (1.7%)	4 (1.6%)	7 (1.9%)	0 (0.0%)	1 (1.6%)
Community committee	83 (11.5%)	35 (14.3%)	38 (10.5%)	0 (0.0%)	10 (16.4%)
Public	99 (13.8%)	20 (8.2%)	59 (16.3%)	11 (21.2%)	9 (14.8%)
Trust in government *n* (%)					
Not trust	151 (21.0%)	62 (25.4%)	60 (16.6%)	8 (15.4%)	21 (34.4%)
Low trust	389 (54.1%)	129 (52.9%)	195 (53.9%)	29 (55.8%)	36 (59.0%)
Medium trust	133 (18.5%)	37 (15.2%)	84 (23.2%)	8 (15.4%)	4 (6.6%)
High trust	46 (6.4%)	16 (6.6%)	23 (6.4%)	7 (13.5%)	0 (0.0%)

3.1.1. Socio-Demographic Characteristics of Respondents

Of the 719 respondents, 244 (33.9%) were from Changjiang District, 362 (50.3%) from Zhushan District, 52 (7.2%) from Leping County, and 61 (8.5%) from Fuliang County (Table 2). Among the 719 respondents, the percentage of male respondents and female respondents were close, accounting for 46.5% and 53.5%, respectively. Respondents aged from 21 to 50 years old were the majority, with the percentage of 21–35 years old and 35–50 years old were 37.8% and 35.5%, respectively. With regard to the education level, most of the respondents were primary school and below (38.0%) and middle school (42.7%), whereas the percentage of high school, bachelor's degree, and master's degree were only 13.4%, 5.7%, and 0.3%, respectively. Six occupation types were classified, 42.4% and 19.6% of the respondents were self-employed person and worked in company, respectively. Most of respondents earned less than 5000 Chinese Yuan (CNY) every month, income less than 2000 CNY per month accounted for 45.1% and monthly income between 2001 and 5000 CNY were majority, accounting for 49.1%. Only 5.8% earned more than 5000 CNY per month. The specific ratios of the four districts/counties were similar to those of the entire city of Jingdezhen.

3.1.2. Other Important Impact Factors

In this study, when considering the characteristics of Jingdezhen City and the objectives of the study, the flood experience, flood knowledge education, flood protection responsibility, and trust in government were considered as other important factors influencing the public flood risk perception (as shown in Table 3).

Most studies revealed that the flood experience can increase flood risk perception and people with recent flood experience would acquire good knowledge of flood and do well in flood mitigation [9,28]. Most of the respondents in Jingdezhen City were influenced by the flood every year, with 47.5% of the respondents experienced at least one flood every year and only 37.3% of respondents experienced less than one flood every two years. Meanwhile, about 25.4% of respondents in Changjiang District and

15.7% of respondents in Zhushan District were seriously affected by the flood, experiencing more than 2 times flood every year.

It has been found the flood knowledge had a positive correlation with the flood risk perception [45]. Medium level was the largest percentage (28.0%) and only 19.6% of respondents had received many flood knowledge education sessions in Jingdezhen. Besides, 26.3% of respondents never received flood knowledge education. Among the four areas, the respondents in Zhushan District received more knowledge education with 31.5% chosen the answer of "Medium" and 27.9% selected the answer of "Many".

Flood protection responsibility, in many studies, was found to reflect the degree of responsibility to which a person took protection behaviors [46]. In this study, most of the respondents thought the government and flood management experts should take the responsibilities to protect the public from the flood disaster, accounting for 48.4% and 24.6%. In contrast, only 13.9% of respondents believed that they themselves should also be responsible for flood protection and disaster mitigation.

Many findings had showed that people with higher degrees of trust in government perceive lower consequences of disaster and tend to prepare less [47]. When asked about their trust in the local government for protecting them from floods, respondents showed low trust levels in Jingdezhen City, with "low trust" answers accounting for 54.1%, followed by not trust (21.0%), medium trust (18.5%), and high trust (6.4%). In addition, Leping County had the largest percentage of high trust in the government (13.5%), followed by Changjiang District (6.6%), Zhushan District (6.4%).

3.2. Public Flood Risk Perception in Four Districts in Jingdezhen

The public flood risk perception was compared among four districts. Here, the p value was less than 0.05 (Table 4), which showed that the public flood risk perceptions among four districts were statistically significant. Therefore, the comparison of scores between the four districts afterwards was credible and reliable.

Table 4. The descriptive statistic of the flood risk perception of case areas.

Residents	Mean	N	Std. Deviation	Std. Error	p Value
Changjiang District	3.24	244	0.842	0.054	
Zhushan District	3.12	362	0.867	0.046	0.000 *
Leping County	2.65	52	1.027	0.142	
Fuliang County	3.11	61	0.686	0.088	
Total	3.13	719	0.867	0.032	

Note: * with statistical difference ($p < 0.05$).

When asked about the flood impact, the respondents in Jingdezhen City showed different flood risk perception level. As shown in Figure 2, more than half the respondents selected "medium impact", accounting for 50.3%, and they thought the flood affected their daily life and work but not serious. And 27.1% of the respondents thought that they were largely affected by the flood, causing inconvenience to life and some loss of property, sometimes unable to work or shutdown. Besides, 4.3% of the respondents felt strongly affected by the flood, and they thought their lives and work were greatly affected and sometimes life-threatening property-damaging. On the contrary, "somewhat impact" and "no impact" recorded a small percentage. In specific, 13.5% of respondents thought they were affected by the flood, but this situation only lasted a very short time and 4.7% of the respondents never felt affected by the flood.

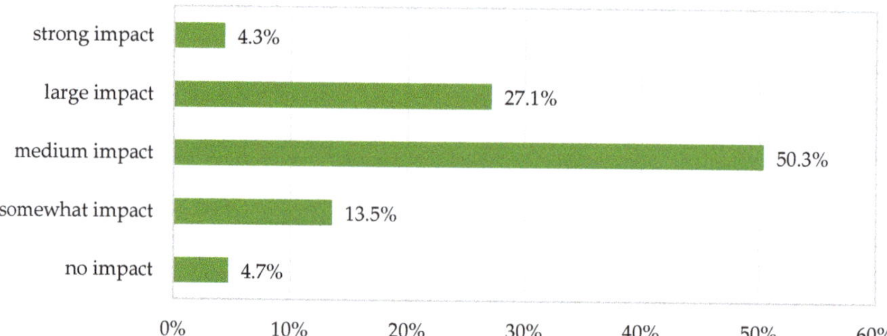

Figure 2. Distribution of the responses regarding the flood impact, the measurement of the flood risk perception.

The specific difference of flood risk perception was compared among four districts in Jingdezhen City. As shown in Table 4, the mean score of flood risk perception for the respondents in the Jingdezhen City was 3.13. This represented the level of flood risk perception among the Jingdezhen City was not high, close to medium level. Among the four districts, Changjiang District had the highest score of flood risk perception, which was 3.24, 0.11 higher than the whole city average level. In the actual investigation, this may be because the area is the city center, where the wealth and population density are large, and the damage that is caused by the disaster was serious in the past. Meanwhile, some streets and plazas got flooded every year, resulting in a high flood risk perception among the respondents in Changjiang District. The lowest score of flood risk perception was 2.65 in Leping County. This may be due to the respondents in Leping County thought the flood was not a serious disaster. The possible reason is 71.2% of the respondents in Leping County experienced less than one flood every two years (Table 3). With less likelihood of future flood as well as the less flood experience, people in Leping County might believe that flood risk had lower importance. In addition, the flood risk perception in Zhushan District and Fuliang County were similar, being 3.12 and 3.11, respectively.

Overall, these results showed that the respondents in the whole Jingdezhen city had a medium level of flood risk perception, and there were large differences among four districts with the respondents in Changjiang District having the highest flood risk perception while the respondents in Leping County had the lowest.

3.3. Impact Factors of Public Flood Risk Perception

ANOVA was conducted to examine and analyze the links between the impact factors and public flood risk perception in Jingdezhen City (as shown in Table 5). In this study, the impact factors were related to:

1. individual socio-demographic characteristics and
2. other four important factors. 95% confidence interval was used. When the $p < 0.05$, it means the impact factor would influence the flood risk perception.

The ANOVA results showed that the impact factors, such as district, gender, age, education level, income per month, flood experience, flood knowledge education, flood protection responsibility and trust in the government have significant relationships with the flood risk perception. Only the occupation factor has insignificant relationship with $p > 0.05$. Then, Post Hoc Tests was conducted to find the flood risk perception differences between different groups under the same impact factor among all respondents (Table 6).

Table 5. Analysis of variance (ANOVA) results of impact factors and public flood risk perception.

Variables	*p* Value	Variables	*p* Value
District	0.000 *	Income per month	0.000 *
Gender	0.000 *	Flood experience	0.000 *
Age	0.000 *	Flood knowledge education	0.001 *
Education level	0.000 *	Flood protection responsibility	0.030 *
Occupation	0.994	Trust in government	0.000 *

Note: * with statistical difference ($p < 0.05$).

Table 6. Multiple Comparisons for factors affecting public flood risk perception.

Variable	Mean Difference	Std. Error	Sig.	95% Confidence Interval Lower Bound	95% Confidence Interval Upper Bound
District (vs. Leping)					
Changjiang	0.584 *	0.131	0.000 *	0.33	0.84
Zhushan	0.470 *	0.127	0.000 *	0.22	0.72
Fuliang	0.461 *	0.162	0.005 *	0.14	0.78
Gender (vs. Male)					
Female	0.228 *	0.064	0.000 *	0.101	0.354
Age (vs. 51–70 years)					
16–20 years	−0.851 *	0.140	0.000 *	−1.13	−0.58
21–35 years	−0.555 *	0.089	0.000 *	−0.73	−0.38
36–50 years	−0.304 *	0.090	0.001 *	−0.48	−0.13
≥ 71 years	−0.004	0.235	0.987	−0.47	0.46
Education level (vs. Bachelor)					
Primary School and below	0.608 *	0.143	0.000 *	0.33	0.89
Middle school	0.512 *	0.142	0.000 *	0.23	0.79
High school	0.324 *	0.160	0.043 *	0.01	0.64
Master and above	0.366	0.620	0.555	−0.85	1.58
Occupation (vs. Work in company)					
Work in government	0.034	0.149	0.819	−0.26	0.33
Self-employed person	0.052	0.089	0.561	−0.12	0.23
Student	0.009	0.161	0.956	−0.31	0.32
Retired person	0.029	0.123	0.814	−0.21	0.27
Others	0.007	0.110	0.950	−0.21	0.22
Income per month (vs. ≤¥2000)					
¥2001–5000	−0.356 *	0.065	0.000 *	−0.48	−0.23
¥5001–8000	−0.472 *	0.149	0.002 *	−0.76	−0.18
≥¥8000	−0.833 *	0.349	0.017 *	−1.52	−0.15
Flood experience (vs. Less than once every two years)					
Once every two years	−0.265 *	0.095	0.005 *	0.08	0.45
1–2 times every year	−0.501 *	0.077	0.000 *	0.35	0.65
More than 2 times every year	0.576 *	0.089	0.000 *	0.40	0.75
Flood Knowledge Education (vs. Many)					
Never	−0.362 *	0.096	0.000 *	−0.55	−0.17
Few	−0.330 *	0.096	0.001 *	−0.52	−0.14
Medium	−0.264 *	0.094	0.005 *	−0.45	−0.08
Protection responsibility (vs. Public)					
Government	−0.262 *	0.098	0.008 *	−0.45	−0.07
Flood management experts	−0.303 *	0.108	0.005 *	−0.52	−0.09
Company	−0.270 *	0.264	0.307	−0.79	0.25
Community committee	−0.173 *	0.129	0.180	−0.43	0.08
Trust in government (vs. Not Trust)					
Low Trust	−0.237 *	0.082	0.004 *	0.08	0.40
Medium Trust	−0.395 *	0.102	0.000 *	−0.60	−0.19
High Trust	−0.445 *	0.145	0.002 *	−0.73	−0.16

Note: * with statistical difference ($p < 0.05$). The mean difference among gender using independent T test because of only two group exist, male and female.

As far as the district was concerned, the above analysis has found that there was a statistically significant difference between the flood risk perceptions in four districts in Jingdezhen City (see Tables 4 and 5). Compared with other three districts, the respondents in Changjiang District perceived the highest flood risk. Besides, the respondents in Leping County showed the lowest flood risk perception, and this may be because they did not think the flood risk was very important and they believed that the likelihood of future flood in this area was very low.

With regard to gender, the female respondents had higher flood risk perception than the male respondents ($p < 0.05$). This is because women had lower socioeconomic status than men and were more vulnerable when facing the floods, which caused women to be more willing to seek flood information, pay more attention to property losses, and more likely to take self-protection measures in advance. Therefore, female perceived higher flood risk than male. This result was in line with the findings in most published studies [4,48].

In this study, there was a significant positive correlation between age and flood risk perception. In general, the older the respondent, the higher the flood risk perception level. Among the age groups, the respondents aged 51–70 years old had the highest flood risk perception comparing with other age groups younger than 51 years old. This may be because the respondents aged 50–71 years old experienced many historical serious floods and more likely to take the responsibilities in family safety. So, they perceived higher flood risk than other age groups. These results were similar to the findings in most published studies [4,48], although some studies thought that age negatively influenced the flood risk perception [43].

It can be seen from Table 6 that the respondents with higher education level had lower flood risk perception. The respondents with bachelor's degree had lower flood risk perception than those with primary school degree, middle school degree, or high school degree. Meanwhile, there was no big difference between bachelor and master or above. Relevant studies also confirmed the significance of education for risk perception with negative correlation [16]. Ho et al. thought that people with higher education level had lower risk perception because highly educated people were more likely to better understand the flood information and government flood mitigation actions, and thus might feel a higher degree of controllability over a disaster [7].

With regard to the occupation factor, in this study, it had no statistically significant relation to the flood risk perception of the respondents. But, from the empirical data, the self-employed respondents had the highest flood risk perception. This may be because, for the self-employed respondents, the damage caused by the flood needs to be borne by themselves. In the relevant studies, Arnaud et al. revealed that, when discussed about the flood risk reduction, the factor of occupation was never significant [49].

With regard to the monthly income of the respondents, in general, higher educated people had higher income and thus their relation to the flood risk perception was similar. This study found that the respondents with lower income per month showed higher level of risk perception for flood. Compared with the respondents whose average monthly income was less than ¥2000, other three groups had poorer flood risk perception ($p < 0.05$ for all). This result was similar to the findings in some previous studies [43]. For example, Kellens et al. reviewed many relevant studies and found that there was a negative correlation between income and risk perception [50,51], though statistical significance was often absent [7,16,52].

The flood experience of the respondents had been found to be positively correlated with the flood risk perception and it was highly significant in this study, i.e., those with more flood experiences had higher flood risk perception than those with less flood experiences. The respondents experienced more than 2 times flood every year had the highest flood risk perception compared to other three groups. This was because people with more flood experience had more knowledge and better understanding of historical floods, and they were more likely to seek flood information and take measure to protect themselves. In fact, in our field research, it was discovered that the residents who were often affected by disasters made small flood control facilities in front of their homes, stored food during the flood season, and established mutual assistance agreements between neighbors. These results were also confirmed by the studies of Pagneux et al. [53] and Kellens W et al. [4].

With regard to the flood knowledge education, the flood knowledge is generally found strongly related to the feeling of security. Individuals with little knowledge of the causes of floods had lower flood risk perception [9]. In this study, it has a similar result that the individuals with more flood knowledge education showed higher flood risk perception. Compared with the respondents who received more flood knowledge education, the other three group showed lower flood risk perception. But the difference between the three groups ("Never", "Few" and "Medium") were small, while the gaps between the group of "Many" and other three groups were large. This means that the level of public flood risk perception would be improved only after a certain amount of flood knowledge education. Thus, the government need to adhere to more flood knowledge education.

The view of flood protection responsibilities has been found that can influence the public flood risk perception [33]. In this study, the respondents who believed themselves should be responsible for the flood protection showed higher flood risk perception than other groups. The reason may be that people who feel responsible for taking protective actions usually doubt the effectiveness of 'public' protective measures [34], thus, they perceived higher flood risk perception and preferred to take self-protection measures. Besides, the difference between other groups were not big. These results also reflect the fact that raising public responsibility for flood protection is very helpful for flood mitigation and risk management.

With regard to the trust in the government, it was found to be negatively related to the flood risk perception in this study (see Table 6). The respondents held the "Very low" trust in the government perceived highest flood risk than other three groups ($p < 0.05$ for all). The reason may be that trust in the government is represented by the trust of the government, experts, and the mass media [26]. The high level of trust showed that the respondents believe that the government can cope with flood hazards and do not need to do too much preparedness themselves. People who have lower trust in the government do not believe that the government can issue early flood warning and timely rescue. Instead, they choose to actively understand flood knowledge, seek flood information, and take measures to protect themselves. These results also were confirmed in most published studies [27,54].

4. Conclusions

This study analyzed the public flood risk perception in Jingdezhen City and explored the impact factors on flood risk perception. Results of ANOVA showed that many impact factors were strongly correlated with the flood risk perception. These include the socio-demographic characteristics of the respondents, previous flood experience, flood knowledge education, flood protection responsibility, and trust in government. The key findings are:

1. the flood risk perceptions of most respondents were in the medium level, which accounted for 50.3% of the total respondents. Only 4.3% of the total respondents recorded high level of flood risk perception and 4.7% of total respondents thought they were not affected by flood;

2. the respondents in the four districts of Jingdezhen City had different levels of flood risk perception, among which the respondents in Changjiang District had the highest level of perception due to this district is the city center and the flood-prone area. The female respondents had higher flood risk perception than male. The respondents with older age, more flood experience, more flood knowledge, lower income per month, less education, and less trusted in government showed higher flood risk perception. The occupation variable did not significantly influence the public flood risk perception, but the self-employed person had higher flood risk perception than other groups from the empirical data. The respondents who thought themselves should also be responsible for flood protection showed higher flood risk perception; and,

3. these results of this study were consistent with the findings in previous studies with few contrasts only.

But, the limitation of the study should also be taken into consideration. First, due to the financial and time limitation, only one subdistrict in Leping County and two downtowns in Fuliang District were chosen to carry out the survey. This may influence the results in these two districts. Second, we used a single measure of risk perception that some researchers might not consider to be optimal. Other research tends to use a set of items (e.g., impact, awareness, likelihood and fear) to measure the flood risk perception. Third, although we told respondents that the survey was anonymous, some respondents were still cautious and avoided choosing extreme answers about the government, which could cause some uncertainty in the results. Despite this, we believe that our analysis will help decision makers to develop effective flood risk communication strategies and flood risk reduction policies. First, it can help decision-makers to grasp the difference between the flood risk perceived by residents and the real flood risk, make reasonable risk regulation according to the characteristics of the people, and reduce the irrational behavior caused by this risk perception deviation. For example, too high-risk perception could lead to social panic while too low risk perceptions could lead to negative flood mitigation. Second, this study can be used to understand which groups have a higher risk perception, and the government can specifically encourage and promote flood prevention products such as flood insurance to reduce flood losses. Third, these results about the "flood protection responsibility" and "trust in government" can provide references for local government to strengthen the relationship between the government and the public because the responsibility and trust are very important for policy promotion and implementation.

Future work will focus on how flood risk perception affects people's flood mitigation behavior; conduct a more detailed survey of flood risk perception for specific vulnerable populations, such as students, females, and farmers; highlight the dominant role of the government in flood risk mitigation in small and medium-sized cities, and refine and select more variables related to government; and, analyze and compare the changes in flood risk perception before and after flood disasters. Finally, our findings suggest that the local government should actively promote the government's credibility and enhance the trust between residents and the government. Although this may partially reduce the flood risk perception of residents, higher trust is conducive to the implementation of the government's disaster reduction policy, and also enables residents to cooperate with the government's disaster reduction work. At the same time, it is important to strengthen flood knowledge education and training for residents based on their socio-demographic characteristics to improve the public flood risk perception and the ability of self-protect. Moreover, the government should reconsider their disaster emergency drills, making it simple and understandable, and more operational.

Author Contributions: Zhiqiang Wang was responsible for literature search, survey design, data collection, data analysis and he also wrote the initial draft of the manuscript. Huimin Wang, Jing Huang principally conceived the idea for the study and provided financial support. Jinle Kang was responsible for field interviews and data collection. Dawei Han contributed to the figures and the revision of English and style.

Funding: This research was funded by Jiangxi Wuxikou Integrated Flood Management Project (Grant number JDZ-WXK-ZX-9), National Natural Science Foundation of China (Grant number 71601070), Postgraduate Research & Practice Innovation Program of Jiangsu Province of China (Grant number KYCX17_0517) and China Scholarship Council (Grant number 201706710094).

Acknowledgments: We would like to express our gratitude to the Jingdezhen district office and street office for the authorization and support to carry out the study. Finally, we thank all the investigators and respondents who participated in the survey.

Conflicts of Interest: The authors declare no conflicts of interest.

References

1. Bradford, R.A.; O'Sullivan, J.J.; van der Craats, I.M.; Krywkow, J.; Rotko, P.; Aaltonen, J.; Bonaiuto, M.; De Dominicis, S.; Waylen, K.; Schelfaut, K. Risk perception—Issues for flood management in Europe. *Nat. Hazards Earth Syst. Sci.* **2012**, *12*, 2299–2309. [CrossRef]

2. Liu, D.L.; Li, Y.; Shen, X.; Xie, Y.L.; Zhang, Y.L. Flood risk perception of rural households in western mountainous regions of Henan Province, China. *Int. J. Disaster Risk Reduct.* **2018**, *27*, 155–160. [CrossRef]

3. Mackay, A. Climate change 2007: Impacts, adaptation and vulnerability. Contribution of working group II to the fourth assessment report of the intergovernmental panel on climate change. *J. Environ. Qual.* **2008**, *37*, 2407. [CrossRef]

4. Kellens, W.; Zaalberg, R.; Neutens, T.; Vanneuville, W.; De Maeyer, P. An analysis of the public perception of flood risk on the Belgian coast. *Risk Anal.* **2011**, *31*, 1055–1068. [CrossRef] [PubMed]

5. Peacock, W.G.; Brody, S.D.; Highfield, W. Hurricane risk perceptions among Florida's single family homeowners. *Landsc. Urban Plan.* **2005**, *73*, 120–135. [CrossRef]

6. Grothmann, T.; Reusswig, F. People at risk of flooding: Why some residents take precautionary action while others do not. *Nat. Hazards* **2006**, *38*, 101–120. [CrossRef]

7. Ho, M.C.; Shaw, D.; Lin, S.Y.; Chiu, Y.C. How do disaster characteristics influence risk perception? *Risk Anal.* **2008**, *28*, 635–643. [CrossRef] [PubMed]

8. Pidgeon, N. Risk assessment, risk values and the social science programme: Why we do need risk perception research. *Reliab. Eng. Syst. Saf.* **1998**, *59*, 5–15. [CrossRef]

9. Botzen, W.J.W.; Aerts, J.; van den Bergh, J. Dependence of flood risk perceptions on socioeconomic and objective risk factors. *Water Resour. Res.* **2009**, *45*, 15. [CrossRef]

10. White, G.F. *Human Adjustment to Floods: A Geographical Approach to the Flood Problem in the United States*; University of Chicago: Chicago, IL, USA, 1945.

11. Brilly, M.; Polic, M. Public perception of flood risks, flood forecasting and mitigation. *Nat. Hazards Earth Syst. Sci.* **2005**, *5*, 345–355. [CrossRef]

12. Bird, D.K. The use of questionnaires for acquiring information on public perception of natural hazards and risk mitigation—A review of current knowledge and practice. *Nat. Hazards Earth Syst. Sci.* **2009**, *9*, 1307–1325. [CrossRef]

13. Starr, C. Social benefit versus technological risk. *Science* **1969**, *165*, 1232–1238. [CrossRef] [PubMed]

14. Sjöberg, L.; Moen, B.; Rundmo, T. *Explaining Risk Perception. An Evaluation of the Psychometric Paradigm in Risk Perception Research*; Norwegian University of Science and Technology, Department of Psychology: Trondheim, Norway, 2004; ISBN 82-7892-024.

15. Fischhoff, B.; Slovic, P.; Lichtenstein, S.; Read, S.; Combs, B. How safe is safe enough? A psychometric study of attitudes towards technological risks and benefits. *Policy Sci.* **1978**, *9*, 127–152. [CrossRef]

16. Qasim, S.; Khan, A.N.; Shrestha, R.P.; Qasim, M. Risk perception of the people in the flood prone Khyber Pukhthunkhwa province of Pakistan. *Int. J. Disaster Risk Reduct.* **2015**, *14*, 373–378. [CrossRef]

17. Diakakis, M.; Priskos, G.; Skordoulis, M. Public perception of flood risk in flash flood prone areas of Eastern Mediterranean: The case of Attica Region in Greece. *Int. J. Disaster Risk Reduct.* **2018**, *28*, 404–413. [CrossRef]

18. Mabuku, M.P.; Senzanje, A.; Mudhara, M.; Jewitt, G.; Mulwafu, W. Rural households' flood preparedness and social determinants in Mwandi district of Zambia and Eastern Zambezi region of Namibia. *Int. J. Disaster Risk Reduct.* **2018**, *28*, 284–297. [CrossRef]

19. Ullah, R.; Shivakoti, G.P.; Ali, G. Factors effecting farmers' risk attitude and risk perceptions: The case of Khyber Pakhtunkhwa, Pakistan. *Int. J. Disaster Risk Reduct.* **2015**, *13*, 151–157. [CrossRef]

20. Luo, X.F.; Lone, T.; Jiang, S.Y.; Li, R.R.; Berends, P. A study of farmers' flood perceptions based on the entropy method: An application from Jianghan Plain, China. *Disasters* **2016**, *40*, 573–588. [CrossRef] [PubMed]

21. Bosschaart, A.; Kuiper, W.; van der Schee, J.; Schoonenboom, J. The role of knowledge in students' flood-risk perception. *Nat. Hazards* **2013**, *69*, 1661–1680. [CrossRef]

22. Armas, I.; Avram, E. Perception of flood risk in Danube Delta, Romania. *Nat. Hazards* **2009**, *50*, 269–287. [CrossRef]

23. Lo, A.Y. The role of social norms in climate adaptation: Mediating risk perception and flood insurance purchase. *Glob. Environ. Chang.* **2013**, *23*, 1249–1257. [CrossRef]

24. Salvati, P.; Bianchi, C.; Fiorucci, F.; Giostrella, P.; Marchesini, I.; Guzzetti, F. Perception of flood and landslide risk in Italy: A preliminary analysis. *Nat. Hazards Earth Syst. Sci.* **2014**, *14*, 2589–2603. [CrossRef]

25. Spitalar, M.; Gourley, J.J.; Lutoff, C.; Kirstetter, P.E.; Brilly, M.; Carr, N. Analysis of flash flood parameters and human impacts in the us from 2006 to 2012. *J. Hydrol.* **2014**, *519*, 863–870. [CrossRef]

26. Lin, S.Y.; Shaw, D.G.; Ho, M.C. Why are flood and landslide victims less willing to take mitigation measures than the public? *Nat. Hazards* **2008**, *44*, 305–314. [CrossRef]

27. Hung, H.C. The attitude towards flood insurance purchase when respondents' preferences are uncertain: A fuzzy approach. *J. Risk Res.* **2009**, *12*, 239–258. [CrossRef]

28. Terpstra, T.; Lindell, M.K. Citizens' perceptions of flood hazard adjustments: An application of the protective action decision model. *Environ. Behav.* **2013**, *45*, 993–1018. [CrossRef]

29. Knuth, D.; Kehl, D.; Hulse, L.; Schmidt, S. Risk perception, experience, and objective risk: A cross-national study with European emergency survivors. *Risk Anal.* **2014**, *34*, 1286–1298. [CrossRef] [PubMed]

30. Wachinger, G.; Renn, O.; Begg, C.; Kuhlicke, C. The risk perception paradox-implications for governance and communication of natural hazards. *Risk Anal.* **2013**, *33*, 1049–1065. [CrossRef] [PubMed]

31. Thieken, A.H.; Kreibich, H.; Muller, M.; Merz, B. Coping with floods: Preparedness, response and recovery of flood-affected residents in Germany in 2002. *Hydrol. Sci. J.* **2007**, *52*, 1016–1037. [CrossRef]

32. Lopez-Marrero, T.; Yarnal, B. Putting adaptive capacity into the context of people's lives: A case study of two flood-prone communities in Puerto Rico. *Nat. Hazards* **2010**, *52*, 277–297. [CrossRef]

33. Becker, G.; Aerts, J.; Huitema, D. Influence of flood risk perception and other factors on risk-reducing behaviour: A survey of municipalities along the R hine. *J. Flood Risk Manag.* **2014**, *7*, 16–30. [CrossRef]

34. Birkholz, S.; Muro, M.; Jeffrey, P.; Smith, H.M. Rethinking the relationship between flood risk perception and flood management. *Sci. Total Environ.* **2014**, *478*, 12–20. [CrossRef] [PubMed]

35. Martens, T.; Garrelts, H.; Grunenberg, H.; Lange, H. Taking the heterogeneity of citizens into account: Flood risk communication in coastal cities—A case study of Bremen. *Nat. Hazards Earth Syst. Sci.* **2009**, *9*, 1931–1940. [CrossRef]

36. Ryan, B. Information seeking in a flood. *Disaster Prev. Manag.* **2013**, *22*, 229–242. [CrossRef]

37. Baan, P.J.; Klijn, F. Flood risk perception and implications for flood risk management in The Netherlands. *Int. J. River Basin Manag.* **2004**, *2*, 113–122. [CrossRef]

38. Petrolia, D.R.; Landry, C.E.; Coble, K.H. Risk preferences, risk perceptions, and flood insurance. *Land Econ.* **2013**, *89*, 227–245. [CrossRef]

39. Tversky, A.; Kahneman, D. Judgment under uncertainty: Heuristics and biases. *Science* **1974**, *185*, 1124–1131. [CrossRef] [PubMed]

40. Slovic, P. Perception of risk. *Science* **1987**, *236*, 280–285. [CrossRef] [PubMed]

41. Gotham, K.F.; Campanella, R.; Lauve-Moon, K.; Powers, B. Hazard experience, geophysical vulnerability, and flood risk perceptions in a postdisaster city, the case of New Orleans. *Risk Anal.* **2018**, *38*, 345–356. [CrossRef] [PubMed]

42. Horney, J.A.; MacDonald, P.D.M.; Van Willigen, M.; Berke, P.R.; Kaufman, J.S. Individual actual or perceived property flood risk: Did it predict evacuation from Hurricane Isabel in North Carolina, 2003? *Risk Anal.* **2010**, *30*, 501–511. [CrossRef] [PubMed]

43. Kellens, W.; Terpstra, T.; De Maeyer, P. Perception and communication of flood risks: A systematic review of empirical research. *Risk Anal.* **2013**, *33*, 24–49. [CrossRef] [PubMed]

44. Peng, Y.; Zhu, X.T.; Zhang, F.Y.; Huang, L.; Xue, J.B.; Xu, Y.L. Farmers' risk perception of concentrated rural settlement development after the 5.12 Sichuan earthquake. *Habitat Int.* **2018**, *71*, 169–176. [CrossRef]

45. Raaijmakers, R.; Krywkow, J.; van der Veen, A. Flood risk perceptions and spatial multi-criteria analysis: An exploratory research for hazard mitigation. *Nat. Hazards* **2008**, *46*, 307–322. [CrossRef]

46. Lara, A.; Sauri, D.; Ribas, A.; Pavon, D. Social perceptions of floods and flood management in a Mediterranean area (Costa Brava, Spain). *Nat. Hazards Earth Syst. Sci.* **2010**, *10*, 2081–2091. [CrossRef]

47. Han, Z.Q.; Lu, X.L.; Horhager, E.I.; Yan, J.B. The effects of trust in government on earthquake survivors' risk perception and preparedness in China. *Nat. Hazards* **2017**, *86*, 437–452. [CrossRef]

48. Kung, Y.W.; Chen, S.H. Perception of earthquake risk in Taiwan: Effects of gender and past earthquake experience. *Risk Anal.* **2012**, *32*, 1535–1546. [CrossRef] [PubMed]

49. Reynaud, A.; Nguyen, M.H. Valuing flood risk reductions. *Environ. Model. Assess.* **2016**, *21*, 603–617. [CrossRef]

50. Lindell, M.K.; Hwang, S.N. Households' perceived personal risk and responses in a multihazard environment. *Risk Anal.* **2008**, *28*, 539–556. [CrossRef] [PubMed]

51. Zhang, Y.; Hwang, S.N.; Lindell, M.K. Hazard proximity or risk perception? Evaluating effects of natural and technological hazards on housing values. *Environ. Behav.* **2010**, *42*, 597–624. [CrossRef]

52. Ling, F.H.; Tamura, M.; Yasuhara, K.; Ajima, K.; Van Trinh, C. Reducing flood risks in rural households: Survey of perception and adaptation in the Mekong delta. *Clim. Chang.* **2015**, *132*, 209–222. [CrossRef]

53. Pagneux, E.; Gisladottir, G.; Jonsdottir, S. Public perception of flood hazard and flood risk in Iceland: A case study in a watershed prone to ice-jam floods. *Nat. Hazards* **2011**, *58*, 269–287. [CrossRef]

54. Terpstra, T. Emotions, trust, and perceived risk: Affective and cognitive routes to flood preparedness behavior. *Risk Anal.* **2011**, *31*, 1658–1675. [CrossRef] [PubMed]

MDPI

St. Alban-Anlage 66

4052 Basel

Switzerland

Tel. +41 61 683 77 34

Fax +41 61 302 89 18

www.mdpi.com

Water Editorial Office

E-mail: water@mdpi.com

www.mdpi.com/journal/water

www.ingramcontent.com/pod-product-compliance
Lightning Source LLC
LaVergne TN
LVHW071026170726
843515LV00006B/1212